A BUR OAK ORIGINAL

Wayne I. Anderson

Iowa's Geological Past

Three Billion Years of Earth History

University of Iowa Press Ψ Iowa City

University of Iowa Press, Iowa City 52242

Copyright © 1998 by the University of Iowa Press

Printed in the United States of America

http://www.uiowa.edu/~uipress

Printed on acid-free paper

Library of Congress
Cataloging-in-Publication Data
Anderson, Wayne I., 1935–

Iowa's geological past: three billion years of earth
history / by Wayne I. Anderson.

p.   cm. — (A bur oak original)

Includes bibliographical references and index.

ISBN 0-87745-639-9, ISBN 0-87745-640-2 (pbk.)

1. Geology — Iowa.   I. Title.   II. Series.

QE111.A582   1998

557.77 — dc21   98-38373

98 99 00 01 02 C 5 4 3 2 1
98 99 00 01 02 P 5 4 3 2 1

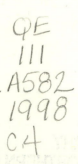

This book is dedicated to William M. Furnish, who first sparked my interest in geology, and Brian F. Glenister, who encouraged me to write on the geology of Iowa; to my wife, Jan, and our four children, Greg, Karen, Jennifer, and Mark; and to the many geologists who have studied Iowa's past with patience and perseverance.

The geologist who undertakes to investigate the vast prairie country of the Mississippi Valley must be provided with no common share of patience and perseverance. He must be content to travel for half a day together without seeing aught but a rich, black soil, covered, as far as the eye can reach, even down to the very edge of the small streams, with a thick and high growth of prairie grass, with perhaps a faint outline of timber cutting the distant horizon. He must be prepared to wade swamps, to ford streams waist deep, or, in times of freshets, to plunge in and breast the current. He must not shrink beneath a broiling sun, without even a bush to cast a faint shadow over an occasional resting place. He must think himself fortunate if he can reach, at night, a few scattered oaks to plenish his fire, and boil his camp kettle; and he may consider it a special instance of good luck, if, in return, he can catch a glimpse of a rock exposure once or twice a day.

Pioneering midwestern geologist David D. Owen, writing in 1852 on the challenges of doing geology in the Shell Rock drainage basin in north-central Iowa

Contents

Preface

This book is written for a varied audience — interested citizen, engineer, teacher, college student, and anyone who is curious about Iowa's geologic past. Although not written specifically for the professional geologist, I believe that geologists will find the publication useful as a general overview of the geology of the state. In writing the book, I have assumed that the reader has some prior knowledge of geology — approximately equivalent to a basic secondary school course in earth science or to an introductory college geology course.

My approach here is stratigraphic and chronological. I start at the beginning of Iowa's geologic record and proceed to tell the geologic story of the state, one rock formation at a time — from the Precambrian to the Quaternary. I follow the format of my earlier book on Iowa geology, *Geology of Iowa: Over Two Billion Years of Change*, published in 1983 by Iowa State University Press. However, this volume is more than a revision. The bulk of the text has been rewritten, and most of the figures are new.

Much of what is presented here has been summarized from other published works. An attempt has been made to keep things simple, although I know from experience that the geologic history of an area never turns out to be simple. Interesting yes, but never simple.

Change has characterized Iowa's geologic past. In fact, the only constant over the long course of geologic time is change. Iowa's geologic resources and natural areas are products of that change. During the Pleistocene Ice Age, a short time ago geologically speaking, Iowa endured multiple episodes of continental glaciation. Prior to that, dozens of seas came and went, leaving behind limestone beds with rich fossil records. Lush coal swamps, salty lagoons, briny basins, vast alluvial plains, ancient rifts, and old Precambrian mountain belts all left their mark. Even the rare extraterrestrial event left its signature; the impact crater and shocked rocks under Manson testify to that.

Overall, Iowa's geologic column records an extraordinary transformation — more than 3 billion years of change. The hills of Iowa are not everlasting when viewed with the perspective of geologic time. Neither is our knowledge of geology fixed for eternity. Geology, like science overall, is by its very nature ever changing. Surely new discoveries and fresh insights will require a rewriting of some of Iowa's geologic history at some point in the future. So take this summary of Iowa geology for what it is: a snapshot of what geologists currently know about the state's fascinating geological past.

I would like to acknowledge the support of the University of Northern Iowa (UNI). They granted me a professional development leave to pursue library and field study on the geology of Iowa during the fall semester of 1995. The Graduate College of UNI provided partial financial support for the project. In addition, the Department of Earth Science at UNI and the staff of the Iowa Department of Natural Resources, Geological Survey Bureau, gave assistance and advice at various times. Nancy Howland, Amy Freiberg, and Johnathan Alexander of the UNI Department of Earth Science provided valuable help. Thanks to the wonders of modern computers, I was able to prepare this manuscript for publication, in spite of my marginal typing skills. Neither I nor my high school typing teacher could have foreseen such a possibility when I struggled through one semester of typing at Keokuk Senior High School in the early 1950s.

I appreciate the constructive suggestions of the following geologists who reviewed portions of the book: Raymond R. Anderson, E. Arthur Bettis III, Bill J. Bunker, Mary R. Howes, Greg A. Ludvigson, Robert M. McKay, Jean C. Prior, and Brian J. Witzke. Several of the paleogeographic reconstructions are based on the work of Brian J. Witzke and were originally published in "Palaeoclimatic Constraints for Palaeozoic Palaeolatitudes of Laurentia and Euramerica," *Geological Society (London) Memoir* 12.

Iowa's Geological Past

1 The Geologic Setting of Iowa

The rocks in question, therefore, so far as relates to Iowa, are nothing more than the consolidated sands and muds of old sea bottoms preserving for our inspection samples of the life that occupied the seas at the time each successive bed was in process of accumulation. Iowa has passed more time under the ocean than as dry land.

>   Preface to "Report on Geology of Iowa's Counties"
>   (1906; author not indicated but probably Samuel Calvin)

Iowa is very very old — as old as the hills and older. So old, in truth, is this fair land that no matter at what period the story is begun, whole eternities of time stretch back to ages still more remote.

>   John Briggs (1920) describing Iowa's antiquity

Iowa has a rich record of the geologic past, particularly in the form of sedimentary rocks. These strata represent deposits of scores of ancient seas that inundated the heartland of the North American continent. However, the state's rock record is not confined to marine deposits alone. The coal beds of southern Iowa are products of coastal swamps and deltas. Former streams left telltale signatures in the sandstones of Ledges, Dolliver Memorial, and Wildcat Den state parks. Fossil amphibians document the presence of freshwater habitats in 340-million-year-old rocks near Delta in southeastern Iowa. And the sedimentary rock record of Iowa lies on older rocks of even greater variety.

These older rocks, assigned to the Precambrian division of the geologic column, consist of a wide assortment of rock types that are representative of the three major rock families (igneous, sedimentary, and metamorphic) that constitute our planet. Over most of the state, the rock record is covered and obscured by unconsolidated sediments that were laid down primarily by wind, streams, or glaciers. A geological cross section (fig. 1.1) illustrates the relationship of the unconsolidated material (labeled Pleistocene) to the underlying sedimentary units (labeled Pennsylvanian, Mississippian, Devonian, Silurian, Ordovician, and Cambrian).

Iowa lies in a part of the United States known as the Stable Interior. In this region, the rocks are generally flat-lying sedimentary rocks that show little or no deformation. In fact, at any one location in Iowa, the rocks typically appear to be horizontal — as flat as a pancake. On a regional scale, however, the rocks of the

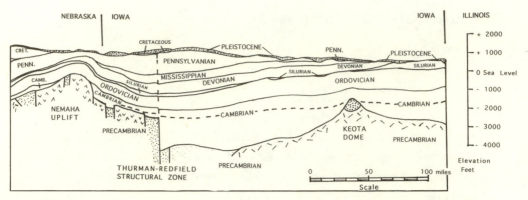

1.1 A geological cross section from near Clinton in eastern Iowa to south of Council Bluffs in southwestern Iowa. As shown, Iowa has a substantial record of layered sedimentary rocks. These strata rest on older Precambrian rocks of varied composition and are generally covered by unconsolidated sediments of Pleistocene age that were laid down by wind, streams, and glacial ice. It is common practice to use different horizontal and vertical scales in order to portray geologic cross sections such as shown here. This produces an exaggeration of the vertical dimension. The vertical exaggeration of this cross section is 85 times. Adapted from Bunker and Witzke 1988.

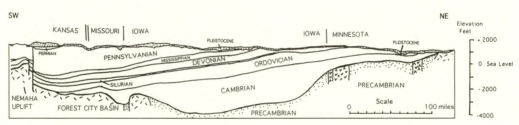

1.2 Iowa's Paleozoic bedrock is inclined gently downward to the southwest. The vertical exaggeration of this cross section is 85 times. Adapted from Bunker and Witzke 1988.

state display a slight inclination, or tilt. A cross-sectional slice from northeast to southwest (fig. 1.2) shows that Iowa's sedimentary rocks sag downward to form a basin. This structure, termed the Forest City Basin, extends into Missouri, Kansas, and Nebraska. In places, the sedimentary rocks of the Stable Interior arch upward to form geologic structures such as the Keota Dome and the Nemaha Uplift (figs. 1.1 and 1.2). A few hundred barrels of petroleum were obtained from rocks of the Keota Dome in Washington County as a result of drilling explorations in 1963 and 1985. To date, this meager recovery represents the bulk of the oil recovered in Iowa.

Figure 1.3 shows the distribution of the bedrock units of Iowa. Throughout most of the state the bedrock is covered by unconsolidated deposits such as glacial till, sand gravel, and loess (wind-blown silt). If these unconsolidated materials were removed, it would be possible to see the rock units as they are displayed on the bedrock map (fig. 1.3). Where the unconsolidated deposits are thin

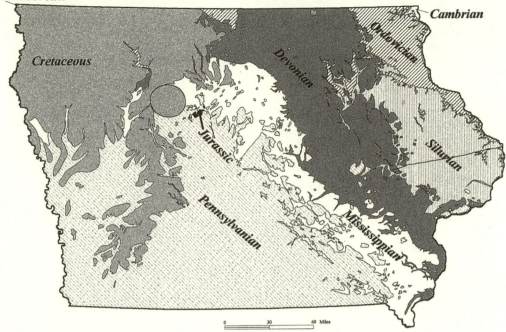

Precambrian

Cambrian

Cretaceous

Ordovician

Devonian

Jurassic

Silurian

Pennsylvanian

Mississippian

0   30   60 Miles

1.3 The distribution of Iowa's bedrock is arranged in roughly parallel belts from northeast to southwest. Cambrian strata are exposed in northeastern Iowa, followed to the southwest by belts of Ordovician, Silurian, Devonian, Mississippian, and Pennsylvanian strata. Cretaceous rocks rest unconformably on older Paleozoic bedrock in western Iowa. Small areas of Jurassic bedrock occur near Fort Dodge. Iowa's oldest exposed bedrock, in the northwestern corner of the state, is of Precambrian age. The nearly circular configuration in northwestern Iowa locates the Manson Impact Structure, and the dark line in eastern Iowa depicts the Plum River Fault Zone. Information provided by Geological Survey Bureau, Iowa City, 1998.

or absent, the bedrock can be viewed and studied in roadcuts, stream valleys, and quarries.

The bedrock map of Iowa shows rocks of several geologic ages, including the Precambrian, Cambrian, Ordovician, Silurian, Devonian, Mississippian, Pennsylvanian, Jurassic, and Cretaceous. This reflects an interesting but somewhat incomplete record of Earth history. Most of these rock systems are only partially represented in the state. Other systems of the standard geologic column, such as the Permian and Triassic systems, have not been recognized in Iowa, although they are well documented in states to the west. Small areas of Tertiary strata are known from western Iowa, but these are too small to depict on the bedrock map.

The distribution pattern of the Cambrian, Ordovician, Silurian, Devonian, Mississippian, and Pennsylvanian rock systems is the result of these strata being warped and tilted and then truncated by erosion. The Cretaceous System was deposited on the eroded surfaces of these older rock units throughout much of the western part of the state (fig. 1.3).

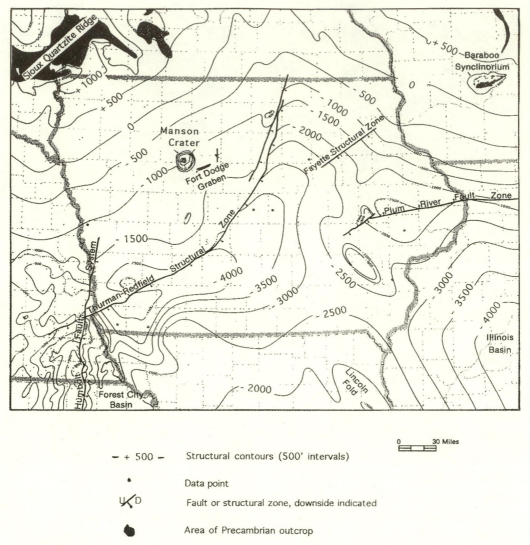

- + 500 -   Structural contours (500' intervals)

•   Data point

U⤬D   Fault or structural zone, downside indicated

●   Area of Precambrian outcrop

1.4 The configuration of the Precambrian surface in Iowa and adjoining states. Also shown are major faults affecting the Precambrian surface in Iowa. Courtesy of Bill J. Bunker, Geological Survey Bureau, Iowa City.

Faults are breaks in rocks along which movement has taken place; geologically speaking, Iowa has few faults. Of the major faults mapped in the state, none are known to present a seismic risk.

Faults offset the Precambrian surface in several regions of the state, including southwestern Iowa, central Iowa, the Fort Dodge area, near Fayette in north-eastern Iowa, and along the Plum River Fault Zone in eastern Iowa (fig. 1.4). Studies of the Plum River Fault Zone suggest that it last had significant offset in the middle of the Paleozoic Era, about 300 million years ago. The bedrock faults of Iowa are buried and lack topographic expression, and the subdued topography overlying areas of known faulting indicates that no significant movements have taken place recently.

Although Iowa is not characterized by earthquake activity, the effects of earthquakes are sometimes felt in the state. The earthquake that proved disastrous to Anchorage, Alaska, on March 27, 1964, produced a shock sufficient to knock a seismometer at Loras College in Dubuque out of service.

Some of the most violent earthquake activity in historic time affected the New Madrid area in southeastern Missouri and adjacent Kentucky and Tennessee during 1811 and 1812. Rated as an 8.8 magnitude earthquake on the Richter scale, the New Madrid earthquake of February 7, 1812, occurred when the Precambrian basement rocks, located several miles below the surface, snapped suddenly and released enormous amounts of energy. Knox and Stewart (1995) describe some of the spectacular events that affected the Mississippi River and its floodplain in the vicinity of New Madrid during the four days immediately following the big earthquake. Boats were thrown from the river; trees were destroyed; the Mississippi flowed backward for a stretch; flooding occurred on the Kentucky side of the river; Reelfoot Lake formed in adjacent Tennessee; and two steep rapids appeared in the Mississippi's channel.

The major earthquakes along the New Madrid Fault Zone during 1811 and 1812 certainly would have caused a stir in Iowa if the state had been heavily settled at the time. Human population has increased significantly since 1811–1812, and the New Madrid Fault Zone lies within 250 miles of the densely populated cities of Memphis, Little Rock, St. Louis, Nashville, Louisville, and Birmingham. Future earthquakes of the magnitude of the 1811–1812 shocks would bring immense damage to this region.

Predictions of a major earthquake along the New Madrid Fault Zone for December 3, 1990, received considerable media attention and aroused some anxiety among the general public. The earthquake forecast, however, was made by a person with limited knowledge of geology or geophysics. Although the prognostication turned out to be faulty in more ways than one, this episode did result in increased public interest in earthquakes and in an upturn in sales of earthquake insurance in Missouri and neighboring states, including southeastern Iowa.

Earthquake prediction can be a relatively inexact science, even as practiced by professional earth scientists. What are the odds of another large earthquake along the New Madrid Fault Zone? According to Anderson and VanDorpe (1991), scientists estimate there is a 90 percent chance that an earthquake of Richter magnitude 6.0 will occur by the year 2040. Estimated recurrence intervals for larger earthquakes, comparable in magnitude to the 1811–1812 events, vary from approximately 175 years to greater than 700 years.

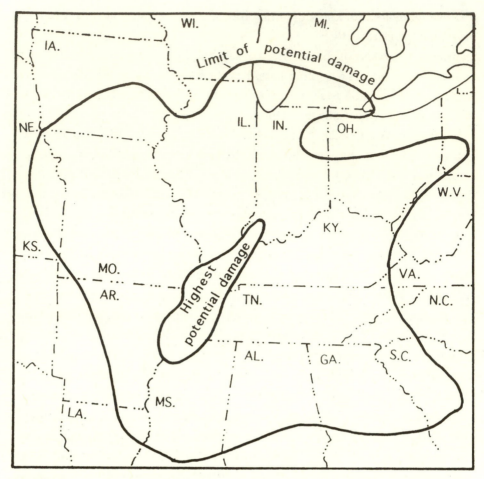

1.5 The potential area affected by a major earthquake (6.5 on the Richter scale) along the New Madrid Fault Zone. Vibrations from such an earthquake would be felt over most of southern Iowa, and Iowans in four southeastern counties would likely experience trembling buildings, cracked windows, and movement of unstable objects. Adapted from Anderson and VanDorpe 1991; based on maps in Atkinson 1989.

The estimated effects of a 6.5 Richter-magnitude earthquake are shown in figure 1.5. According to this interpretation, Iowans in the southeastern corner of the state would experience cracked windows, broken dishes, and trembling buildings. The effects would decrease to the north, but people throughout the southern half of the state would definitely know they had been in an earthquake.

In addition to the New Madrid Fault Zone, another region of seismic activity is located close to Iowa's borders. The Nemaha Uplift, located in Nebraska and Kansas (figs. 1.1 and 1.2), and the associated Humboldt Fault System, which extends into southwestern Iowa (fig. 1.4), experience numerous tiny earthquakes. The frequent release of small amounts of energy through these minor earthquakes prevents stresses building to dangerous levels, however, and southwestern Iowa is considered to be a safe area as far as seismic activity is concerned.

Geologic Time

The rock systems shown on the bedrock map of Iowa (fig. 1.3) represent deposits throughout immense intervals of geologic time. The magnitude of geologic time is so great that it is difficult for most people to imagine or comprehend. Geologic time is measured in millions of years rather than by the hours, days, weeks, months, years, and centuries to which humans are accustomed. In order to examine Iowa's geologic past, one must attempt to fathom the dimension of geologic time, for the state's geologic history was hundreds of millions of years in the making. Figure 1.6 shows the standard geologic time scale. Note that the divisions of geologic time are not of equal duration. The geologic column, on which the geologic time scale is based, was established before methods for absolute dating of rocks were discovered. The rock systems contained in the geologic column were placed in a relative chronology based on the law of superposition, a rather obvious principle that can be verified by observations of recent sediments. The principle contends that if layered rocks are relatively undisturbed, then older rocks are found at the bottom of a sequence and are overlain by progressively younger rocks.

Iowa's oldest rock record is of Precambrian age and includes a complex igneous and metamorphic terrane that is approximately 3 billion years old. Paleozoic rocks constitute most of Iowa's rock record. The oldest of these Paleozoic rocks were deposited approximately 530 million years ago; the youngest formed nearly 320 million years ago. Iowa also has a partial record of Mesozoic strata consisting of rocks deposited approximately 160 to 65 million years ago. The Cenozoic record of the state is limited primarily to deposits of the Pleistocene and Holocene (Recent) epochs of geologic time. However, some areas of Tertiary deposits have been documented in the western part of the state. Iowa's Pleistocene and Holocene deposits give the state a significant and relatively complete record of the last 1.65 million years of geologic history.

The dimension of geologic time can be understood better if we compare the geologic time scale (4.6 billion years) to a calendar year (365 days). If geologic time were scaled on the basis of a calendar year, most of the calendar year (January 1–mid-November) would be represented by Precambrian time (about 4 billion years in duration). Paleozoic rocks (545–245 million years old) would have accumulated during about one month (November 19–December 13). Mesozoic time (245–65 million years ago) would be represented by two weeks (December 13–December 27), and Cenozoic time (the last 65 million years of geologic time) would be represented by the last five days of the calendar year. The Pleistocene Ice Age (approximately 1.65 million years to 10,500 years ago) would be represented by only three hours. The portion of the geologic time scale assigned to the Holocene Epoch is about 10,500 years in duration, or a mere 1.1 minutes of the calendar year.

		(PERIODS)	(Epochs)

The geologic time scale diagram:

			(Epochs)
		(PERIODS)	**Recent** (the last 10,500 years)
PHANEROZOIC	**CENO-ZOIC**	**QUATERNARY** (0-1.65)	Pleistocene
		TERTIARY (1.65-65)	Pliocene
			Miocene
	MESOZOIC ERA	**CRETACEOUS** (65-146)	Oligocene
		JURASSIC (146-208)	Eocene
		TRIASSIC (208-245)	
	PALEOZOIC ERA	**PERMIAN** (245-290)	Paleocene
		PENNSYLVANIAN (290-323)	
		MISSISSIPPIAN (323-363)	
		DEVONIAN (363-409)	
		SILURIAN (409-439)	
		ORDOVICIAN (439-510)	
		CAMBRIAN (510-545)	
PRECAMBRIAN (.545-4.6 billion years)			

1.6 The geologic time scale. The numbers represent the ages of periods in millions of years. The Quaternary Period consists of the last 1.65 million years of geologic time and is represented by the Pleistocene and Holocene (Recent) epochs. The Holocene Epoch consists of the last 10,500 years, a dimension that is too small to plot to scale on this diagram. The Precambrian, commonly divided into the Proterozoic and Archean eons, spans the interval of geologic time from .545 to 4.6 billion years ago. Adapted from Stanley 1993.

Another way to try to comprehend the enormous dimension of geologic time is to scale some known distance in proportion to the geologic time scale. Let's do that for the 300-mile distance across Iowa on Interstate 80 between Davenport and Omaha. Precambrian time represents the first 264 miles of our trip. Paleozoic time, known as the Age of Invertebrates, is equivalent to approximately 20 miles, and the Mesozoic Era, the Age of Reptiles, spans only 12 miles. The Cenozoic Era, the Age of Mammals, is confined to the last 4 miles of our journey. The Pleistocene and Holocene epochs correspond to the final 565 feet of our trip, as we cross the Missouri River bridge from Council Bluffs to Omaha.

Iowa's Rock Record

Iowa's exposed rock record is composed almost entirely of sedimentary rocks. These layered rocks are as revealing to the geologist as archival documents are to the historian. Abundant clues (fossil remains, mineral grains, ripple marks, and so forth) in the layered rocks enable geologists to read much of Iowa's past history from the rock record.

Not all of the geological record is present in the state, however, for certain rocks were never deposited here. Other strata were deposited but later were eroded away. These gaps in Iowa's rock record are represented by surfaces that geologists call unconformities. Significant unconformities in Iowa occur at the base of the Cambrian, within the Ordovician, at the base of the Devonian, between the Mississippian and Pennsylvanian, at the base of the Cretaceous, and below the late Cenozoic glacial record. There are no Permian or Triassic rocks exposed in Iowa, and the Jurassic record is very limited. Iowa received marine deposits when shallow seas advanced (transgressed) over the continental interior during Cambrian, Ordovician, Silurian, Devonian, Mississippian, Pennsylvanian, Jurassic, and Cretaceous times. The deposits of Pennsylvanian age were cyclic in nature and consist of both marine and nonmarine beds. Figure 1.7 provides a graphic view of the state's record of ancient marine strata. Note that there are significant gaps in the rock record.

The record of deposition within the continental interior (craton) is characterized by deposits of widespread transgressive seas separated by major unconformities. These unconformity-bounded packages of sedimentary strata were called sequences by L. L. Sloss, a renowned North American stratigrapher. Sloss (1963) recognized six major sequences for North America and named them for Native American tribes. This terminology is shown in figure 1.8.

The use of tribal names for sequences followed a practice that was initiated when the original geologic column was established in Europe in the 1800s. For example, the Ordovician System was named for the Ordovices, an ancient Welsh tribe that was the last in Britain to submit to Roman rule. The Silurian System owes its name to the Silures, an ancient tribe that once inhabited Wales. Other

1.7 A generalized geologic time scale showing when inland seas covered the heartland of North America. Iowa's record of ancient marine deposits is particularly good for the Paleozoic Era. Adapted from Witzke 1989.

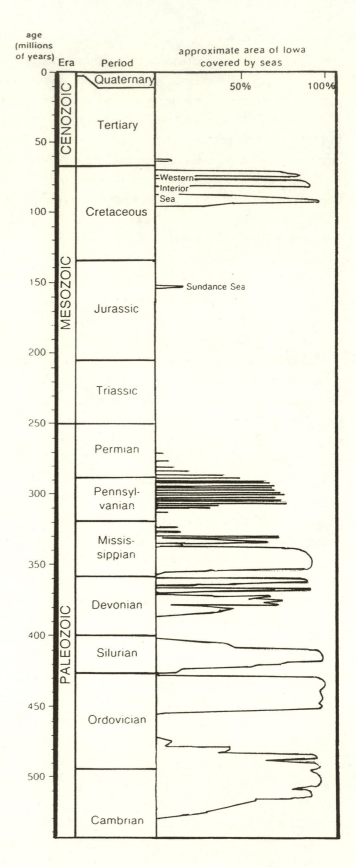

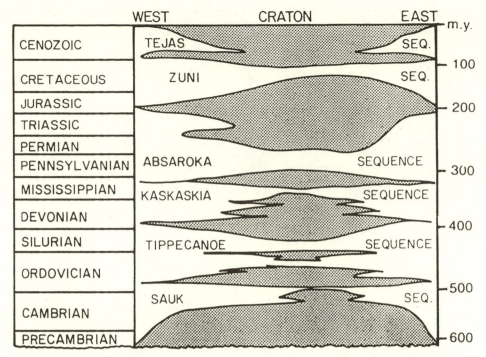

	WEST	CRATON	EAST	m.y.
CENOZOIC	TEJAS		SEQ.	
CRETACEOUS	ZUNI		SEQ.	— 100
JURASSIC				— 200
TRIASSIC				
PERMIAN				
PENNSYLVANIAN	ABSAROKA		SEQUENCE	— 300
MISSISSIPPIAN	KASKASKIA		SEQUENCE	
DEVONIAN				— 400
SILURIAN	TIPPECANOE		SEQUENCE	
ORDOVICIAN				
CAMBRIAN	SAUK		SEQ.	— 500
PRECAMBRIAN				— 600

1.8 The terminology for the six major unconformity-bounded packages of sedimentary strata, called sequences. The unconformities mark huge gaps in the rock record. Note that the rock record is more complete at the margins of the continent and is less complete in the continental interior (craton). Geologic ages in terms of millions of years ago are shown on the right. The irregular line at the base of the diagram signifies that only part of the Precambrian is shown. Adapted from Bunker et al. 1985; based on Sloss 1963.

names from the standard geologic column were derived from geographic localities, such as Cambrian for the Roman name for Wales (Cambria) and Devonian for the region of Devonshire, England.

The concept of sequences is useful for relating Iowa's Paleozoic record to that of surrounding states. The transgressions (advances) and regressions (retreats) of the seas that produced Iowa's sedimentary record laid down similar deposits in adjacent states. Many of these widespread marine transgressions and regressions across the North American continent were probably produced by worldwide changes in sea level.

What causes sea-level fluctuations on a global scale? Several suggestions have been offered. The rise and fall of ocean ridges is one possibility. The uplift and later erosion of mountain belts along continent-ocean boundaries can also produce worldwide changes in sea level. In addition, widespread glaciation can lead to a drop in sea level as some of the water of the oceans is converted to continental ice sheets. Subsequently, the melting of glacial ice produces a worldwide rise in sea level.

As we have seen, Iowa's rock column is somewhat incomplete. It has numerous

SEQUENCE	ERA	SYSTEM		EPOCH/SERIES/STAGE	WESTERN IOWA; EASTERN NEBRASKA; NORTHWESTERN MISSOURI	EASTERN IOWA; SOUTHEASTERN MINNESOTA
	MES.	TR.	U			
			L+M			
ABSAROKA		PERMIAN	U	Ochoan		
				Guadalupian		
			L	Leonardian		
				Wolfcampian	Chase / Council Grove / Admire Gp.	
		PENN.	U	Virgilian	Waubaunsee/Shaw-nee/Douglas	
				Missourian	Lansing/Kans.City/Pleas	
			M	Desmoinesian	Marmaton Gp. / Cherokee Gp.	Cherokee Gp.
				Atokan		Caseyville Fm.
			L	Morrowan		
KASKASKIA		MISS.	U	Chesterian		
			M	Meramecian	Ste. Genevieve/St. Louis/Spergen/Warsaw	Ste. Genevieve/St. Louis Spergen/Warsaw
				Osagean	Keokuk/Burlington	Keokuk Fm. / Burlington Fm.
			L	Kinderhookian	Gilmore City Fm. / Hampton Gp.	Hampton Fm. / North Hill Gp.
	PALEOZOIC	DEVONIAN	U	Famennian	"Maple Mill" Sh.	Yellow Spring Gp.
				Frasnian	"Lime Creek" — Cedar Valley undiff.	Lime Creek Sh. / Cedar Valley
			M	Givetian	Wapsipinicon	Wapsipinicon
				Eifelian		Spillville-Otis
			L			
TIPPECANOE		SILUR.	U	Pridolian		
				Ludlovian		Gower Fm.
			L	Wenlockian	Sil. undiff.	Scotch Grove Fm.
				Llandoverian		Hopkinton/Blanding fms. / T.d Morts-Mosalem
		ORDOVICIAN	U	Richmondian	Maquoketa Fm.	Maquoketa Fm.
				Maysvillian	Galena Gp.	Galena Gp.
				Edenian		
				Sher./Kirk./Rockland.	Decorah Sh.	Decorah Sh.
			M	Blackriveran	Platteville Fm. / Glenwood Fm.	Platteville Fm. / Glenwood Fm.
				Chazyan	St. Peter Ss.	St. Peter Ss.
				Whiterockian		
			L		Shakopee Fm. / Oneota Fm.	Shakopee Fm. / Oneota Fm.
SAUK		CAMBRIAN	U	Trempealeauan	Jordan/St. Lawrence fms.	Jordan/St. Lawrence
				Franconian	"Davis" Fm.	Lone Rock/Franconia / Wonewoc Fm.
				Dresbachian	Bonneterre Fm. / Mt. Simon Ss.	Eau Claire Fm. / Mt. Simon Ss.
			M			
			L			

1.9 A generalized correlation diagram of the Paleozoic stratigraphic units of Iowa and adjacent states. Four of the six Sloss (1963) sequences are shown on the left, and Iowa's principal rock units (formations or groups) are displayed in the two columns on the right. Note gaps in Iowa's stratigraphic record, shown by slanted pattern. Adapted from Bunker et al. 1988.

gaps in Mesozoic and Cenozoic times (figs. 1.7 and 1.8). Nevertheless, the record is sufficient to provide a good account of the state's geologic past. In fact, Iowa has a record of some strata that are missing in the well-known and exceptionally exposed layers of the Grand Canyon of Arizona. Iowa has an excellent record of both Ordovician and Silurian strata; the Grand Canyon has no record of either of these major rock systems. In addition, whereas the Grand Canyon reveals only a partial record of Devonian strata, Iowa's Devonian record is rich and varied. Thousands of people marveled at the abundantly fossiliferous Devonian rocks that were revealed by erosion below the spillway at Coralville Lake as a result of widespread flooding in Iowa during the summer of 1993. Floodwaters scoured out tens of feet of glacial deposits to expose a remarkable expanse of ancient seafloor, exceptionally preserved in the local 375-million-year-old bedrock.

The rock layers exposed below the spillway of Coralville Lake are assigned to the Little Cedar Formation. To geologists, formations represent the pages of the geologic past, and Iowa's rock record is made up of scores of formations, each recording a story in stone. The Pennsylvanian System alone is comprised of more than seventy formations. Geologists define formations on the basis of composition, not on the basis of geologic age. However, the geologic age of formations can often be determined through the use of index fossils or by radiometric dating. Formations will be discussed in greater detail in subsequent chapters, and selected formations will be examined more closely to illustrate specific geologic principles and concepts. Two or more contiguous or associated formations are often assigned to a formal rock unit called a group. Figure 1.9 shows the principal Paleozoic rock units (formations or groups) of Iowa and adjacent states.

Geographic Setting

Some of the iron-bearing minerals in lavas are magnetized by the earth's magnetic field as they cool. Under favorable conditions, this magnetic alignment remains locked in rocks and can be utilized to plot the position of ancient magnetic poles. If it is assumed that the positions of ancient magnetic poles were approximately the same as the ancient geographic poles, then paleogeographic reconstructions can be drawn showing the latitudes of ancient continents. Latitudinal settings can also be inferred from the distribution in the rock record of climate indicators, such as coal beds, selected plant fossils, coral reefs, and evaporite deposits.

Several reconstructions of the position of ancient continents have been attempted (fig. 1.10). Note that North America is shown in equatorial or low-latitudinal positions in the geologic past. In fact, much of Iowa's rock record accumulated in warm, shallow seas in tropical and subtropical settings when the North American continent was much closer to the equator than it is at present.

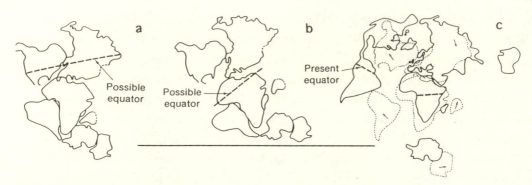

1.10 The position of continents in the geologic past: (a) Late Paleozoic, approximately 250 million years ago; (b) Jurassic, approximately 150 million years ago; and (c) end of Cretaceous, approximately 65 million years ago, shown by dotted pattern, and present continental positions, shown by solid lines. Adapted from Dietz and Holden 1970.

Paleogeographic reconstructions depicting possible positions of the equator are shown for specific geologic time intervals in subsequent chapters.

During Paleozoic time, the continents were close together, part of one giant supercontinent. Later, in Mesozoic time, the continents broke apart and started their slow drift and rotation to their present positions (fig. 1.10).

Interpreting the Past

An interpretation of the past on the basis of the rock record is founded primarily on the principle of uniformitarianism. This principle, simply stated, contends that the present is the key to the past. This means that by studying modern depositional features and geologic processes, insights can be gained on how to interpret the rock record of the past. Although the modern version of uniformitarianism holds that the fundamental laws and principles of science are invariant over the course of geologic time, it does recognize that rates of intensity of geologic processes do vary. For example, glaciation was a dominant process in Iowa during the Pleistocene Ice Age, but glaciers have been absent from the state for over 10,000 years.

Iowa's past has been interesting and varied. Looking back we see a setting far different than the gently rolling farm fields of present-day Iowa. In the days before Iowa was settled by Euro-Americans, prairie grasslands and scattered stands of hardwood trees were the dominant elements of Iowa's landscape. Earlier, during the Pleistocene Ice Age, Iowa was alternately covered by continental glaciers, tundra, and expanses of spruce.

If we go back beyond the Pleistocene to Mesozoic time, we find the state covered with shallow seas, partially similar to the modern Gulf of Mexico. Still earlier in Mesozoic time, Iowa's environment resembled the saline lagoons of the present-day Persian Gulf.

Ancient Iowa in Paleozoic time experienced a variety of coastal plain and shallow marine environments similar to those found along the modern coasts of Texas, Louisiana, and Florida. Coastal swamps and deltas, somewhat analogous to those of present-day Louisiana, prevailed during Late Paleozoic time, when Iowa's coal deposits were formed.

Iowa's most ancient rock record, the Precambrian, is harder to decipher; still, the record can be interpreted in general terms. During the later part of Precambrian time, ancient Iowa was located along a major rift-valley system, something like that of present-day Africa. Dark-colored igneous rocks formed along the rift system as lava flows and shallow intrusives, and streams deposited huge thicknesses of sands and muds within the ancient rift valley. Earlier, the ingredients of Iowa's oldest exposed bedrock accumulated on the floor of a sandy basin, some 1.6 billion years ago. Still earlier in Precambrian time, Iowa was part of ancient mountain belts in which granite, granite gneiss, and other igneous and metamorphic rocks were formed well below the Earth's surface. The details of Iowa's Precambrian history are covered in the following chapter.

The quartz is of a close grain and exceedingly hard, eliciting the most brilliant sparks from steel, and in most places where its surface is exposed to the sun and air it is highly polished beyond any results that could have been produced by diluvial action, being perfectly glazed as if by ignition.

> George Catlin (1840) describing Iowa's oldest exposed bedrock, the Sioux Quartzite

Terranes older than those of the Paleozoic age occupy in Iowa a very small surface area. . . . It seems all sufficient merely to note the existence of these rocks in the extreme northwestern corner of the state. Yet these very rocks now appear to have a history longer, more complicated, and more vicissitudinous than that of any other terrane represented within our borders.

> Charles Keyes (1919) on Iowa geology

The lengthy interval of time between the formation of the Earth and the appearance of an abundant record of hard-shelled invertebrates about 545 million years ago is known as the Precambrian. In Iowa, Precambrian rocks are exposed only in the northwestern corner of the state (fig. 2.1). Rocks of Precambrian age occur over a large area in the heart of the North American continent (the craton) and are particularly well exposed in the Canadian Shield (fig. 2.2). These rocks are primarily of igneous and metamorphic origins and formed by crystallization of hot silicate melts or alteration (metamorphism) of other rocks. Such rocks typically form well below the Earth's surface.

Most of the Precambrian rocks of the North American craton have a dense, nonporous texture and are often referred to as the Precambrian basement complex. The Precambrian basement complex served as the foundation on which the Paleozoic rocks of the Midwest were deposited.

At various times during the last 545 million years, the craton was covered by shallow inland seas or by low-lying lagoons and swamps. In such settings, the sedimentary rock record of the midwestern states accumulated, and these sedimentary rocks now cover the Precambrian rocks over much of the interior of the North American continent.

2.1 The Sioux Quartzite in Gitchie Manitou State Preserve, Iowa's oldest exposed bedrock. Photo by the author.

Figure 2.3 shows the areas of Precambrian rock exposures in Iowa and adjacent states. Precambrian rocks are well exposed throughout much of Canada and in the Canadian Shield area of northern Minnesota, northern Wisconsin, and the Upper Peninsula of Michigan. In South Dakota, the Precambrian is exposed in the core of the Black Hills Uplift and in a small area in the southeastern part of the state. Thousands of tourists view Precambrian bedrock each summer in South Dakota when they visit Mount Rushmore. In Minnesota, in addition to the exposures in the Canadian Shield, the Precambrian is exposed in the southwestern corner of the state and along the Minnesota River valley. In Missouri, Precambrian rocks are found in the St. Francois Mountains, representing the structural center of the Ozark Dome. Those who vacation in Minnesota and Wisconsin are familiar with Precambrian rocks from exposures in the Mesabi Range, the North Shore country of Lake Superior, the Wisconsin Highlands, and the Baraboo Range and Devils Lake area. Important mineral deposits such as ores of iron and copper are produced from the Precambrian rocks of neighboring states.

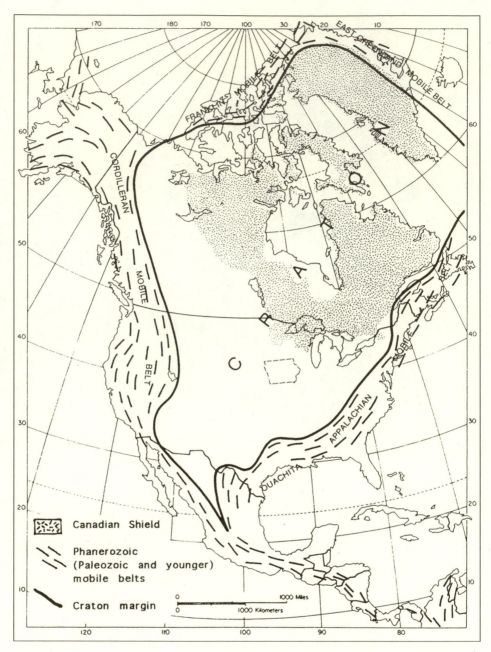

2.2 The craton and the Canadian Shield area of North America. Phanerozoic (Paleozoic and younger) mobile belts are also shown. Adapted from Dott and Batten 1971 and Anderson 1983.

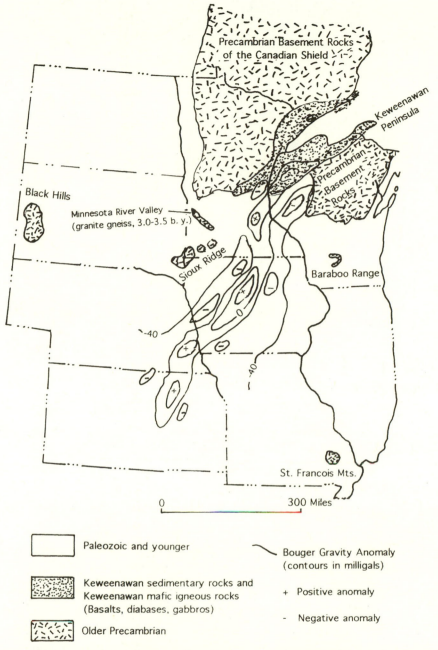

2.3 The Precambrian rock exposures in Iowa and adjacent states. The Midcontinent Gravity Anomaly and its relationship to the Keweenawan rocks of the Lake Superior region are also shown. Dense rocks of the Keweenawan type underlie the region of high-gravity values in central Iowa. Adapted from Dott and Batten 1971 and Anderson 1983.

2.4 The time frame for the Precambrian rock record of Iowa.

Age	Rock Record
Late Proterozoic	No rock record in Iowa
Middle Proterozoic	Upper Red Clastic Sequence
	Lower Red Clastic Sequence
	Thor Volcanic Group
	Finnegan Gabbro (1.02 billion years old)
	Eischeid Gabbro of the "MacKenzie Dike Swarm" (1.28 billion years old);
	Spencer Granite (1.37 billion years old);
	Quimby Granite and Manson Granite (1.43 billion years old);
	Green Island Granite (1.49 billion years old)
Early Proterozoic	"Baraboo-Interval Quartzites" (Sioux Quartzite, Washington County Quartzite, and Cedar Rapids Quartzite, 1.64–1.76 billion years old);
	Hull Keratophyre and Matlock Keratophyre (1.76 billion years old);
	Harris Granite and Hawarden Granite (1.80 billion years old);
	"Manson Gneiss" (1.85 billion years old);
	LeMars Gneiss (2.06 billion years old);
	Matlock Banded-Iron Formation
Archean	Lyon County Gneiss (2.52 billion years old);
	Otter Creek Layered Mafic Complex (2.89 billion years old)

Time Frame

In order to establish a time frame for the Precambrian in Iowa, we will need to rely on the knowledge gained from examining exposed Precambrian rocks in the surrounding states because Iowa's Precambrian rock record is almost entirely buried under younger rocks. Information from studies of the Precambrian of South Dakota, Minnesota, Wisconsin, and northern Michigan is particularly helpful. Figure 2.4 is a general summary of Iowa's Precambrian rock record and the geologic events that took place in Iowa during Precambrian time. Several of these events have been dated by radiometric means (fig. 2.5), so it is possible to place them in a time frame.

Radiometric ages always have a plus and minus factor of a few percentage points. One reason for this range of possible ages is the slow rate of transformation of radioactive "parent" elements to stable "daughter" elements. A second factor is that only minute amounts of material are involved in these transformations, making reproducible measurements difficult. Other difficulties arise if samples have been metamorphosed or reheated. Even with these potential problems, radiometric dating is a helpful and essential tool for deciphering Precambrian events. Because this book is intended primarily to serve a general rather than a specialized audience, I will often report radiometric ages that are in the middle of the range of accepted values in order to keep the text as straightforward as possible. Remember, however, that these dates actually represent ranges. Although radiometric ages are sometimes referred to as absolute ages,

2.5 Modern radiometric dates from the Precambrian rocks of Iowa. An asterisk indicates that rocks sampled from the Manson area were deformed by impact, so these dates are not considered to be reliable. Courtesy of Raymond R. Anderson, Geological Survey Bureau, Iowa City.

Rock Unit	Type of Rock	Sample Location by County	Age in Billions of Years
Finnegan Gabbro	Gabbro	Guthrie	1.02
Eischeid Gabbro	Gabbro	Carroll	1.28
Spencer Granite	Granite	Clay	1.37
Quimby	Granite	Cherokee	1.43
Manson*	Granite	Pocahontas	1.43
Green Island Granite	Granite	Jackson	1.49
Wheeler "Granite" = Hull Keratophyre; Matlock Keratophyre	Keratophyre	Lyon	1.76
Harris Granite	Granite	Osceola	1.80
"Manson Gneiss"*	Gneiss	Pocahontas	1.85
Le Mars Gneiss	Quartz-feldspar gneiss	Plymouth	2.06
Lyon County Gneiss	Quartz monzodiorite	Lyon	2.52
Otter Creek Layered Mafic Complex	Mafic igneous rock	Lyon	2.89

this means only that the dates are given in terms of years rather than as relative ages that are older or younger than some other event.

Some of the oldest rocks in North America are found along the Minnesota River valley in southwestern Minnesota, only 60 to 90 miles from Iowa's northern border. These rocks, the Morton and Montevideo gneisses, are placed in the Early Precambrian (Archean) by the Minnesota Geological Survey. Inclusions in the Montevideo Gneiss have been dated as approximately 3.8 billion years old, and the bulk of the gneiss is about 2.7 billion years old. Radiometric dates suggest that the Morton Gneiss formed nearly 3.5 billion years ago.

Iowa's oldest radiometrically dated bedrock is the Otter Creek Layered Mafic Complex. Recovered by drilling in northwestern Iowa, the rocks of the Otter Creek Layered Mafic Complex are about 2.9 billion years old and are associated with Archean rocks of the Superior Province, some of the oldest bedrock in North America. The dates and chronology of other Precambrian rocks in Iowa are shown in figure 2.4. The following time divisions are used to divide the Precambrian of the Midwest: Archean (older than 2.5 billion years), Early Proterozoic (1.6–2.5 billion years old), and Middle Proterozoic (0.9–1.6 billion years old). Rocks of Late Proterozoic (0.9–0.55 billion years old) age are unknown in Iowa.

Before looking at specific examples of Iowa's Precambrian history, a discussion of the theory of plate tectonics will be helpful. This widely accepted theory, considered by many to be one of geology's major contributions to modern scientific thought, provides a broad framework that is helpful to explain the origin of continents, ocean basins, and mountain belts through geologic time.

Plate Tectonics and Iowa's Precambrian Record

The geologic history of Iowa's Precambrian can best be understood in the framework of modern plate tectonics. The theory of plate tectonics evolved from a variety of separate geological studies and observations in the 1950s and 1960s, and crucial evidence has come from studies of the ocean floors since the 1960s.

It is generally well known that the major earthquakes of the world occur in specific zones and that volcanism and earthquake activity are frequently linked. In general, the location of major earthquake zones and volcanoes serves to outline the boundaries between plates.

Plates can be thought of as huge slabs of the Earth's crust and uppermost mantle (lithosphere) that move as discrete blocks of rigid rock. A typical continental plate is about 60 miles thick. Underlying the plates is a plasticlike zone with a thickness of approximately 120 miles. This zone, termed the asthenosphere, provides a pliable layer on which the plates move. Below this level, the Earth again behaves as a rigid body. Plates move at the rate of a few inches per year, about the rate of growth of a human fingernail. The major plates of today's world are shown in figure 2.6. The positions of plate boundaries in the geologic past varied significantly from those of today. It will be helpful to examine a generalized cross-sectional view (fig. 2.7) of the upper part of the Earth to see the location of the asthenosphere and lithosphere.

The crust under the continents differs significantly from the crust under the oceans. Not only is the continental crust thicker than oceanic crust, it is composed of different material. Continental crust is granitic in composition and "lighter" (lower in density) than oceanic crust. Whereas continental crust is close to the density of granitic rocks (density of 2.7 grams per cubic centimeter), the density of oceanic crust is like that of basalt (density of 2.9 grams per cubic centimeter). Basalt, a fine-grained, dark-colored igneous rock, is the major constituent of the ocean crust.

The mantle directly underlies the Earth's crust and is composed of even denser rocks. Called ultramafic rocks, these mantle components reach densities of 3.3 grams per centimeter. Ultramafic rocks are composed almost entirely of dark-colored minerals of so-called mafic composition. The word "mafic" is derived from "ma" (magnesium) and "fic" (ferric, or iron). Basalt is considered to be a mafic rock because it is composed of appreciable amounts of dark-colored minerals that contain magnesium and iron. In addition to bearing mafic minerals, basalt also possesses large amounts of the common mineral feldspar. Ultramafic rocks, however, contain little or no feldspar; they are made up almost exclusively of mafic minerals. Their composition is similar to that of stony meteorites.

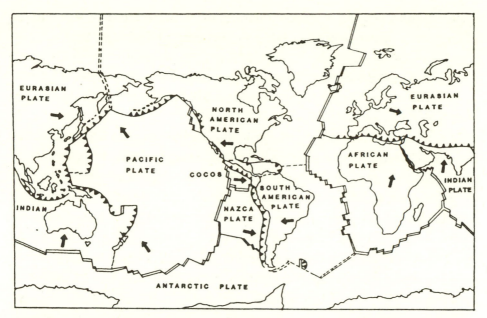

2.6 The major tectonic plates of the world. Three types of plate boundaries are shown: midoceanic ridges (double lines); transform-fault boundaries (single lines that offset the midoceanic ridges); and subduction zones (barbed patterns). The teeth point in the direction of subduction of the descending plate. In addition to the major plates, there are a number of smaller plates, such as the Juan de Fuca Plate (located adjacent to the north-western United States), that are active today. From LaBerge 1994.

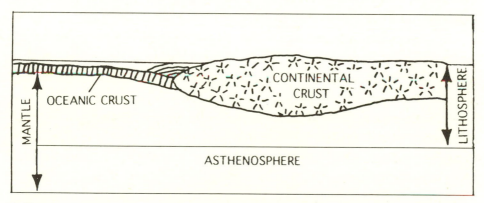

2.7 A generalized cross section of the Earth's crust and upper mantle. The oceanic crust is typically basaltic in composition and averages about 6 miles in thickness. The continental crust is much thicker (up to 25 miles in thickness) and is composed of rocks of granitic character. The lithosphere, or plate, consists of the crust and the upper part of the mantle. These rigid slabs of lithosphere rest on a plasticlike zone in the upper mantle known as the asthenosphere. Adapted from LaBerge 1994.

Mafic rocks are much less common in the continental crust than in the oceanic crust. The composition of the continental crust is described as felsic ("fel" = feldspar and "sic" = silicic). The continental crust contains large amounts of quartz and light-colored feldspar minerals, and they give the continental crust a composition that is significantly higher in aluminum and silicon than that of the oceanic crust.

Dating of rocks and sediments obtained from modern ocean floors demonstrates that today's ocean basins are relatively young, geologically speaking. The modern Atlantic started to form in Mesozoic time. Little, if any, of the modern ocean floor is older than 200 million years. Ocean floors older than Mesozoic age have apparently been destroyed at convergent plate boundaries. In these locations, oceanic plates are subducted, melted, and converted to igneous rocks. In some cases, portions of old ocean floors have been added to continents in structurally complex mountain belts such as the Appalachians, Alps, and Cordilleran systems.

The youngest parts of modern ocean basins are found along the crests of midocean ridges. For example, active volcanism is currently forming new basaltic ocean crust along the Mid-Atlantic Ridge at Iceland, the Azores, and elsewhere. The hypothesis that new ocean crust forms along midocean ridges and that old ocean crust is consumed in deep-sea trenches (subduction zones) was formulated in 1962 by Harry Hess, a noted American geologist. Although Hess's pronouncement did not immediately revolutionize geologic thinking, it was an important contribution to the modern theory of plate tectonics.

Today, geologists recognize three types of plate boundaries: plates that spread apart (diverge), plates that push together (converge), and plates that grind or slide past other plates (transform-fault motion). Figure 2.8 illustrates the three major types of plate boundaries.

At divergent plate boundaries, new crust forms when basalt is generated in the asthenosphere and rises toward the surface to fill the gap between widening plates. Most divergent plate boundaries are located within ocean basins, and the result is an outpouring (extrusion) of new basaltic lava on the ocean floor. The lavas cool on contact with seawater, and the resulting chemical reactions alter the composition of the basalts to some degree. In addition, characteristic features called pillows are formed. Pillow basalts, the product of submarine lava flows, are very common on modern ocean floors. They form when red-hot lava flows into a body of water and is cooled quickly. This cooling produces a rind around the margin of the lava flow. Pressure within the lava flow builds and produces cracks in the rind, and molten material is extruded through the cracks. Although contact with water again chills the outer portion of these new extrusions, the interior of the flow remains hot and molten. Continued extrusions enlarge the pillows. Gene

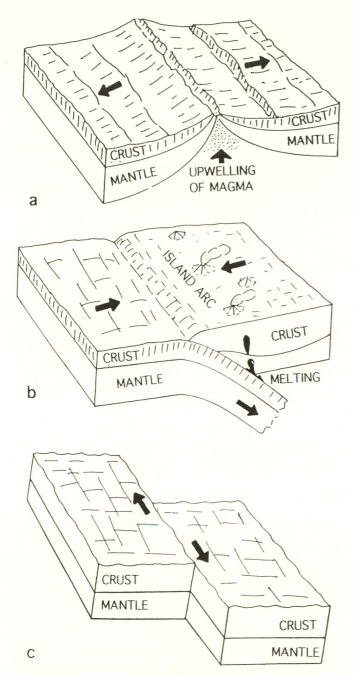

a

b

c

2.8 The three major types of plate boundaries. (a) A divergent plate boundary, where two plates pull apart. New oceanic crust forms at divergent plate margins, such as the present-day Mid-Atlantic Ridge. Divergent boundaries also form under continents, producing rift features such as found today in East Africa. (b) A convergent plate boundary, where two plates move toward each other. Oceanic crust is consumed at convergent plate boundaries in oceanic trenches and subduction zones, such as those located near Japan. (c) Plates also grind or slide past each other with lateral offset. The San Andreas Fault of California is a modern example of such a plate boundary. Adapted from LaBerge 1994.

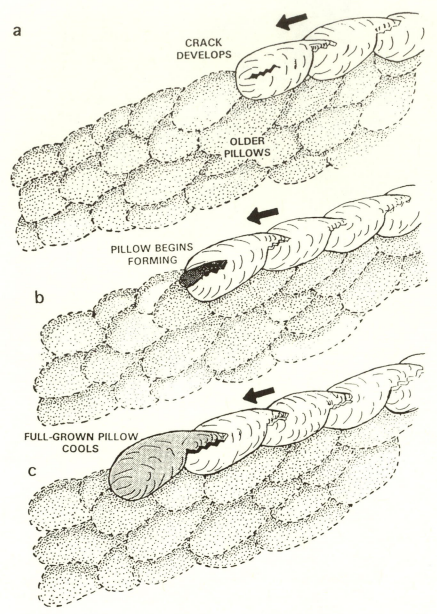

2.9 The process of forming pillow basalt. (a) A rind develops along the margin of a submarine lava flow, and a crack forms in the rind. (b) Molten material oozes from the crack, and a new pillow is formed as lava drains. (c) The pillow cools completely and is preserved in the rock record as pillow basalt. From LaBerge 1994.

LaBerge (1994) compares this process to blowing up a balloon. The amount of internal pressure within a lava flow probably determines the ultimate size that a pillow reaches. Figure 2.9 depicts the process of formation of pillow basalt.

Ancient pillow basalts are common in the Archean terranes of northern Minnesota and northern Michigan and provide evidence that these areas were once ocean floors. The ocean basalts in these ancient terranes have been metamorphosed and altered to a rock called greenstone. Greenstone terranes also consist of other deformed and metamorphosed igneous and sedimentary rocks.

The present-day Mid-Atlantic Ridge is a well-documented divergent plate boundary in an oceanic setting, but divergent plate boundaries can also form under continents. The rift-valley system in eastern Africa is located along a divergent plate boundary, and the Red Sea in northern Africa, where the Arabian Peninsula is being ripped from the continent by tectonic forces, is considered to be an ocean basin in the process of formation. Iowa's Precambrian rocks contain ample evidence of an ancient continental rift that nearly tore the ancestral North American continent apart.

Although new crust is still forming at divergent plate boundaries such as the Mid-Atlantic Ridge, geologists see no evidence that the planet is getting larger. The growth of new crust is compensated by the destruction of older crust at convergent plate boundaries. In places where two plates converge, the denser plate usually subducts (sinks), and as it descends to deeper levels it is heated and melted. The melt produced from the descending ocean plate and the sediments that cover it produces magma (a molten silicate melt). If this magma is extruded or erupted onto the surface of the Earth, explosive volcanic rocks are the usual result. Slow cooling of the magma at depth, however, produces masses of granite, or related igneous rocks, in the form of batholiths or plutons.

A present-day example of a convergent plate boundary is found along the west coast of South America where oceanic crust of the Nazca Plate is being subducted under the South American continent. Earthquakes, igneous activity, and mountain building characterize this boundary. In Mid-Cenozoic time, the mighty Himalaya Mountains formed by convergent processes. Through time, a former ocean basin located between the subcontinent of India and the Eurasian Plate was destroyed in a subduction zone. This resulted in two continental plates coming together at a convergent boundary. Because both the subcontinent of India and the Eurasian continent are made of low-density granitic crust, neither plate subducted. Instead, the result was a thickening of the continental crust, along with compression, uplift, and mountain building.

The recognition of extensive areas of igneous rocks, metamorphic rocks, and complex structural features in the Precambrian rocks of Iowa and neighboring states suggests that several episodes of mountain building occurred in the Midwest in the distant past. Therefore, it appears that the geologic history of this region was influenced significantly by processes associated with active plate boundaries. The junction where two continental terranes are welded together by plate-tectonic processes is called a suture. Several tectonic sutures are shown in figure 2.10.

The Midcontinent Rift System (1.0–1.1 billion years old) is the youngest of the Precambrian regions in Iowa. Regionally, the Midcontinent Rift cuts across five

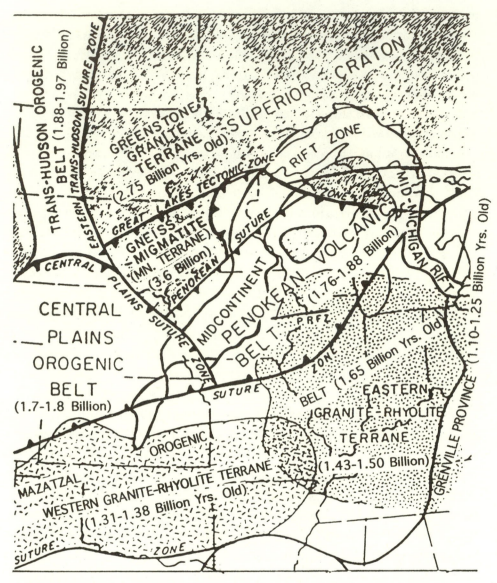

2.10 The Precambrian terranes of central North America. Barbed patterns represent tectonic sutures where continental terranes have been welded together. Adapted from Anderson 1990d.

older Precambrian terranes. Four of these older terranes are represented in Iowa's subsurface (fig. 2.10).

The oldest of these terranes is the Minnesota Terrane, at least 3.6 billion years old. Composed of highly deformed igneous and metamorphic rocks, this terrane underlies the northwestern corner of Iowa. Gneiss (a banded metamorphic rock) and migmatite (a complex rock composed of mixtures of igneous and metamorphic materials) are the main components of the Minnesota Terrane.

The Penokean Orogen (also known as the Penokean Volcanic Belt or Wisconsin Magmatic Zone) underlies much of Wisconsin and Iowa. This complex terrane represents a 1.76-to-1.88-billion-year-old mountain belt with abundant volcanic and plutonic rocks. In addition, sedimentary rocks are found that apparently formed on unstable seafloors near island arcs.

The Central Plains Orogenic Belt (1.70–1.80 billion years old) is found in the Precambrian of the southwestern corner of Iowa. The rocks of this belt, comparable in part with the age of the Penokean Orogen, suggest a complicated past involving igneous and metamorphic processes in an island arc setting.

Granites and rhyolites (a light-colored, fine-grained igneous rock) are key constituents of the Eastern Granite-Rhyolite Province (1.43–1.50 billion years old). This terrane underlies most of Illinois and the eastern part of Iowa (fig. 2.10).

In addition to convergent and divergent boundaries, plates also meet at transform boundaries. At transform boundaries, large blocks of the outer Earth grind past each other laterally. California has current examples of this type of junction, including the famous San Andreas Fault Zone. At transform boundaries, earthquakes are common, and faulting produces areas of crushed and sheared rock. These crushed and sheared zones are sometimes recognized in the ancient rock record.

Not all igneous activity takes place at plate boundaries. For example, the Hawaiian Islands are thought to be positioned over a hot spot, a small area of heating in the Earth's crust caused by a thermal plume rising from the mantle. The islands of this volcanic chain get progressively older toward the northwest, a pattern that is explainable if the ocean plate of which the Hawaiian Islands are a part moved northwestward over a fixed mantle plume through geologic time. The volcanism at Yellowstone National Park has also been explained by hot-spot, or mantle-plume, processes. Mantle-plume activity probably played an important role in the formation of some of the igneous rocks of Precambrian time as well. The Middle Proterozoic granites in Iowa's subsurface may have formed from magma sources within the continental crust that were not related to plate boundaries.

As we have seen, plate tectonics provides a broad framework to help explain the origin of Precambrian rocks and terranes. In addition, remote-sensing techniques have produced a variety of geophysical data that have proven to be helpful in understanding the buried Precambrian record of the state. A brief review of some of these geophysical techniques is in order before we examine the Precambrian rock record of Iowa.

Geophysical Measurements and Iowa's Precambrian Record

Geologists prefer to make their interpretations from direct examinations of the rock record. At times, however, it is necessary to rely on remote-sensing techniques, which enable scientists to measure the Earth's properties from a distance. These geophysical techniques involve the measurement of sound waves, electrical currents, radioactivity, and magnetic and gravitational fields. Such measurements have been useful in determining the physical properties of rock bodies in Iowa's subsurface. Of these techniques, magnetic and gravity surveys have proven to be the most helpful in deciphering the history of Iowa's Precambrian rocks.

Gravity surveying is used to measure the acceleration of the Earth's gravity. Gravitational attraction, or pull, varies slightly from one location to another because of lateral differences in density of underlying rock bodies. These density differences are related to the composition of the rock bodies, so gravity data can be used to map rock types in a general way. It is essential, however, to have some direct data, such as samples of rocks, in order to tie these indirect measurements to the actual rock record. In general, positive gravity anomalies indicate the presence of dense rock bodies, such as basalt or gabbro, and negative gravity anomalies signify the presence of lower-density materials, such as sandstone or shale. Geophysical measurements have been particularly useful in mapping and modeling the Midcontinent Rift System under central Iowa.

Midcontinent Rift System

The Midcontinent Rift System is a 1.0-to-1.1-billion-year-old tear in the continental crust that nearly ripped the early North American continent apart. This gigantic rift stretches from eastern Lake Superior country to beneath the plains of southern Kansas. Rocks produced during the formation of the Midcontinent Rift System are exposed in the Lake Superior area of Canada, Minnesota, Wisconsin, and northern Michigan. The rock record of the rift system is particularly well exposed in the Keweenaw Peninsula of Upper Michigan, and the name Keweenawan Supergroup is used in the classification of the igneous and sedimentary rocks that formed during the long history of the Midcontinent Rift System.

Rocks associated with the rift include extrusive basalts and intrusive rocks of related composition. In addition, a thick sequence of sandstones, siltstones, and shales is known. Because these clastic sedimentary rocks are reddish, they are sometimes referred to as red clastics.

In Iowa, the Precambrian rocks of the Midcontinent Rift are buried under 1,200 to 5,500 feet of Phanerozoic rocks (strata of Cambrian and younger ages), but their presence has been detected by sensitive gravity measurements. The dense basalts (2.9 grams per cubic centimeter) and the lower-density red clastics (2.0–2.65 grams per cubic centimeter) produce patterns of positive and negative gravity anomalies that trace the path of the Midcontinent Rift System under Iowa. This is one of the most pronounced gravity anomalies in North America. Because the Midcontinent Rift is deeply buried in Iowa, few wells have been drilled to its level. Much of our knowledge of the extent and characteristics of the rift system has come from interpretations of gravity, aeromagnetic, and seismic data.

The initial development of the Midcontinent Rift System probably began about 1.1 billion years ago when tensile stresses pulled the Earth's crust apart, allowing hot basaltic lava to ooze upward along fractures from the Earth's upper mantle. Large crustal blocks (grabens) subsided along the zone of rifting, and molten basalt moved along fractures and flowed into the grabens. In this manner, tens of thousands of cubic miles of basaltic melts were emplaced during the formation of the Midcontinent Rift System.

As volcanism waned and finally ceased, the grabens continued to sink and apparently dragged down adjacent areas, too. Surface drainage that was diverted into the descending rift system deposited a thick sequence of sand and mud. In Iowa, these deposits are known as the Lower Red Clastic Sequence. At various times during deposition of the Lower Red Clastic Sequence, lakes occupied the subsiding axial region of the rift system. Although these ancient lakes were probably somewhat similar to those of the modern-day East African Rift System, they contained less complex life forms. The lakes of the Precambrian rifts supported only bacteria and algae, the dominant types of life during the Earth's early history. The altered remains of this early life produced hydrocarbons within the rift sequence of northern Michigan and contributed carbonaceous residues to Iowa's Lower Red Clastic Sequence.

About 1 billion years ago, subsidence of the Midcontinent Rift halted abruptly when compressive forces pushed the graben of the central rift upward to form an uplifted block known as a horst. Sedimentary rocks of the Lower Red Clastic Sequence were eroded from the uplifted area and redeposited in adjacent alluvial basins. This episode of deposition produced the Upper Red Clastic Sequence of Iowa's Precambrian record.

A prominent feature of the Midcontinent Rift System under Iowa is a mammoth block of basalt that follows the axis of the rift across the entire state, from south-central Minnesota to southeastern Nebraska. This enormous block, the so-called Iowa Horst, measures from 20 to 40 miles in width and was faulted upward more than 30,000 feet during the history of the rift formation. Over time, most of the sedimentary cover was stripped from this uplifted block of dense basaltic rock. Today, the location of this feature in Iowa's subsurface leaves a signature in the form of a large positive gravity anomaly. The Iowa Horst is bordered on the east and west by basins that are filled with thick deposits of sandstone, siltstone, and shale. These basins cover an area of approximately 150,000 square miles in Iowa's subsurface and contain over 35,000 cubic miles of red clastic sedimentary rocks that produce a pattern of negative gravity anomalies.

The Precambrian Record in Iowa's Deepest Well

One of the thickest sections of Precambrian sedimentary rocks anywhere along the Midcontinent Rift System is found under Carroll County, Iowa. This stratigraphic record was sampled in 1987 when the Pan American Petroleum Company, a subsidiary of Amoco Production Company, completed drilling a well on the farm of M. G. Eischeid near Halbur. After 208 days of drilling, the Eischeid well reached a depth of 17,851 feet and was by far the deepest well ever drilled in the state. Previously, the deepest well in Iowa was to a depth of 5,305 feet. This was only the second well to explore the petroleum potential of the Midcontinent Rift System; Texaco completed a "dry hole" in the rift system under Kansas in 1985.

Amoco spent nearly $20 million on the exploration project, starting with an initial investment of approximately $10 million to acquire and process a variety of seismic and other geophysical data. After reviewing the data, Amoco leased over 800,000 acres of Iowa farmland at a cost of nearly $5 million. The actual drilling costs in Carroll County were estimated to be $4.8 million. Although Amoco's Eischeid well was plugged and abandoned shortly after its completion, a wealth of scientific information was obtained from this exploration effort.

Perhaps the most remarkable thing about the Eischeid well is its depth. Not only is the Eischeid well the deepest well in Iowa, it is one of the deepest wells in the entire midcontinent region. The Eischeid well encountered 2,802 feet of Phanerozoic sedimentary rocks and sediments overlying the Precambrian record (fig. 2.11). Within the Precambrian, 14,898 feet of clastic rocks of the Keweenawan Supergroup were discovered. Included within this interval are 813 feet of clastic rocks, designated as Unit H, that initially were placed above the Keweenawan Supergroup and below the Cambrian-age Mt. Simon Formation. The sequence of Keweenawan clastics in the Eischeid well is the thickest ever sampled along the Midcontinent Rift. These rocks rest on a mafic (gabbro) intrusive.

Brian Witzke (1990a) divided the rocks of the Keweenawan Supergroup into the Upper Red Clastic Sequence and the Lower Red Clastic Sequence. The Upper Sequence (including Unit H), encountered at well depths from 2,802 to 10,510 feet, consists of sandstones, siltstones, and shales. These rocks are interpreted as ancient fluvial deposits. The term "fluvial," as used to describe the environment of deposition here, includes a variety of settings associated with streams and floodplains, including channels, ponds and lakes, and overbank sedimentation on floodplains during times of flooding.

The Lower Sequence, found between depths of 10,510 and 17,700 feet, is also primarily of fluvial origin. Sandstones, siltstones, and shales are the common rocks of the Lower Sequence. Some of the deposits of the Lower Sequence, however, differ from those of the Upper Sequence. Of particular interest are black shales, which may have served as source beds for the generation of petroleum (fig. 2.11). Traces of gases (methane and ethane) were detected in certain intervals

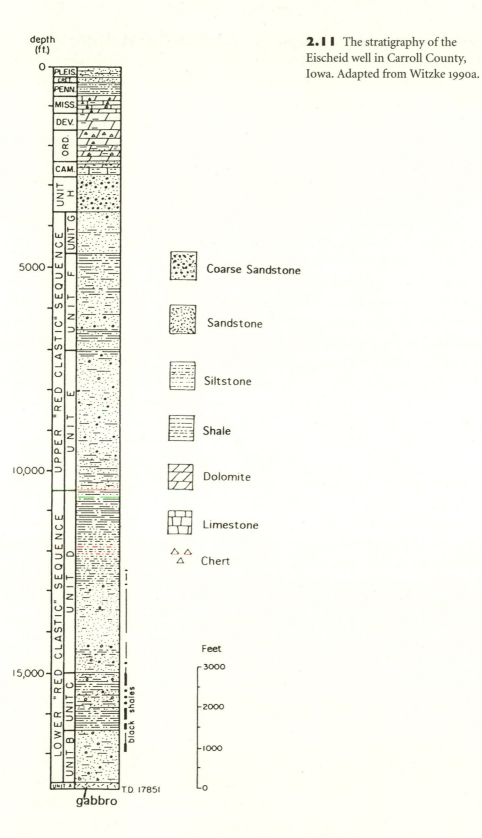

2.11 The stratigraphy of the Eischeid well in Carroll County, Iowa. Adapted from Witzke 1990a.

depth (ft.)

Coarse Sandstone

Sandstone

Siltstone

Shale

Dolomite

Limestone

Chert

Feet

in the Lower Sequence, and black residues were reported. These black residues may represent relict hydrocarbons.

An organic-rich interval within the Lower Sequence, referred to as Unit C, appears to have been deposited in standing water. Could these deposits be the remains of an ancient lake that occupied a portion of the Midcontinent Rift? Several geologists believe that is the case. The black coloration in Unit C is primarily the product of disseminated organic carbon. Unit C is similar in age and appearance to the Nonesuch Formation of the Lake Superior area, a formation that drips oil in a copper mine in northern Michigan.

What about the petroleum potential of the Precambrian rocks encountered in the Eischeid well? Oil shows were not reported in any of the rocks, but trace amounts of methane and ethane were detected in three rock units within the Lower Red Clastic Sequence. In addition, black residues, suggestive of past oil movement, were noted at three levels in the Lower Red Clastic Sequence. It appears that petroleum may have been generated in rocks of the Lower Red Clastic Sequence during Late Precambrian time, but it probably migrated elsewhere. Findings from the Eischeid well have not ruled out the possibility of finding oil in the Midcontinent Rift System in the future. Oil exploration and deep drilling, however, are very expensive ventures. Consequently, exploration companies are reluctant to pursue additional deep drilling in Iowa.

The basal rock body encountered in the Eischeid well is a mafic igneous intrusive of gabbroic composition. Gabbro is similar to basalt in composition, but it contains larger crystals, which suggests that it cooled below the surface of the Earth. Dated at 1.28 billion years old, the gabbro is considerably older than igneous rocks associated with the Keweenawan Supergroup (0.9–1.1 billion years old). The gabbro encountered in the Eischeid well is similar in age to the mafic intrusives of the Mackenzie Dike Swarm of Canada (fig. 2.4).

As has been noted, the Eischeid drill hole revealed nearly 3 miles of Iowa's Precambrian record, adding much new knowledge about the state's geologic history. Although not as deep, other drill holes into Iowa's Precambrian have been equally revealing. The Quimby drill hole near Cherokee recovered 1.43-billion-year-old granite, and a series of drill holes near Matlock sampled a variety of Precambrian rocks, including 2.9-billion-year-old rocks of the Otter Creek Layered Mafic Complex. A brief discussion of these significant Precambrian sites follows.

The Quimby Drill Hole

The Quimby drill hole was selected and drilled for scientific information rather than for information on potential economic mineral wealth. Several positive magnetic anomalies were mapped in northwestern Iowa based on data from aeromagnetic surveys. These anomalies played an important role in the selection

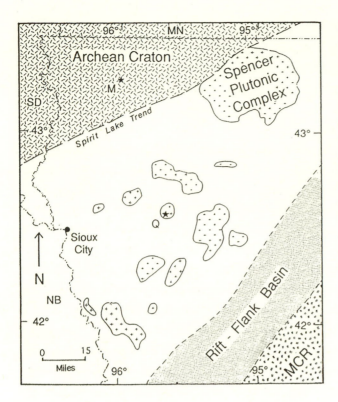

2.12 The location and pattern of anomalies (+ pattern) near Quimby (Q), Iowa. Granite, 1.43 billion years in age, was sampled from a drillhole (★) near Quimby. The other areas shown by the + pattern, including the Spencer Plutonic Complex, are inferred to represent Middle Proterozoic granites, too. The Archean Craton represents Iowa's oldest bedrock and includes 2.9-billion-year-old rocks recovered from drillholes (★) near Matlock (M). The unshaded area between the Archean Craton and the Rift-Flank Basin of the Midcontinent Rift (MCR) represents Early Proterozoic crustal rocks. From Van Schmus et al. 1989.

of a site for the drill hole, and prior to drilling Raymond R. Anderson modeled the magnetic anomalies based on magnetic properties of Wisconsin granites. The granitic rocks in Wisconsin were thought to be similar in age and physical properties to the Precambrian basement rocks in the subsurface under northwestern Iowa. Anderson's modeling suggested that granites were present beneath Quimby and that they could be reached by drilling.

Drilling centered on a positive magnetic anomaly located near Quimby in Cherokee County (fig. 2.12). It was hypothesized that the positive magnetic anomaly pattern was produced by magnetite-bearing granites in the subsurface. Magnetite, a magnetic iron-oxide mineral, is an important source of iron. Drilling confirmed the hypothesis when magnetite-bearing granite was found at a depth of approximately 2,000 feet. Nearly 330 feet of granitic core was recovered, and dating of a granite sample from 92 feet below the top of the Precambrian surface yielded an age of 1.43 billion years old.

The granite appears to be part of a continent-wide episode of granite and rhyolite formation. The intrusions were probably derived from the melting of preexisting continental crust; they are not associated with compressional mountain building. Granitic bodies such as these have been termed anorogenic (post-mountain building).

Detailed geochemical analyses were completed on a sample of the Quimby core taken 312 feet below the top of the Precambrian surface. The results of these studies suggest that the lower basement granites under Quimby were derived from Early Proterozoic crust at the time of the Penokean Orogeny, some 1.86 billion years ago.

The Matlock Cores

In 1963, the New Jersey Zinc Company conducted a core drilling program in search of iron ore in northwestern Iowa. Drilling centered around a strong magnetic anomaly in Lyon and Sioux counties, near the town of Matlock. The Matlock anomaly is one of several strongly positive magnetic anomalies that form a linear trend across northeastern Nebraska, southeastern South Dakota, northwestern Iowa, and southwestern Minnesota (fig. 2.13). The trace of the anomalies in Iowa lies just north of the Spirit Lake Trend, which marks the boundary between ancient Archean rocks and younger Proterozoic rocks. Three papers (Van Schmus and Wallin 1991; Windom et al. 1991a, 1991b) detailed the findings from the Matlock drill cores. Key points from these sources follow.

Rocks from a dark-colored igneous rock body, called the Otter Creek Layered Mafic Complex, were found in ten of the twelve holes drilled. This rock body is composed of layers of ultramafic and mafic igneous rocks; it may represent a large sill (tabular-shaped igneous body emplaced parallel to existing rocks). Analyses from the Otter Creek Layered Mafic Complex yielded an age of 2.9 billion years old. To date, this is the oldest rock unit documented in the state.

Samples of iron formation and dark-colored dike rock (lamprophyre) were also recovered from one of the Matlock cores. Both the iron formation and the lamprophyre display evidence of having been metamorphosed at high temperatures. Initially, the iron formation was interpreted as a block that had been engulfed by the magma that gave rise to the Otter Creek Layered Mafic Complex. According to this interpretation, the iron formation is older than the Otter Creek complex and, therefore, is Iowa's oldest rock unit. Raymond R. Anderson (personal communication, 1997) has revised this interpretation, and the Matlock Banded-Iron Formation is currently assigned (fig. 2.4) to the Lower Proterozoic (1.6–2.5 billion years old).

Iron formations are composed of both iron-oxide and silicon-dioxide minerals; these deposits are well known from the Precambrian of Minnesota, where they are mined for iron ore. Iron formations probably represent ancient marine deposits. Although iron formations are common in the Precambrian record, they are essentially unknown elsewhere in the geologic column. Apparently, conditions favorable for the deposition of iron formations were fairly common in Precambrian time but were quite rare later in geologic time. Bacteria may have played a role in the deposition of iron-oxide minerals found in iron formations, and volcanism probably contributed silica to the sites of deposition. Economic

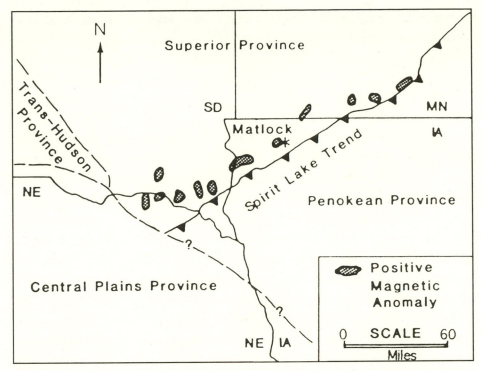

2.13 The location and pattern of anomalies near Matlock, Iowa. The Otter Creek Layered Mafic Complex, sampled from drillholes (★) near Matlock, is Iowa's oldest known bedrock, 2.9 billion years old. From Windom et al. 1991a.

factors prompted the drilling at Matlock, but the small quantity of iron formations recovered did not justify further exploration.

The Matlock cores provided additional insight into Iowa's Precambrian history when pyroclastic rocks were recovered. Pyroclastic rocks are formed from ash and other ejecta from volcanoes. Originally logged as rhyolite, these pyroclastic rocks are now interpreted as flow breccias (rocks composed of angular fragments) and tuffs (consolidated ash deposits). The pyroclastic rocks lie unconformably on the older Otter Creek Layered Mafic Complex, and chunks of Otter Creek mafic rocks are found incorporated in the younger pyroclastics.

Chemical analyses show that the pyroclastic rocks are depleted in some elements but are relatively rich in sodium. The proper name for such an igneous rock is keratophyre. Radiometric dating of the Matlock Keratophyre gave an age of 1.76 billion years old, similar to the age of felsic igneous rocks in Wisconsin.

The Matlock drilling confirms earlier evidence from a well at Hull that keratophyres underlie the Sioux Quartzite in northwestern Iowa. In fact, the Sioux Quartzite lies directly on keratophyre in the Matlock cores. In the Matlock area, the Sioux Quartzite contains a basal conglomerate with rounded clasts of banded iron formation and what appear to be clasts of keratophyre and clasts of layered mafic rock. Because the beds of the Sioux Quartzite were deposited on the keratophyre, we know that the base of the quartzite is younger than the keratophyre, or younger than 1.76 billion years old.

Sioux Quartzite

In 1870, Charles White applied the name Sioux Quartzite to beds of red and pink quartzite exposed along the banks of the Big Sioux River in extreme northwestern Iowa. This area, which represents the designated type area for the Sioux Quartzite, was dedicated as a state geological, natural, and archaeological preserve in 1969. Its name, Gitchie Manitou State Preserve, comes from the Sioux name for Great Spirit. The formation is also exposed on private property a few miles east of the preserve. Although the Gitchie Manitou area was overridden by at least half a dozen continental glaciers, no evidence of glacial grooves or striations remain. Instead, the highly polished surface of the quartzite at Gitchie Manitou State Preserve probably reflects the work of wind-driven clay, silt, and sand.

More extensive outcrops of the Sioux Quartzite occur in adjacent South Dakota and Minnesota, and a small area of exposure occurs in northeastern Nebraska (fig. 2.14). The Sioux Quartzite is still quarried extensively near Sioux Falls, South Dakota. At one time, the formation was locally called the Sioux Falls granite from exposures in the vicinity of Sioux Falls. An unincorporated village in Lyon County in northwestern Iowa still bears the name Granite. These were misnomers, however, as the Sioux Quartzite is composed almost exclusively of tightly cemented grains of quartz. Quartzite is not at all related to granite — an igneous rock that cooled from a melt and has interlocking crystals of quartz, feldspar, and other silicate minerals.

Although the area of the Sioux Quartzite in South Dakota, Minnesota, Nebraska, and Iowa has been called the Sioux Uplift, there is no structural or stratigraphic evidence to support this usage. Rather, geological observations suggest that this area of durable Sioux Quartzite formed a positive topographic feature because it is more resistant to erosion than the surrounding bedrock. A better name for this topographic prominence is Sioux Ridge.

The Sioux Quartzite in Gitchie Manitou State Preserve consists predominantly of fine- to medium-grained quartz sand grains that have been tightly cemented by silicon dioxide. The cement has filled the spaces between the grains so thoroughly that the quartzite has little or no porosity. In addition, the cement has grown around and over the sand grains in a process called secondary growth. As a result of this process, the sand grains and the silicon dioxide cement are now so tightly bound to each other that the rock typically breaks across the grains rather than around them.

In some areas outside of Gitchie Manitou State Preserve, the Sioux Quartzite is poorly cemented. Rocks of the formation in these areas are somewhat friable —

2.14 The distribution of the Sioux Quartzite and related Early Proterozoic quartzites of the Baraboo interval. From Anderson 1987b.

2.15 Native Americans in a quarry in the Sioux Quartzite near Pipestone, Minnesota. This photo taken by Samuel Calvin in the early 1900s captures the open, treeless expanse that early French explorers called the Coteau de Prairies. Courtesy of the Calvin Photographic Collection, Department of Geology, University of Iowa.

the rocks are crumbly, and individual grains can be broken from the cementing agent rather easily.

Conglomerate and mudstone are found in the Sioux Quartzite in Minnesota. Red mudstones from the Sioux Quartzite in southwestern Minnesota have been used as sources of pipestone by Native Americans for centuries (fig. 2.15). Pipe-

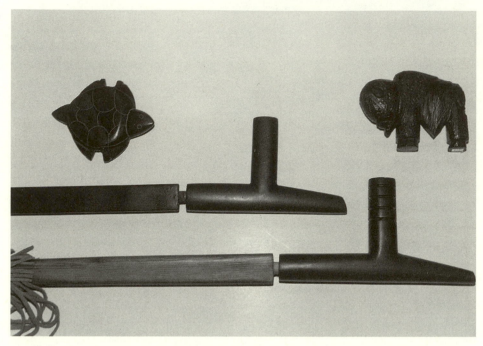

2.16 Ceremonial pipes and figurines such as these are carved from red mudstone (pipestone) that occurs within the Sioux Quartzite at Pipestone National Monument, Minnesota. Photo by the author.

stone National Monument at Pipestone, Minnesota, located 35 miles north of the Iowa border, gives recognition to this use of rocks from the Sioux Quartzite.

The town of Pipestone took its name from the place where Native Americans quarried a soft rock (pipestone) for hundreds of years to make figurines and ceremonial pipes (fig. 2.16). Pipestone is sometimes referred to as catlinite, a name that was coined to honor George Catlin (1796–1872), an American painter of Native Americans. Catlinite was and is considered to be a spiritual item by members of the Sioux Nation and other Plains tribes. It was a highly prized trade good, and Native American groups have long attached spiritual significance to the pipestone region.

According to Murray (1968), the catlinite pipes made by the Dakota (Sioux) were famous throughout the Great Plains and west to the Columbia River Basin. On the western edge of the Great Plains, a well-made catlinite pipe was worth a horse in trade. The Dakota tell of the gift of a sacred pipe from a supernatural power early in their history. To this day, one tribe of the Dakota maintains and preserves, under close security, the presumed original pipe.

Geologists once considered the mudstones at Pipestone to be low-grade metamorphic rocks because of the reported presence of a talclike mineral called pyrophyllite. Later studies have failed to confirm the presence of minerals of un-equivocal metamorphic origin, however.

The Sioux Quartzite varies from pink to reddish in exposures in Iowa. In adjacent South Dakota and Minnesota, the quartzite varies from light gray to orange, red, purple, and dark gray. The color of the quartzite is produced by a thin coating of iron oxide around the sand grains; differences in color are caused by slight variations in the composition and abundance on the iron-oxide coating. The iron-oxide coating constitutes only a small amount of the rocks — less than 1 percent.

Sedimentary structures such as cross-bedding, ripple marks, and mud cracks occur in the Sioux Quartzite. These structures, along with information based on composition, texture, and regional distribution, have been used to interpret the environment of deposition for the formation. There are no fossils in the Sioux Quartzite to help determine its origin.

The formation may be as thick as 7,800 feet, and the quartzite deposits appear to be widespread laterally. The sand and mud that now make up the beds of the Sioux Quartzite were probably derived from a landmass somewhere north of the present outcrop area of the formation. The Sioux Quartzite has been correlated with formations in Wisconsin known as the Baraboo, Barron, Flambeau, Waterloo, and Rib Mountain quartzites. Several of these Late Precambrian quartzites have been grouped together as the Baraboo Interval quartzites (fig. 2.14). These quartzites were laid down along the stable margin of an old continental nucleus about 1.64 to 1.76 billion years ago. Quartzites belonging to this interval are known in the subsurface under Cedar Rapids and Keota in eastern Iowa. The Baraboo Quartzite has experienced structural deformation and some metamorphism, whereas the Sioux Quartzite is relatively undeformed and unmetamorphosed.

The absolute or radiometric dating that gave the 1.64-to-1.76-billion-year-old age for the Baraboo Interval quartzites came from samples in Wisconsin. Igneous rocks underlying the Sioux Quartzite in Iowa have also yielded radiometric dates.

Traditionally, the Sioux Quartzite had been interpreted as a shallow marine deposit, laid down close to the shoreline (Van Schmus 1979; Van Schmus et al. 1975). However, in 1984, geologists at the University of Minnesota, Duluth, proposed that the quartzite beds formed in a mixed environmental setting, with the lower portion of the formation reflecting stream deposition and the upper part of the unit indicating deposition in either tidal estuaries or on a shallow marine shelf (Ojakangas and Weber 1984).

Interesting bedding features with alternately reversed directions of inclination (herringbone cross-bedding) are present in the Sioux Quartzite in Minnesota and South Dakota. Depositional features of this nature are excellent indicators of tidal currents and are formed in shoreline settings where the direction of transport of bottom sediments is reversed when the tides change direction.

In 1986, geologists at the Minnesota Geological Survey (Southwick et al. 1986) interpreted the environment of deposition of the Sioux Quartzite to be entirely fluvial (stream deposited) and proposed that the formation was laid down by braided streams that were confined to fault-bounded basins. Braided rivers are characterized by major channels that split into numerous smaller interlacing channels that flow around sand bars and islands. In most cases, a braided steam has more sediment than the stream can transport. Southwick et al. (1986) observed that the particle-size distribution became finer (smaller in size) upward in the vertical sequence of quartzite deposits. This observation suggested to them that stream gradients (slopes) decreased and that stream flow became more sluggish during deposition of the Sioux Quartzite.

Although these differing interpretations for the environment of deposition of the Sioux Quartzite appear to be in conflict, this quartzite interval may actually reflect regional environmental settings that grade laterally from fluvial, through estuarine and tidal, to shallow marine. Regardless of its origin, the Sioux Quartzite is a significant rock unit in Iowa because it is the oldest exposed bedrock in the state.

The Sioux Quartzite rests unconformably on older Proterozoic and Archean rocks. In Africa and Canada, Early Proterozoic quartzites that rest unconformably on deeply eroded igneous and metamorphic terranes contain economic deposits of uranium and gold. To date, no commercial concentrations of either of these mineral resources have been discovered in the Sioux Quartzite in Iowa and surrounding states.

The Sioux Quartzite has served as a source for building stone, road aggregate, and railroad ballast. A number of buildings in Iowa, South Dakota, and Minnesota have been constructed from this tough and resistant material. Crushed quartzite from the Sioux Formation, used in asphalt pavement in southeastern South Dakota and southwestern Minnesota, gives the roads a pink or reddish appearance.

Small lumps of Sioux Quartzite were found with the remains of an ancient seagoing reptile (mosasaur) in the chalk beds of western Kansas. This occurrence suggests that a Mesozoic reptile gulped down quartzite pebbles on the shallow seafloor adjacent to bedrock exposures of the Sioux Quartzite near northwestern Iowa, southeastern South Dakota, southwestern Minnesota, or northeastern Nebraska. The presence of the quartzite gizzard stones in western Kansas, 400 miles from their probable source, gives some indication of the distance that mosasaurs ranged.

The Precambrian at Manson

In a 12-square-mile area along the border of Pocahontas and Calhoun counties near Manson in north-central Iowa, the Precambrian basement is elevated

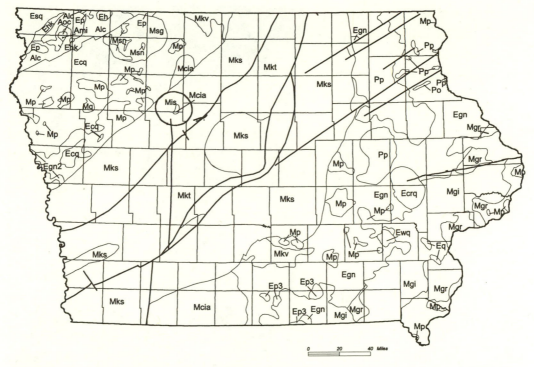

2.17 Iowa's principal Precambrian rock units. Legend: Mis, Manson Impact Structure; Mks, Keweenawan Clastic Sedimentary Rocks; Mkt, Keweenawan Thor Volcanic Group; Mkv, Keweenawan Volcanic Rocks; Mp, Eastern Granite-Rhyolite Terrane Granitic Plutons; Mq, Quimby Granite; Mgi, Green Island Granitic Plutons; Mcia, Central Iowa Arch Granites; Msn, Spencer Norite; Msg, Spencer Granitic Pluton; Mgr, Dominantly Rhyolite with Granitic Plutons; Ep3, Mazatzal Interval Granitic Plutons; Eq, Baraboo Interval Quartzite; Esq, Sioux Quartzite; Ecrq, Cedar Rapids Quartzite; Ewq, Washington County Quartzite; Egn2, Central Plains Interval Gneiss and Granite Terrane; Ehk, Hull Keratophyre; Ep, Penokean Interval Post-Orogenic Granitic Plutons; Eh, Harris Granite; Egn, Penokean Interval Orogenic Gneiss and Granite Terrane; Ecq, Camp Quest Gneiss; Alc, Lyon County Gneiss; Aoc, Otter Creek Layered Mafic Complex; Ami, Matlock Banded Iron Formation; Pp, Northeast Iowa Plutonic Complex; Po, Osborne Mafic Complex. Information provided by Geological Survey Bureau, Iowa City, 1998.

and occurs directly beneath glacial deposits at a depth of approximately 100 feet. This area of uplifted and deformed rocks is represented in figure 2.17 by a nearly circular pattern.

Broken and crushed rocks have been recovered from several wells that penetrate the Precambrian near Manson. The Paleozoic and Mesozoic rocks are also deformed. This area of unusual geology was originally termed the Manson Disturbed Area by geologists. Today, the area is accepted as an impact structure.

Impacts by meteorites and other extraterrestrial bodies typically deform the rocks and minerals of the impact area and produce a variety of shock-formed features. Some of these characteristic features have been documented from rocks recovered from drill holes at Manson. Chief among these is shocked

2.18 Shocked quartz with planar deformation features (PDFs) from the Crystalline Clast Breccia in the subsurface at Manson, Iowa. The shocked quartz grain shown here was broken free from Precambrian granitic source rock. Photographed through crossed polarizers. Width of field of view is .025 inch. Courtesy of Raymond R. Anderson, Geological Survey Bureau, Iowa City.

quartz grains with multiple sets of planar deformation features (PDFs) (fig. 2.18). Many of the rocks at Manson also display distinctive broken or brecciated textures (fig. 2.19).

Seismic refraction investigations, detailed gravity surveys, and aeromagnetic surveys have been conducted in the Manson area. These studies have helped delineate the configuration of the Manson Impact Structure. The Manson crater is twenty times larger in diameter than the well-known Meteor Crater located between Flagstaff and Winslow, Arizona. Indeed, the Manson crater is currently ranked as the third largest impact crater in the continental United States. A larger crater in Montana, ranked as number one, is sliced and deformed by mountain-building processes. The crater ranked as number two is offshore, under Chesapeake Bay. So, it can truthfully be proclaimed that the Manson crater is the largest intact, onshore meteorite crater in the continental United States.

Dating of the Manson Impact Structure gives an age of 73.8 million years old. This places the impact in Late Cretaceous time, before the demise of the last of the dinosaurs. I will return to the fascinating story of the Manson Impact Structure when Iowa's Mesozoic history is discussed in chapter 9.

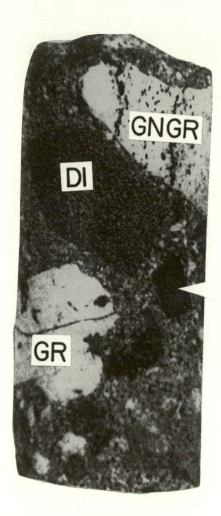

2.19 Brecciated Precambrian rocks from a core at Manson, Iowa. The core, 1.5 inches in diameter, contains fragments of gneissoid granite (GNGR), granite (GR), and diabase (DI) in a matrix of dark-colored microbreccia. From Hoppin and Dryden 1958.

A variety of Precambrian rocks have been recovered from drilling projects in the Manson area. The oldest is garnet-bearing gneiss, part of an island arc sequence accreted to the early North American craton about 1.85 billion years ago. The gneiss was intruded by granite approximately 1.43 billion years ago. Both of these older rock units have been cut by dark-colored diabase dikes, part of the Keweenawan rifting events about 1.0 billion years ago. Precambrian sedimentary rocks belonging to the red clastic sequences are also known in the Manson area.

The Precambrian rocks of the Manson area are an excellent source for groundwater. This is unusual because Precambrian rocks typically are dense and lack pore space to store groundwater. The broken and fractured Precambrian rocks of the Manson area, however, provide lots of storage for groundwater. In addition, water from the Precambrian wells is much softer than water from wells in the adjacent sedimentary rocks, which usually contain lime. The city wells at Manson produce very soft water — comparable to rain water, according to local residents.

Additional Investigations
of Iowa's Precambrian Geology

Granite has been recovered from wells that reach the Precambrian in the eastern part of the state in Allamakee, Dubuque, and Clinton counties; in southwestern Iowa in Taylor County; and in northwestern Iowa in Kossuth and Ida counties. Dark-colored igneous rocks such as basalt, diabase, and gabbro have been sampled from wells in central Iowa in Cerro Gordo, Webster, Boone, and Dallas counties. Near Spencer in Clay County, gabbro and the iron-bearing mineral magnetite are known, and gabbro with iron-bearing and titanium-bearing minerals has been documented in northeastern Iowa in Clayton County.

To date, fewer than a hundred drill holes have probed Iowa's Precambrian rock record. Additional drilling will undoubtedly provide more information about the state's Precambrian history. The Precambrian geology of Iowa map (fig. 2.17) is based on drill-hole information and inferences drawn from geophysical data. Twenty-seven units are shown on the map, indicating something of the variety and complexity recorded in Iowa's Precambrian bedrock.

Summary

Iowa's long and varied Precambrian history begins with emplacement of complex igneous and metamorphic rocks nearly 3 billion years ago, followed by mountain-building events that produced the Penokean Orogen, the Central Plains Orogenic Belt, and the Eastern Granite-Rhyolite Province. Iowa's oldest exposed bedrock is the Sioux Quartzite, visible in the northwestern corner of the state. The Sioux Quartzite and other related quartzites were laid down along the stable margin of an old continental nucleus about 1.64 to 1.76 billion years ago. Approximately 1 billion years ago, the early North American continent was nearly ripped apart. Iowa's rock record contains evidence of this event in the form of extrusive basalts, mafic intrusives, and terrestrial clastics of the Midcontinent Rift System. The Precambrian rocks under the Manson area in Pocahontas and Calhoun counties are brecciated and deformed as a result of a Late Cretaceous impact by an extraterrestrial object. The impact produced the Manson crater, which ranks as the third largest impact crater in the continental United States.

Acknowledgments

Raymond R. Anderson, Iowa Department of Natural Resources, Geological Survey Bureau, is recognized as the expert on Iowa's Precambrian rock record. Ray completed his doctoral dissertation on the Precambrian of Iowa, and he continues to study this fascinating and puzzling interval of geologic history. I acknowledge his many contributions to this chapter, and several of his scientific papers are referenced at the end of this book. In addition, investigations by

Karl Seifert and Kenneth Windom of Iowa State University contributed to my understanding of the Precambrian geology of northwestern Iowa. Robert M. McKay, Greg A. Ludvigson, and Brian J. Witzke interpreted the significant Precambrian sedimentary record of the Eischeid well. Gene L. LaBerge's work provided helpful background.

> As to lithological characters the portion of the Saint Croix (Cambrian)
> Sandstone exposed in Iowa is very variable. . . . With few exceptions the
> several beds of this formation (Cambrian System) vary in character laterally,
> so much so, that sections taken quite near together would differ greatly in
> minor details. This whole complex mass of arenaceous strata throughout
> its entire thickness . . . is simply a shore deposit laid down in shallow
> water upon a subsiding sea bottom.
>
> Samuel Calvin (1894) discussing Cambrian strata near Lansing, Iowa

By the beginning of Paleozoic time, approximately 545 million years ago, Iowa
and the rest of the North American craton were part of a low landmass undergoing weathering and erosion. Weathering and erosion continued in the area
into Middle and Late Cambrian time, some 530 million years ago.

This old landmass consisted of a variety of Precambrian rocks, including granite, gneiss, quartzite, rhyolite, tuff, basalt, diabase, gabbro, sandstone, and shale.
Weathering altered most of the original silicate minerals in the bedrock, con-
verting them into a variety of clay minerals. Quartz, a common mineral in much
of the original bedrock, did not undergo any significant chemical change. Since
quartz is very resistant to chemical weathering, it persisted in its original chemical form, silicon dioxide. A few minor constituents, such as durable zircon, tourmaline, and garnet minerals, also survived the long episode of chemical weathering. Some feldspar endured as well.

Thus, the weathering products of the old Precambrian landmass were primarily grains of quartz and feldspar, clay minerals, and minor amounts of stable
silicate minerals. These constituents provided most of the ingredients for Iowa's
Cambrian record. Quartz became the rounded and sorted quartz sands of sandstones (fig. 3.1). Clay minerals became important components of argillaceous
(clay-bearing) siltstones, sandstones, and carbonate rocks.

At the beginning of Cambrian time, the central portion of the North American continent was a low-lying landmass. Seas existed on the continental margins where the Appalachian and Rocky mountains stand today (fig. 3.2). During Cambrian time, these marginal seas received a thick accumulation of marine
sedimentary rocks. As the interior of the continent slowly sank, the seas transgressed over the craton. By Middle to Late Cambrian time, approximately 530 million years ago, the seas reached Iowa. Iowa's Cambrian rock record accumulated
in such an inland, or epeiric, sea. These deposits are part of the Sauk Sequence,

3.1 An exposure of quartz sandstones of the Jordan Formation, Allamakee County, Iowa. Photo by the author.

the oldest of six continent-wide packages of shallow-water marine strata formed by worldwide sea-level fluctuations. Each of the six sequences represents a major transgressive-regressive (T-R) cycle. Figure 3.3 shows the subsurface thickness and the outcrop belt of the Sauk Sequence of rocks.

Iowa's Cambrian record lies unconformably on Precambrian rocks of various ages. The gap between the formation of Iowa's youngest Precambrian rocks and

3.2 An interpretation of Late Cambrian paleogeography. Iowa's Cambrian rock record accumulated on the sandy floor of a vast inland sea. Note the inferred position of the equator (EQ) during Cambrian time. Iowa's Cambrian rocks formed south of the equator in a subtropical setting. Since then, the North American plate has rotated and drifted northward to its present position. Shaded areas represent land, and unshaded areas depict seas. Note also the presence of the Transcontinental Arch and a large area of land in the Canadian Shield. Symbols: fe = oolitic ironstones, o = carbonate ooids (oolites) and coated grains, m = evaporite crystal molds, s = sulfate evaporites (primarily gypsum and anhydrite). From Witzke 1990b.

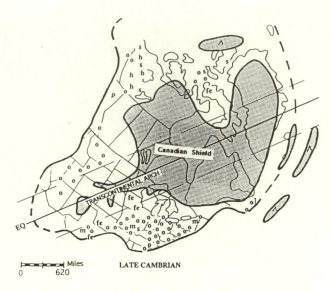

LATE CAMBRIAN

the start of deposition of the Cambrian is at least 370 million years. This represents an immense hiatus for which the state has no record.

West of Iowa (in terms of present geography), in what is now the Great Plains region, Precambrian rocks projected above the Cambrian seas as a large island system, a surface reflection of the Transcontinental Arch (fig. 3.2). The Transcontinental Arch, composed of Precambrian basement rock, supplied sand and other sediment to the shallow seas that covered the interior of the continent. Although the arch influenced sedimentation to some extent during Cambrian time, much of the warping and uplift of the structure probably took place after Cambrian time. The presence of several major unconformities on and adjacent to the Transcontinental Arch suggests a long history of periodic flooding by shallow seas, followed by episodes of uplift and erosion.

Studies of the Cambrian in Wisconsin show that the advance of the sea there was cyclic. Overall, the pattern of advance was interrupted by several intervals of partial regression. The rate of transgression of the sea was only about 10 miles per million years, or about half a foot per year.

In general, we expect that a belt of sand will form near the shoreline of an advancing sea and that finer particles like silt and clay will be swept farther from the shore zone by marine currents and deposited offshore as marine mud. Carbonate rocks (limestone or dolomite) may form seaward from the belt of mud under favorable conditions.

This pattern of sedimentation occurs in a general way in the present-day environments of the northwest Gulf Coast off the shores of Louisiana, Texas, and

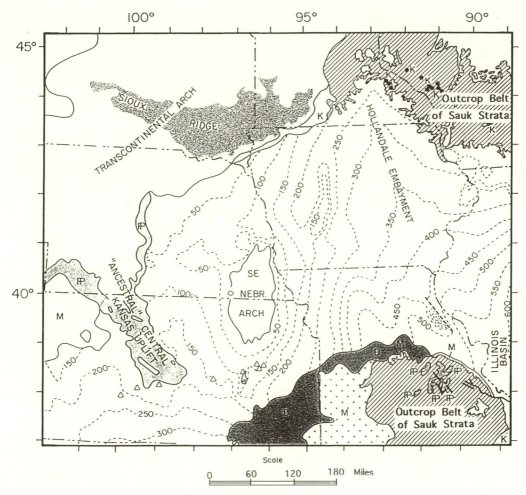

3.3 The subsurface thickness and outcrop belt of rocks of the Sauk Sequence in Iowa and adjacent states. Thickness shown by isopach lines; the contour interval is 50 meters (approximately 165 feet). In Iowa, the Sauk Sequence includes primarily Upper Cambrian and Lower Ordovician rocks. Thickness of Mt. Simon is not included. The diagonal pattern represents areas where the Sauk Sequence is the surface bedrock. Symbols: ℙ = Pennsylvanian, D = Devonian, M = Mississippian, K = Cretaceous. From Bunker and Witzke 1988.

Mexico. Here, beach and shoreline sands grade seaward into alternating sands and muds. The sands and muds, in turn, grade seaward into muds. Farther seaward on the shelf, a few patchy carbonate deposits are found in areas where the mud supply is low. The modern North Sea in Europe and the Middle Atlantic shelf of the United States also provide helpful insights for understanding the ancient Cambrian deposits of the Midwest.

Overall, the environment of deposition of Iowa's Cambrian record was that of a shallow inland sea with well-oxygenated waters and periodic wave agitation. The term "shallow" in geology is commonly used to describe water depth similar to that found on modern marine shelves; this corresponds to depths less than approximately 600 feet. Deep-water environments in geology include those of the continental slopes, continental rises, and abyssal plains. None of these traditional deep-water settings are represented in Iowa's Paleozoic and Mesozoic sedimentary rock record. However, some of these rocks reflect deposition in fairly deep offshore marine settings.

Nature of the Rock Record

Quartzose and feldspathic sandstones make up the bulk of Iowa's Cambrian rock record, and they are even more abundant in the Cambrian of adjacent Minnesota and Wisconsin. The sandy nature of the Cambrian was recognized early. David Dale Owen (1807–1860), a pioneering American geologist, was the first scientist to study the Lower Paleozoic strata of the Upper Mississippi Valley region. Owen included all of the Cambrian rocks in one formation, the Lower Sandstone. Although we now divide Owen's Lower Sandstone into several formations and recognize the presence of siltstone, shale, and dolomite, there is no question that sandstone is the predominant rock type in the Cambrian of the Upper Mississippi Valley.

Several of the Cambrian sandstones of the Midwest have been described as mature sandstones. Mature sandstones are characterized by the presence of stable minerals (chiefly quartz) and the general absence of easily weathered minerals such as feldspar and iron-bearing and magnesium-bearing silicates. Also, mature sandstones are characteristically well sorted — they have grains that fall within a limited size range.

Many of the Cambrian sandstones of the Midwest fit most of these criteria. They contain a high percentage of quartz grains and traces of other very stable minerals, such as zircon, garnet, and tourmaline. However, significant amounts of feldspar are present in some Cambrian sandstones.

Most of the mineral grains in Cambrian rocks originated in the Precambrian rocks of the Canadian Shield and Transcontinental Arch. Later, these mineral grains experienced a long and complex sedimentary history involving weathering, erosion, and transport before they came to rest on the floor of the shallow Cambrian sea.

The roundness of sand grains in Cambrian sandstones is often exceptional, as can be seen by examining sand from the Jordan Formation of northeastern Iowa (fig. 3.4). A very long history of transport and abrasion is suggested by rounding such as this. Although the sand grains last came to rest in a shallow marine environment, the grains were likely rounded by both wind and water during their long history of weathering and transport during Late Precambrian and Early and Middle Cambrian time.

The distribution of the sizes of sand grains in many Cambrian sandstones closely resembles the distribution of sand sizes found in modern marine beach and shelf sands. Both display excellent sorting.

3.4 Rounded and sorted grains of quartz sand from the Jordan Formation of northeastern Iowa. Photo by the author.

Several varieties of cross-bedding are common in the Upper Cambrian sandstones of the Midwest, and stratification features of this type are very useful in interpreting the depositional environments of the rocks. The Jordan Formation displays well-developed cross-bedding in northeastern Iowa. In general, the direction of inclination of the cross strata is in the direction of current movement.

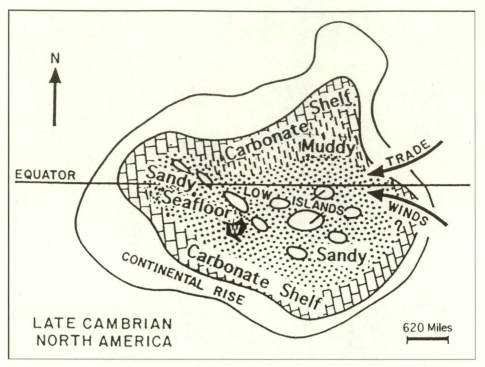

3.5 A generalized facies and paleogeographic setting during Late Cambrian time in North America. The present-day location of the state of Wisconsin (W) is indicated. Iowa, located just above and slightly to the left of Wisconsin, was a sandy marine shelf. Elliptical areas, without pattern, portray the exposed Transcontinental Arch and other land areas. Adapted from Byers and Dott 1995.

Therefore, cross-bedding can be used to reveal the direction of flow of ancient currents if numerous measurements are made over a wide area so that the data can be treated by statistical methods. Studies show that the direction of current movement in the Late Cambrian sea in Iowa was southerly (in reference to present continental positions).

The Cambrian rocks of Iowa formed in a shallow sea that flooded the old Precambrian craton. Quartz-rich sandstone accumulated along the shoreline of the sea and on the wave-agitated seafloor of the inner shelf. Offshore on the shelf, argillaceous and glauconitic sandstones and siltstones formed. Carbonate rocks (limestone or dolomite) formed very thick accumulations on the outer shelf along the edges of the craton. Figure 3.5 shows a generalized facies distribution and paleogeographic reconstruction for Late Cambrian time in North America.

The advance of the Cambrian sea over the craton was not one continuous transgression but instead was marked by fluctuations as the shoreline shifted back toward the continental margins and then readvanced toward the continental interior again. Fluctuations of this nature produced a somewhat repetitious

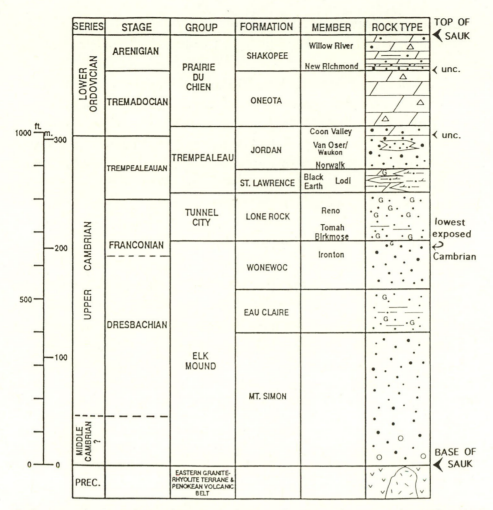

3.6 A stratigraphic column of Upper Cambrian and Lower Ordovician rocks in Allamakee County, northeastern Iowa. Only the Oneota Formation and Waukon Member are named for Iowa localities; all other formation and member names originate from Wisconsin or Minnesota. The large arrows to the right of the rock column mark the positions of major unconformities at the bottom and top of the Sauk Sequence. The smaller arrows indicate unconformities of lesser magnitude. The arrow with the curve denotes the lowest stratigraphic level of exposed Cambrian rock in Iowa; Cambrian units below this stratigraphic position are known only from subsurface information in Iowa. Standard rock symbols are used: dot pattern for sandstone; dot and dash for siltstone; dash for clay or shale; rhombs for dolomite; and triangles for chert. G indicates glauconitic rocks. Adapted from McKay 1993.

rock record in Iowa. Mature quartz sandstones mark the base and top of each transgressive-regressive sequence and represent the deposits of a shoreline or inner-shelf setting. The Mt. Simon, Wonewoc, and upper Jordan formations are examples of such deposits (fig. 3.6). The Eau Claire, Lone Rock, and St. Lawrence formations probably accumulated in deeper water on marine shelves farther from the shoreline.

Cambrian Formations of Iowa

Formations are mappable bodies of rock. They are also the basic stratigraphic divisions used in subdividing the rock column of a given area. The first part of a formation name often originates from the geographic locality where the formation was first studied and described. The second part of the name commonly derives from the composition of the formation, such as in the Jordan Sandstone.

Although Jordan Sandstone is the formal stratigraphic name of the formation, the Jordan Formation is also acceptable, as is the Jordan. Most formations contain more than one kind of rock. For example, the St. Lawrence Formation contains both sandstone and dolomite. Formations of this nature often have the term "Formation" as part of their formal stratigraphic name.

The Iowa Department of Natural Resources, Geological Survey Bureau (hereafter cited as Geological Survey Bureau), recognizes six formations in the Cambrian System of northeastern Iowa (fig. 3.6). Two of these formations are known in Iowa only from subsurface information, and only the upper portion of a third formation is exposed. The other three formations occur along the bluffs and tributaries of the Mississippi River in Clayton and Allamakee counties.

General descriptions of the Cambrian formations of northeastern Iowa and their depositional environments are shown in figure 3.7. Discussions of each of the six formations follow, starting with the Mt. Simon Formation, the basal Cambrian unit in the state.

The Mt. Simon Formation is known only from the subsurface in the state. Named for a locality in Wisconsin, this formation is widespread in the central United States. The name Mt. Simon Formation, or Mt. Simon Sandstone, has been applied to basal Cambrian sandstones that overlie the Precambrian basement in Iowa, Illinois, Indiana, Ohio, Michigan, Minnesota, Wisconsin, and Nebraska. The formation reaches thicknesses of more than 1,000 feet in eastern Iowa, Illinois, Indiana, and Michigan. In western Iowa, however, the Mt. Simon is less than 100 feet thick.

Although the Mt. Simon is generally listed as Late Cambrian in age and therefore assigned to the Upper Cambrian of stratigraphic columns, its age is poorly known. The Eau Claire Formation, which overlies the Mt. Simon, contains index fossils of Late Cambrian age in Wisconsin, but the Mt. Simon contains no diagnostic fossils. It is possible that part of the thick Mt. Simon deposits of the Midwest may be of Middle Cambrian age.

The Eau Claire Formation, also known only from subsurface information in Iowa, consists primarily of glauconitic sandstones, siltstones, and shales. In the subsurface of central and western Iowa, the Eau Claire is assigned to the Eau Claire–Bonneterre interval. Portions of the interval contain glauconitic and argillaceous limestone and dolomite, similar to the Bonneterre Formation of Missouri (fig. 3.8). Fragmented remains of marine fossils are moderately com-

3.7 The Cambrian formations of northeastern Iowa, their compositions, and depositional environments. Adapted from McKay 1993 and Witzke 1990a.

Formation	Rock Types	Depositional Environments
Jordan	Dolomitic sandstone and sandy dolomite, quartz sandstone	Marine shelf and shoreline
St. Lawrence	Silty dolomite, dolomitic siltstone, very fine grained sandstone	Marine shelf, offshore setting
Lone Rock	Argillaceous (muddy) sandstone and siltstone, glauconitic sandstone and siltstone	Marine shelf, offshore setting
Wonewoc	Quartz sandstone	Marine shoreline and shelf
Eau Claire	Glauconitic sandstone, siltstone, and shale	Marine shelf, offshore setting
Mt. Simon	Quartz sandstone	Marine shoreline and inner shelf

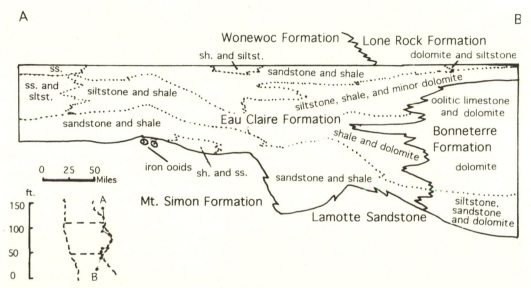

3.8 A stratigraphic cross section of the Eau Claire Formation extending from the western Wisconsin outcrop belt through eastern Iowa into northern Missouri. To the south, clastic rocks (sandstones, siltstones, and shales) of the Eau Claire are replaced primarily by carbonates (limestones and dolomites) of the Bonneterre Formation. Adapted from McKay 1988.

mon in the Eau Claire–Bonneterre beds, including trilobites (early hard-shelled arthropods), inarticulate (unhinged) brachiopods (a type of bishelled invertebrate), hyolithids (tapering shells of possible molluskan origin), echinoderms, rare sponges, algae, and trace fossils (tracks, trails, and burrows).

The Wonewoc Formation is the oldest Cambrian rock unit exposed in Iowa. Only the upper part of the formation (Ironton Member) is visible, however. The Wonewoc is composed of quartz sandstone with well-developed, trough-shaped cross-bedding. The remains of marine invertebrates are present in this formation. Inarticulate brachiopods and trilobites occur in the Wonewoc in Iowa, and gastropods (snails) are known from localities in Wisconsin.

Dott et al. (1986) interpret the environment of deposition of the Wonewoc in Wisconsin to represent wind-formed coastal sand dunes and fluvial sand plains that were covered by an advancing sea. They view the lower part of the formation as eolian (wind-deposited) and fluvial and the upper part of the formation as shallow marine.

The quartz sandstones of the Wonewoc Formation thin to zero thickness and pinch out in west-central Iowa, where the Wonewoc interval consists of marine shales and carbonate rocks. These rocks formed offshore on a marine shelf beyond the area of sand transport and deposition.

The Lone Rock Formation (Franconia Formation of Minnesota usage and Tunnel City of Wisconsin terminology) is composed of glauconitic and feldspathic (feldspar-bearing) sandstones and siltstones. Some layers in the Lone Rock Formation are designated as greensands because they contain an abundance of the green-colored silicate mineral glauconite. Elsewhere, glauconite is mined for fertilizer because of its high potassium content. Glauconite forms only in marine environments, so it is a useful indicator of the marine realm.

Marine fossils such as trilobites, inarticulate brachiopods, and trace fossils occur in the Lone Rock Formation, although they are not abundant or well preserved. The Lone Rock Formation was probably deposited offshore on a marine shelf where storm waves periodically stirred the bottom.

The Lone Rock Formation contains large amounts of clay, and the St. Lawrence Formation is composed of low-porosity, argillaceous dolomite. Thus, both formations are relatively impermeable (incapable of transmitting fluids under pressure). Together, these two formations comprise a confining layer between two of Iowa's important bedrock aquifers: the underlying Dresbach aquifer and the overlying Cambro-Ordovician aquifer. These important aquifers are discussed in more detail in chapter 11.

The St. Lawrence Formation in Allamakee County is divisible into two members: the Lodi Member consists of dolomitic siltstone and fine-grained sandstone; the Black Earth Member is composed of silty dolomite and dolomitic siltstone. Stratification and primary depositional features of the formation include both symmetrical and asymmetrical ripple marks, several varieties of cross-bedding, and beds of intraclast conglomerate (a rock with locally derived clasts larger than approximately 10 inches in diameter). The St. Lawrence Formation formed on a marine shelf in an offshore setting where wave-generated currents transported and deposited sediments. Under storm conditions, strong waves ripped up chunks of lithified seafloor, forming intraclast conglomerates.

The fossil record of the St. Lawrence Formation, although sparse, has been documented in northeastern Iowa and adjacent states. Horizontal burrows

3.9 Nodular bodies and "grapeshot" clusters of quartz sandstone form by the weathering of irregularly cemented areas of sandstone within the Jordan Formation of northeastern Iowa. Paper scale is 1 inch in length. Photo by the author.

occur, as do trilobites and other arthropods, both articulate (hinged) and inarticulate brachiopods, gastropods, hyolithids, conodonts (toothlike phosphatic microfossils), graptolites (carbonaceous remains of complex colonial animals), and algal mats and mounds.

The St. Lawrence and Lone Rock formations are exposed along the base of the bluffs of the Mississippi and its tributaries in Allamakee County in northeastern Iowa. These formations are not very resistant, however, and the exposures are poor and incomplete. A partial section of these formations is exposed at Lansing, Iowa.

Named for the type locality at Jordan, Minnesota, the Jordan Formation is the uppermost formation of Iowa's Cambrian record. It is more resistant to weathering than are the other Cambrian formations in the state, and consequently this unit provides better exposures (fig. 3.1). The Jordan is well exposed along the bluffs of the Mississippi River between McGregor and Marquette in Clayton County. In Allamakee County, the Jordan is visible in roadcuts west of Lansing and along the bluffs of the Upper Iowa River. The Jordan Formation forms steep sandstone cliffs beneath the resistant dolomites of the Ordovician Prairie du Chien Group.

Irregular carbonate-cemented layers in the Jordan give the formation a distinctive appearance (fig. 3.1). Weathered occurrences of these zones of cementation produce nodular and "grapeshot" forms (fig. 3.9). The Jordan Formation

3.10 A generalized stratigraphic column for Cambrian and Ordovician strata exposed in the northern Mississippi Valley. The Sauk Sequence of strata consists of the record from the base of the Mt. Simon Formation through the top of the Prairie du Chien Group. The exact position of the Cambrian-Ordovician boundary relative to the Jordan–Prairie du Chien contact is subject to disagreement. G indicates glauconitic rocks. U-shaped features represent trough-shaped cross-bedding. Adapted from Byers and Dott 1995.

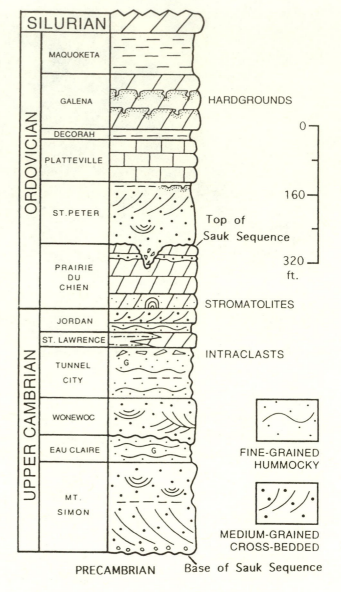

contains well-rounded quartz sand and prominent cross-bedding in northeastern Iowa.

In northeastern Iowa, the Jordan Formation (fig. 3.6) consists of four members (Norwalk, Waukon, Van Oser, and Coon Valley). The basal member (Norwalk) is poorly exposed in Iowa, however. In southwestern Wisconsin, the lower part of the Jordan contains fine-grained sandstones with hummocky cross-stratification (fig. 3.10). Hummocky cross-bedding is considered to be a significant environmental indicator because typically it forms where storm waves move sand in offshore settings, below fair-weather wave base.

The Van Oser Member of the Jordan Formation is visible at the Hanging-Rock locality in Effigy Mounds National Monument, where it consists of fine- to coarse-grained quartz sandstones with distinctive cross-bedding and prominent scour marks. The environment of deposition for the Van Oser Member was a high-energy marine shelf in a nearshore setting.

The uppermost facies of the Jordan Formation, the Coon Valley Member, contains sandstones and sandy dolomites. The stratigraphic assignment of this unit has a history of uncertainty. Does it belong in the Jordan Formation of the Cambrian System, or should it be placed in the overlying Prairie du Chien Group of the Ordovician System? Geologists at the Geological Survey Bureau still include Coon Valley strata in the Jordan Formation, but geologists in Wisconsin and Minnesota have recently moved the unit back into the Prairie du Chien Group.

The Coon Valley Member formed in a shallow marine sea that became progressively shallower. Mud cracks in the upper part of the unit record times when fine-grained sediments were exposed to the atmosphere and to drying conditions in intertidal or supratidal (above tide level) settings.

The Jordan Formation in northeastern Iowa contains few fossils. Stromatolites and vertical burrows occur. Stromatolites are laminated structures produced by sediment trapping, binding, and/or precipitation as a result of the growth and photosynthesis of blue-green algae (cyanobacteria). In addition, molds of brachiopods and a few poorly preserved gastropods have been recovered. Conodont faunas recovered in Wisconsin suggest that the Cambrian-Ordovician boundary falls within the upper part of the Jordan Formation.

Figure 3.6 shows the current convention of the Geological Survey Bureau of placing the Cambrian-Ordovician boundary at the base of the Coon Valley Member of the Jordan Formation. Byers and Dott (1995) present evidence for an erosional contact at the top of the Jordan Formation. According to their interpretation, the Lower Ordovician Prairie du Chien Group lies unconformably on different layers of the truncated Jordan Formation. Although Byers and Dott indicate that the exact position of the Cambrian-Ordovician boundary relative to the Jordan–Prairie du Chien contact is uncertain, they place the contact between the Cambrian and Ordovician systems at the base of the Prairie du Chien (fig. 3.10). In Minnesota, Runkel (1994) also finds evidence for an unconformity at the Cambrian and Ordovician boundary.

Recognition of specific Cambrian formations is more difficult in the subsurface. In central, western, and southeastern Iowa, the formations described previously are difficult to differentiate.

The Jordan Formation is well known in Iowa's subsurface, having been drilled extensively for groundwater. The sandstones of the Jordan Formation have a high percentage of void space (porosity) between the sand grains. In addition, the pore spaces are generally interconnected (permeable) and allow for movement of fluids between grains. The Jordan acts as a giant rock sponge and stores and transmits a large volume of groundwater. This formation is one of the state's

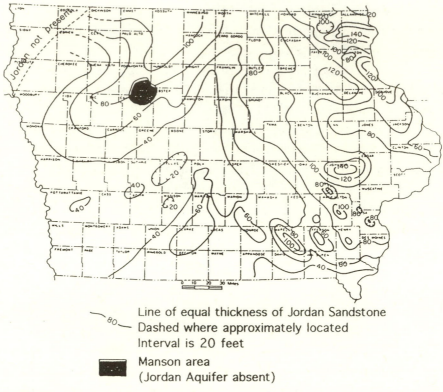

~80— Line of equal thickness of Jordan Sandstone
Dashed where approximately located
Interval is 20 feet

Manson area
(Jordan Aquifer absent)

3.11 The thickness and distribution of the Jordan Sandstone in Iowa. From Horick and Steinhilber 1978.

best aquifers (water-bearing rock body). It is the most widespread source of large yields of good-quality groundwater for cities and industries. More than 150 wells penetrate the Jordan Formation throughout Iowa, and much is known concerning its subsurface thickness and distribution (fig. 3.11). Because of large withdrawals, the water levels in this formation have dropped significantly in several parts of Iowa. For example, water levels in the Jordan have gone down 100 to 200 feet in the Mason City area and approximately 100 feet in Grinnell and Ottumwa. The Jordan Formation and other aquifers are discussed further in chapter 11.

Cambrian Life

Cambrian time was the Age of the Trilobites. Although any earth history book will tell you that, it is not easy to find complete trilobites in Iowa's Cambrian rocks. Trilobites and brachiopods comprise over 90 percent of the Cambrian faunas in other regions, but neither group is particularly well represented in the Cambrian record of Iowa. Preservation of fossils is often poor in the sandy deposits of shoreline and inner-shelf settings where the abrasive action and agitation of marine currents and waves grind and pulverize fossils. Fossil shells are reduced to particles of sand size. Fossils are more common and better preserved in the muddy sediments laid down in offshore settings.

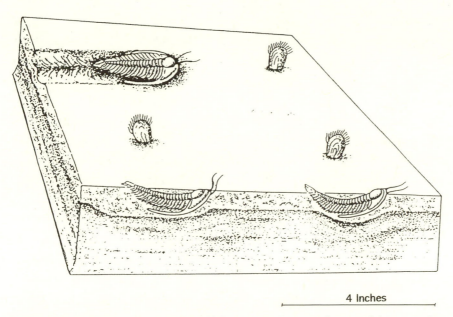

4 Inches

3.12 A generalized reconstruction of life on a Cambrian seafloor. Trilobites are shown crawling (upper left) and burrowing (lower left and right). Inarticulate brachiopods are attached to the seafloor and partially buried in sediment (upper right and center). Adapted from McKerrow 1978.

The St. Lawrence, Lone Rock, and Eau Claire formations were deposited offshore on marine shelves. Fossils can be collected from these formations in Wisconsin where the units are fairly well exposed. In Iowa, there is limited opportunity to collect from these units. The St. Lawrence and Lone Rock formations are poorly exposed in Iowa, and the Eau Claire Formation is known only from the subsurface.

By relying on information gained from studies in Wisconsin and Minnesota, along with limited fossil collections in Iowa, we know that the Cambrian seafloor was home to trilobites and other arthropods, both articulate and inarticulate brachiopods, gastropods and other mollusks, burrowing wormlike creatures, and sticky algal mats. Blue-green algae play a key role in the formation of stromatolites. Stromatolites often display wavy structures and fine laminations, the products of deposition of fine-grained carbonate sediments on the sticky surfaces of algal mats and algal colonies. Stromatolites form today where blue-green algae thrive in shallow, sunlit water. We know that blue-green algae carry on photosynthesis, which requires sunlight. They cannot survive below the photic zone (the region of the ocean with sufficient light for photosynthesis). Although the depth of the photic zone varies, it averages about 260 feet deep in modern oceans. Therefore, the presence of stromatolites in the Cambrian record suggests that the ancient shelves were sufficiently shallow to allow sunlight to penetrate to the seafloor. Figure 3.12 provides a reconstruction of trilobites and inarticulate brachiopods on a Cambrian seafloor.

Summary

During Late Precambrian time and much of Cambrian time, the region that is now Iowa was above the sea undergoing weathering and erosion. No deposits of Early Cambrian age are found in the state. The Mt. Simon Formation may prove to be Middle Cambrian in age, but it has not yet been reliably dated.

By Late Cambrian time, Iowa was flooded by the marine waters of an inland sea that covered much of North America. During Late Cambrian time, the sea shifted back and forth across Iowa, leaving behind a record of shoreline, inner-shelf, and offshore-shelf deposits. By the end of Cambrian time, some 510 million years ago, Iowa was covered by a sunlit sea on a carbonate shelf. The record of carbonate sedimentation in Iowa appears to have continued into Ordovician time. Modern environments in the northwest Gulf of Mexico, the Middle Atlantic shelf, and the North Sea are somewhat comparable to the environments that existed in Iowa during Late Cambrian time.

Conditions for fossilization were not ideal in the Cambrian seas over Iowa, and exposures in the state have not yielded a great variety of fossils. Nevertheless, the basic pattern is well known. The shallow Cambrian seas of ancient Iowa were inhabited by trilobites and other arthropods, articulate and inarticulate brachiopods, gastropods and other mollusks, burrowing wormlike organisms, and sticky algal (cyanobacterial) mats.

Acknowledgments

Robert M. McKay published a guidebook on the Lower Ordovician and Upper Cambrian geology of Allamakee and northern Clayton counties in 1993. I relied on Bob's work for several sections of this chapter, and I acknowledge his helpful review and suggestions. Brian J. Witzke's description of the Eischeid core includes information about Iowa's subsurface Cambrian record in west-central Iowa. A 1987 article by Witzke and McKay presents useful information on Cambrian and Ordovician rock exposures in northeastern Iowa. The scientific papers by C. W. Byers and R. H. Dott Jr. provide significant background on Cambrian exposures and environments in Wisconsin. Similarly, A. C. Runkle's research in Minnesota supplies important insights for a regional understanding of the Cambrian of the Midwest.

> These mural escarpments, exhibiting every variety of form, give to the
> otherwise monotonous character of the landscape in Iowa a varied and
> picturesque appearance. Sometimes they may be seen in the distance, rising
> from the rolling hills of the prairie, like ruined castles, moss-grown under the
> hand of time.
>
> David Dale Owen (1840) describing exposures of Ordovician bedrock
> along the Mississippi River near Dubuque, Iowa

Ordovician rocks are well exposed in northeastern Iowa (fig. 4.1) and indicate
that the state was under the sea for much of Ordovician time. Marine deposition
that started during Late Cambrian time continued largely uninterrupted into
Early Ordovician time. Consequently, the paleogeographic setting of Iowa during Early Ordovician time was nearly identical to that of Late Cambrian time.

Toward the end of Early Ordovician time, the sea completely withdrew from
the interior of the continent. During this time, the continental interior was a low
landmass comprised of Precambrian, Upper Cambrian, and Lower Ordovician
bedrock. The Upper Cambrian and Lower Ordovician rocks were exposed to
weathering and erosion during the remainder of Early Ordovician time, and a
widespread erosion surface was produced. Middle Ordovician rocks were deposited on this erosion surface.

In Middle Ordovician time, the sea advanced inward from the continental
margins to again flood Iowa and adjacent parts of the continental interior. This
major transgression produced the deposits of the Tippecanoe Sequence during
Middle Ordovician through Silurian time. The paleogeography, representative
of Early and Middle Ordovician time, is shown in figure 4.2.

The Transcontinental Arch was a prominent feature during Middle Ordovician time, extending (in terms of modern geography) for more than 1,000 miles
from the Canadian Shield across the Great Plains region, terminating in northern Mexico (fig. 4.3). At various times, this low arch divided the Ordovician seas
into northern and southern segments. Precambrian rocks were exposed on top
of the arch, and Upper Cambrian and Lower Ordovician sedimentary rocks were
present on its flanks (fig. 4.3). The weathered products of these Precambrian,
Upper Cambrian, and Lower Ordovician rocks served as sources of sediment for
the Middle Ordovician seas that invaded Iowa.

During Middle and Late Ordovician time, the eastern margin of the early

4.1 Bold cliffs of the lead-bearing beds (Ordovician Galena Group) along the Mississippi River near Dubuque, Iowa. Drawing by David Dale Owen in 1844. From Owen 1844.

North American continent was located near an active plate boundary, where uplift and volcanism produced the Taconic Mountain System (fig. 4.4). Volcanoes associated with the Taconic Mountains erupted and ejected large quantities of volcanic ash into the atmosphere. Some of this ash settled out of the atmosphere over what is now Iowa and accumulated on the shallow seafloor. Several thin layers of volcanic ash, now altered to the rock bentonite, occur in the Ordovician rocks of northeastern Iowa.

Mud, produced by weathering and erosion of the Taconic Mountains, washed into the inland seas. Marine currents transported and deposited the mud, and it is now part of the Upper Ordovician rock record of the state as shales or argillaceous carbonate rocks.

Nature of the Ordovician Rock Record

The Ordovician System has its type locality in Europe, as do most of the other geologic systems. The duration of time represented by the deposition of the Ordovician System is the Ordovician Period; it is judged to be about 70 million years in length. The preceding Cambrian Period was once thought to be about 60 million years in length, but current studies suggest that it was of shorter duration, perhaps encompassing 35 million years. The Silurian Period, which follows the Ordovician, records some 30 million years of Earth history. Obviously,

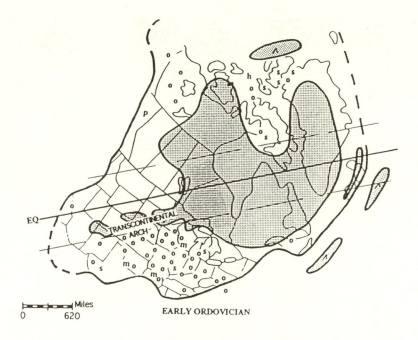

EARLY ORDOVICIAN

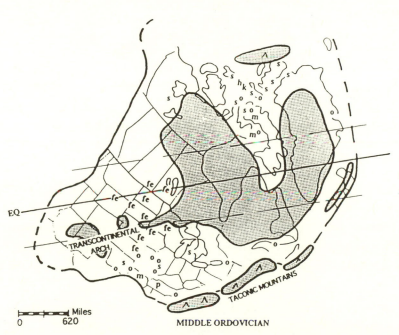

MIDDLE ORDOVICIAN

4.2 The interpreted paleogeography for the Early and Middle Ordovician continent of Laurentia. Land areas are shaded, and areas of marine deposition are unshaded. Symbols: o = carbonate ooids (oolites) and coated grains, p = phosphorites or phosphatic sediments, m = evaporite crystal molds, s = sulfate evaporites (primarily gypsum and anhydrite), h = halite (rock salt), k = potash salts, fe = oolitic ironstones, ∧ = mountainous terrains, EQ = proposed paleoequator. The Early Ordovician reconstruction portrays conditions during deposition of Iowa's Prairie du Chien Group. The Middle Ordovician map represents the setting during deposition of the St. Peter, Glenwood, Platteville, Decorah, and Dunleith strata. From Witzke 1990b.

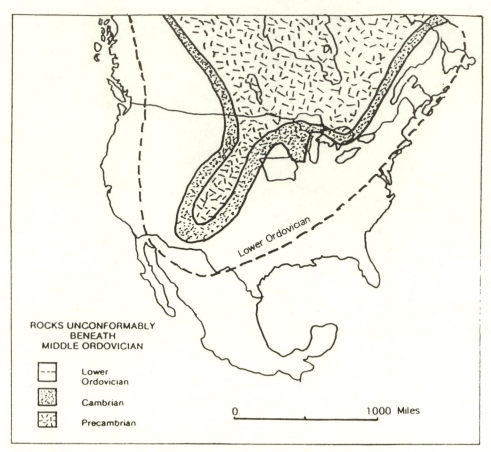

ROCKS UNCONFORMABLY
BENEATH
MIDDLE ORDOVICIAN

Lower Ordovician

Lower
Ordovician

Cambrian

Precambrian

0   1000 Miles

4.3 The distribution of Precambrian, Cambrian, and Lower Ordovician bedrock unconformably beneath Middle Ordovician strata. The sands of the St. Peter Formation were derived from exposed rocks of Precambrian, Cambrian, and Early Ordovician ages. Note the location of the Transcontinental Arch. Adapted from Dott and Batten 1971 and Anderson 1983.

the geologic time periods are not of the same duration. The geologic systems were originally established on the basis of physical characteristics. Later, after radiometric dating had been discovered, it was possible to fix dates in years to the geologic column.

The rocks of the Ordovician System in Iowa are divided into three divisions: lower, middle, and upper. The Lower Ordovician rocks were deposited as sediments during Early Ordovician time; the Middle Ordovician strata originated during Middle Ordovician time; and the Upper Ordovician record accumulated during Late Ordovician time. Iowa has a fairly complete record of Ordovician rocks and one that preserves a variety of compositions and depositional environments (fig. 4.5). Whereas sandstone was a dominant deposit during Cambrian time, carbonate rocks (limestone and dolomite) became increasingly important in Ordovician time and in the subsequent Silurian, Devonian, and Mississippian periods as well.

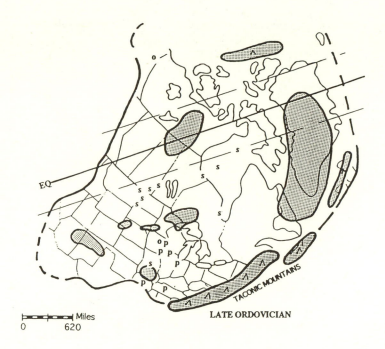

LATE ORDOVICIAN

Miles
0   620

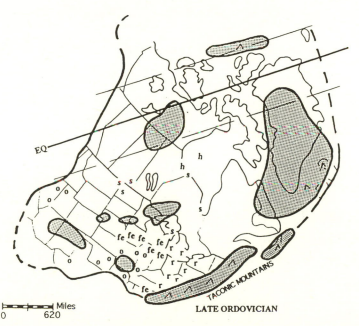

Miles
0   620

LATE ORDOVICIAN

4.4 The interpreted paleogeography for the Late Ordovician continent of Laurentia.
Symbols as in figure 4.2 with the addition of r for redbeds. The upper map is a reconstruction for Late Ordovician time (Edenian-Maysvillian) and shows the general
setting for the deposition of Iowa's Wise Lake, Dubuque, and lower Maquoketa formations.
The lower map portrays the paleogeographic setting of a later interval of Late Ordovician
time (Richmondian) and the depositional setting of the upper Maquoketa Formation.
The Taconic Mountain belt was active in Middle and Late Ordovician time and supplied
volcanic ash that is now represented by bentonite layers in Iowa's Ordovician rock record.
From Witzke 1990b.

4.5 Iowa's Ordovician formations and their depositional environments. Adapted from Delgado 1983; McKay 1993; Witzke 1980, 1990a; Witzke and Glenister 1987a, b; Witzke and Kolata 1988; and Witzke and McKay 1987.

Age	Group or Formation	Depositional Environments
Late Ordovician	Maquoketa Formation	Carbonate shelf and muddy shelf; locally abundant invertebrate life; oxygen levels on the seafloor varied from normal to below-normal levels; some settings were anoxic (without oxygen)
Late Ordovician	Galena Group (Dubuque Formation, Wise Lake Formation)	Normal marine seafloor of shelf depth; deposition primarily below normal wave base; normally, quiet-water conditions prevailed, but occasionally turbulent storms laid down well-washed fossil debris; abundant hardgrounds formed by early submarine cementation, followed by later subsea erosion and/or dissolution during times of very slow sedimentation; abundant and diverse marine life
Middle Ordovician	Galena Group (Dunleith Formation)	Same as above, but also includes deposits of altered volcanic ash (bentonite) from sources in the Taconic Mountains (eastern U.S.) and possibly from the Baltic region of Europe
Middle Ordovician	Galena Group (Decorah Formation)	Carbonate and muddy marine shelf; mud was washed into the sea from the exposed Transcontinental Arch; minor influx of volcanic ash from sources to the east; abundant and diverse invertebrate life; mostly well-oxygenated bottom waters, but minor episodes of low-oxygen conditions on the seafloor
Middle Ordovician	Platteville Formation	Carbonate and muddy marine shelf; abundant and diverse invertebrate life
Middle Ordovician	Glenwood Formation	Muddy shelf with slow rate of sedimentation; includes sandy deposits in southeastern Iowa (Starved Rock Sandstone)
Middle Ordovician	St. Peter Formation	Nearshore and shelf deposits of well-washed and well-sorted quartz sand, deposited along the margin of an advancing sea; may include some reworked stream deposits and coastal dunes

Age	Group or Formation	Depositional Environments
Early Ordovician to Middle Ordovician	Major unconformity between Sauk Sequence (below) and Tippecanoe Sequence (above)	A major episode of erosion; no rock record
Early Ordovician	Prairie du Chien Group (Shakopee Formation)	Carbonate shelf and sandy shoreline setting; the New Richmond Member of the Shakopee Formation formed in a sandy inner shelf or shoreline setting; the upper member (Willow River) formed on a carbonate shelf with algal stromatolites
Early Ordovician	Prairie du Chien Group (Oneota Formation)	Shallow carbonate shelf with algal stromatolites

LOWER ORDOVICIAN

Variety characterizes Iowa's Ordovician rock record. The Lower Ordovician rocks show a shift away from the sandstones that were so typical of Late Cambrian deposition. Carbonate rocks are the dominant rock type of the Lower Ordovician. These carbonate rocks are composed primarily of the mineral dolomite, a calcium-and-magnesium carbonate.

The Lower Ordovician dolomites of the Midwest formed by the process of dolomitization whereby limestone ($CaCO_3$) is changed chemically to dolomite [$CaMg(CO_3)_2$] by reactions involving magnesium-bearing waters. This process involves a structural transformation of the mineral calcite to form dolomite, which differs in crystalline arrangement. The dolomitization process generally destroys or alters the fossil content and original textural features of the rock. Dolomitization can occur in a variety of ways and at various times after deposition of the original calcium carbonate sediments.

Dolomitization affected Iowa's Lower Ordovician rocks in significant ways. It also played a key role in transforming the composition of the Middle and Upper Ordovician strata of the state. The process of dolomitization is discussed in more detail later in this chapter. Dolomitization had a profound effect on Iowa's Silurian, Devonian, and Mississippian rocks, too.

Stromatolites (fig. 4.6) are common in the Lower Ordovician Prairie du Chien Group of northeastern Iowa, indicating that these strata were deposited in a shallow sea. At times, the Early Ordovician environments were probably like modern tidal flats where sediments are exposed to drying when the flats are above water. As the sediments of the exposed tidal flats dry, they shrink to form mud-cracked surfaces. Later, the mud-cracked sediments are stirred up when marine

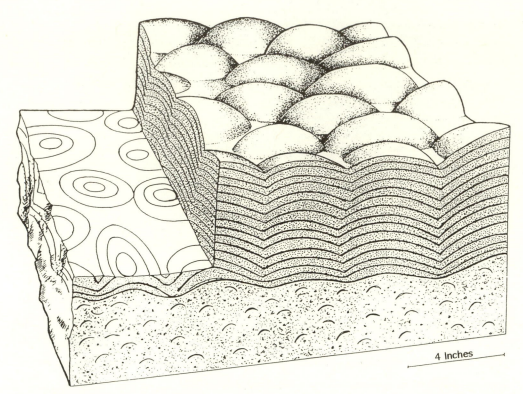

4.6 Stromatolites. Mats and colonies of blue-green algae (cyanobacteria) trap carbonate sediments on their sticky surfaces, forming laminated deposits called stromatolites. Structures similar to these are present in the Lower Ordovician rocks of northeastern Iowa, indicating that shallow, sunlit seas once covered the area. From McKerrow 1978.

waters again wash over the area. Fragments of mud-cracked sediments occur in the Lower Ordovician rock record as clasts in conglomerates (intraclast conglomerates).

Oolites, or ooids, are present in Iowa's Ordovician rock record, and they, too, indicate a very shallow marine setting. Today, oolites form in shallow, warm seas that are nearly saturated with calcium carbonate (lime). Oolites are sand-size bodies with laminations of calcium carbonate; they look like the grains in tapioca pudding. Oolites start to form when calcium carbonate is deposited on the surface of a sand grain, shell fragment, or some other particle. Additional layers of calcium carbonate form on the surface of the oolites as they are rolled and shifted around on the seafloor. In many cases, the end product is a spherical or subspherical grain of sand-size dimension, with concentric laminae of calcium carbonate surrounding a nucleus such as a shell fragment or sand grain. Currently, oolites are forming in the shallow marine (10–20 feet deep) setting of the Grand Bahama Banks where the waters are nearly saturated with lime and constantly agitated by wave action.

At the end of Early Ordovician time, the sea withdrew from Iowa, and Lower Ordovician rocks were exposed to the processes of weathering and erosion. At this time, Iowa was a low coastal plain similar to southern Mississippi or southern Alabama of today.

In Middle Ordovician time, the sea again transgressed across the interior of the continent. Sand and mud were deposited along the margins of the transgressive sea and are now incorporated in the Middle Ordovician rock record as sandstone and shale. The most characteristic deposit of the Middle Ordovician advancing sea was very pure quartz sandstone.

The sources for the quartz sand were the exposed Precambrian rocks of the Transcontinental Arch and Canadian Shield and the Upper Cambrian sandstones that were exposed in a belt fringing the Canadian Shield and Transcontinental Arch (fig. 4.3). Since the Upper Cambrian sandstones are generally described as mature because of their high quartz content and well-rounded and well-sorted nature, it would be appropriate to describe the Middle Ordovician sandstones as supermature. In the process of weathering and transport, unstable minerals were nearly all eliminated from the sand, and the final product was a highly rounded sand composed almost exclusively of quartz.

A blanketlike deposit of such sand, represented by the St. Peter Formation and its equivalents, is recognizable over a wide geographic area of the continental interior from Oklahoma to Minnesota. The St. Peter Formation and its equivalents represent the deposits of a marine shoreline and inner shelf that advanced over the interior of the continent.

As the edge of the Middle Ordovician sea crept slowly northward beyond Iowa, the sandy shelf gave way to more offshore settings where marine muds and carbonate sediments formed. These deposits now constitute the Glenwood, Platteville, Decorah, and Dunleith formations. Carbonate deposition continued in Late Ordovician time, and part of the Galena carbonate sequence (Wise Lake and Dubuque) is assigned to the Upper Ordovician. All of the major groups of marine invertebrate fossils are represented in these formations. Many of the rocks are literally crammed with fossils (fig. 4.7).

The dominant deposits of Middle Ordovician time consist of carbonate minerals, shells, shell fragments, other fossils, and other fossil fragments. The Decorah Formation, however, represents a rather substantial body of shale derived from terrestrial sources in Minnesota and from the emergent Transcontinental Arch. Although Iowa was part of a shallow seafloor receiving carbonate deposits during Middle and Late Ordovician time, the deposition was not without interruption. The sediments on a shallow marine seafloor are subjected to considerable wave action. Storm waves often scour and erode previously deposited

4.7 Shell-rich limestone from Middle Ordovician strata of northeastern Iowa. Many of Iowa's Ordovician limestones are highly fossiliferous, and all of the major groups of shelled invertebrates occur in the Ordovician rock record of the state. Photo by the author.

sediments and transport them seaward on the shelf. In the Galena Group of northeastern Iowa, there are a number of well-washed beds of fossil debris that are interpreted as ancient storm deposits.

In addition to interruption by submarine and tidal erosion, shelf deposits are also disrupted by solution and by the activities of burrowing organisms. A variety of bottom-dwelling creatures burrow and disturb the sediments on modern seafloors. If the rate of sedimentation is slow on the seafloor, a widespread burrowed surface may be produced. Many such surfaces are found in the rocks of the Galena Group in northeastern Iowa. Burrow patterns are considered to be trace fossils, and they provide important information about the activity of ancient organisms.

Rocks of the Middle and Late Ordovician Galena Group in northeastern Iowa display hardgrounds on a grand scale. Hardgrounds, often an inch or so thick, are hardened surfaces produced by submarine cementation. They are often bored by organisms, etched by solution processes, and stained by iron sulfides. In places, they contain attached or incrusting fossils. The presence of hardgrounds implies gaps in sedimentation. David Delgado (1983) estimates that the time represented by hardground surfaces in the Galena Group is 15 to 30 percent of the total time involved in deposition of these strata.

Several thin beds of altered volcanic ash occur in Iowa's Ordovician rocks (fig. 4.5). Such layers have been discovered in the Decorah and Dunleith formations of northeastern Iowa. The volcanic ash, now altered to bentonite, appears as thin layers of gray or orange clay. The mineral composition of these clays substantiates that they represent water-laid volcanic ash deposits. In addition to having a definitive clay-mineral composition, the bentonite beds have a trace-mineral content that is indicative of a volcanic origin, allowing correlation of specific ash layers across vast areas of the eastern United States.

The Upper Ordovician rock record of Iowa is composed primarily of shale, limestone, and dolomite. The shale represents the accumulation of marine mud, much of which came from the Taconic Mountains (fig. 4.4).

Well-preserved graptolites occur in some of the Upper Ordovician shales of Iowa. Graptolites are representatives of a nearly extinct group of colonial organisms that possess a proteinlike composition. According to some paleontologists, graptolites are represented today by living pterobranchs. Graptolites fall into the phylum Hemichordata — somewhat transitional between the invertebrate and vertebrate worlds. Graptolites evolved rapidly and many varieties floated; they make excellent index fossils.

The skeletal framework of graptolites was composed of protein-based compounds in contrast to the calcium carbonate compositions typical of common colonial organisms such as corals and bryozoans. Graptolites were highly susceptible to destruction by scavenging organisms that inhabited shallow seafloors. Proteinaceous substances are also unstable in oxidizing environments. Given that most shallow seafloors possess scavenging organisms and are characterized by the presence of some dissolved oxygen, it follows that graptolites are rare in the deposits of shallow inland seas. The preservation of graptolites is favored by oxygen-deficient conditions (reducing environments) on the seafloor.

Iowa's Upper Ordovician shales contain an abundance of carbonaceous material, which is another indication of oxygen-deficient conditions in the depositional environment. The carbon content is so high in some of the shales that they will burn when ignited. Iron pyrite (FeS_2) is also found in the Upper Ordovician shales; it, too, suggests an oxygen-poor environment.

The marine shelves of Late Ordovician time experienced episodes when the bottom waters contained little or no dissolved oxygen. Some of the sedimentary rocks that now constitute the Maquoketa Formation of Iowa were deposited in such oxygen-poor environments.

Phosphatic rocks, composed of the mineral apatite (calcium phosphate), are found in moderate abundance in the Upper Ordovician rocks of Iowa. Phosphatic layers occur in the lower part of the Maquoketa Formation in the form of pebbles, pellets, ooids, and phosphatic fossils. In general, the deposition of marine phosphate is favored in areas of upwelling. Water in these areas is high in nutrients and supports a large population of floating organisms. The floating organisms, mainly phytoplankton, incorporate phosphate into their tissues during growth. When the organisms die, their remains settle to the seafloor and increase the phosphate content of the bottom waters, creating conditions whereby calcium phosphate may form. Calcium phosphate forms either as a direct precipitate or as a replacement of calcium carbonate sediments.

Generally, there is insufficient oxygen available on the seafloor to oxidize all of the organic matter that settles below areas of upwelling waters and high organic productivity. This results in the development of reducing environments on the seafloor, where preservation of organic matter and formation of iron sulfides are favored. For this reason, it is common to find an association of iron sulfides and organic matter with marine phosphate beds. Such an association is found in the phosphatic beds of the Maquoketa Formation in eastern Iowa.

In Dubuque County, the basal phosphate bed of the Maquoketa Formation is approximately 2 feet thick. The dolomite beds in the lower Maquoketa are also phosphatic in the vicinity of Dubuque, where several feet of highly phosphatic dolomite have been reported.

In addition to the occurrence of phosphatic rocks in the Maquoketa Formation, minor amounts of phosphate are known from some of the other Ordovician formations of Iowa. For example, phosphate is found in small amounts as nodules, pellets, and phosphatic fossil remains in the Glenwood, Platteville, and Decorah formations.

Ordovician Strata and Models to Explain Dolomitization

The two main models for dolomitization are the mixing zone model (fig. 4.8) and the evaporative reflux model (fig. 4.9). Widespread regional dolomitization of Ordovician rocks in Iowa is not completely explained by either model, but, of the two, the mixing zone model seems more promising. Dolomitization is still poorly understood, so no single model accounts for all aspects of this important but somewhat mysterious process that changes limestone into dolomite.

The mixing zone occurs at the interface of the land and sea (fig. 4.8) where potable groundwater under the land mixes with salt water that saturates the pores of sediments under the sea. If fresh groundwater is mixed with seawater in a ratio of approximately nineteen parts of freshwater to one part of salt water, the resulting mixture is undersaturated with respect to calcite ($CaCO_3$) but saturated with respect to dolomite [$CaMg(CO_3)_2$]. This zone of mixed fresh groundwater and normal seawater serves as a dolomitizing fluid. As a shallow sea advances over the continent, the mixing zone moves landward. Likewise, when the sea retreats, the zone of dolomitization also migrates seaward. In theory, the mixing

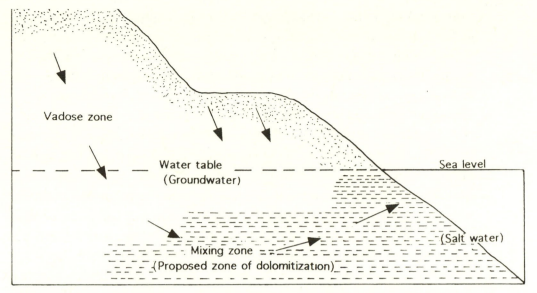

4.8 A schematic cross section showing the mixing zone model for dolomitization. The zone of mixed groundwater and saltwater serves as a magnesium-bearing fluid and provides an environment where limestone is converted to dolomite. Adapted from Land 1973.

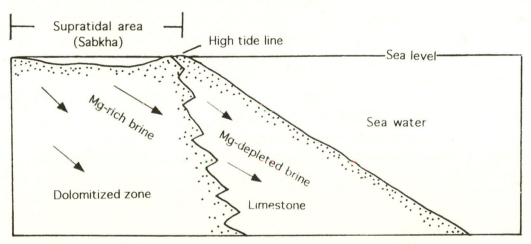

4.9 A schematic cross section showing the evaporative reflux model for dolomitization. Gypsum is precipitated in shallow surface waters, removing calcium and sulphate ions from the water. The remaining fluid is relatively enriched in magnesium ions, and this magnesium-rich brine infiltrates the subsurface layers, where it reacts with limestone deposits to form dolomite. Adapted from Ehlers and Blatt 1982.

zone model is useful in explaining laterally extensive areas of dolomitization associated within transgressive-regressive sequences of carbonate rocks.

In the case of the mixing zone model, the initial composition of the limestone beds that are dolomitized reflects a normal marine origin. This is in line with observations of the Ordovician rocks of Iowa. These Ordovician carbonate rocks carry a typical marine fauna and appear to be the products of deposition under normal marine salinities.

High rates of evaporation and the deposition of evaporites such as gypsum ($CaSO_4 \cdot H_2O$) are associated with the evaporative reflux model (fig. 4.9). Such processes occur today in the Persian Gulf, where broad coastal flats called sabkhas are subjected to periodic flooding by seawater during times of storms. Evaporation on the flooded sabkhas increases the salinity of the waters two- to threefold, causing the deposition of gypsum. The precipitation of gypsum removes calcium and sulfate ions from the water and leaves the remaining brines relatively enriched in magnesium ions. These dense magnesium-rich brines sink downward and soak the underlying deposits of limestone. The calcium carbonate in the limestones reacts with the magnesium-laden fluids to produce dolomite, a calcium-and-magnesium carbonate. This transformation may take a few centuries to complete.

With the evaporative reflux model, one expects to find an association of dolomites and evaporite deposits (gypsum) in the rock record. We do not see this association in the Ordovician strata of Iowa. Evaporite deposits are known in Iowa's subsurface in strata of the Devonian and Mississippian systems, however, and the evaporative reflux mechanism may have played a role in the conversion of limestone to dolomite in these rocks.

Ordovician Stratigraphic Units

Ordovician formations are well exposed in northeastern Iowa and contribute significantly to the beautiful scenery along the bluffs of the Upper Iowa, Turkey, Volga, and Mississippi rivers. Exposures of Ordovician rocks primarily occur in Winneshiek, Allamakee, Clayton, Fayette, Dubuque, and Jackson counties. Ordovician strata form the bedrock in several other areas in the state as well (fig. 4.10). In these areas, however, the Ordovician rocks are covered by surficial deposits. The Ordovician System occurs throughout most of the state in the subsurface.

David Dale Owen was the first person to make a scientific study of Ordovician and other Lower Paleozoic strata of the Upper Mississippi Valley region. Owen, with a corps of assistants, visited Iowa and adjacent Wisconsin and Minnesota in 1839 and again in the late 1840s to investigate the geology, geography, and natural history of the region. Owen's charge was to evaluate the resources of the region for the General Land Office.

Geologic maps made by Owen's group were the first for the region, and the rock units shown on these maps are still readily recognizable. The rock units were established primarily on the basis of rock types and the nature of the fossils they contained. Subsequent geologists proposed more formal formation and member names for these mappable bodies of rock. The stratigraphic units

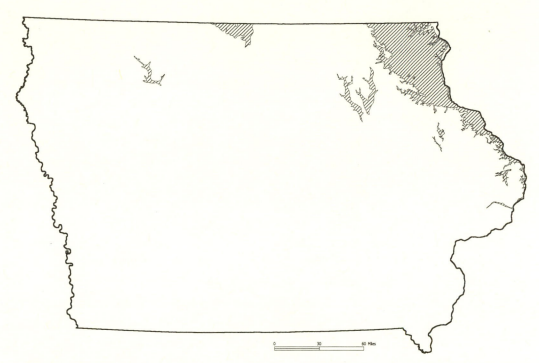

4.10 The distribution of Ordovician bedrock in Iowa. Information provided by Geological Survey Bureau, Iowa City, 1998.

mapped and described by the Owen Survey include, in ascending order: Lower Sandstone, Lower Magnesian Limestone, St. Peter Sandstone, St. Peter's Shell Limestone, and Upper Magnesian Limestone.

These rock bodies are still recognizable. The Lower Sandstone represents the Cambrian formations. The Lower Magnesian Limestone is now known as the Prairie du Chien Group (Ordovician). The St. Peter Sandstone, a distinctive quartz arenite (sandstone), is still referred to by the same name. The Glenwood Formation, of current stratigraphic usage, is poorly exposed in northeastern Iowa; it is positioned stratigraphically between the St. Peter Sandstone and the St. Peter's Shell Limestone. The St. Peter's Shell Limestone is known today as the Platteville and Decorah formations. The Upper Magnesian Limestone is represented by most of the Galena Group, the carbonate members of the Maquoketa Formation, and strata of the Silurian System. Owen termed the lower portion of the Upper Magnesian Limestone (part of the Galena Group of today's usage) the "lead-bearing beds."

A generalized stratigraphic column is shown in figure 4.11. The column displays the terminology and thicknesses for key strata of the Ordovician System in northeastern Iowa. The environments of deposition of the Ordovician formations of Iowa are summarized in figure 4.5. Figure 4.12 presents qualitative sea-level curves and inferred depositional settings for rocks of Middle and Late Ordovician ages. Discussions of major Ordovician stratigraphic units follow.

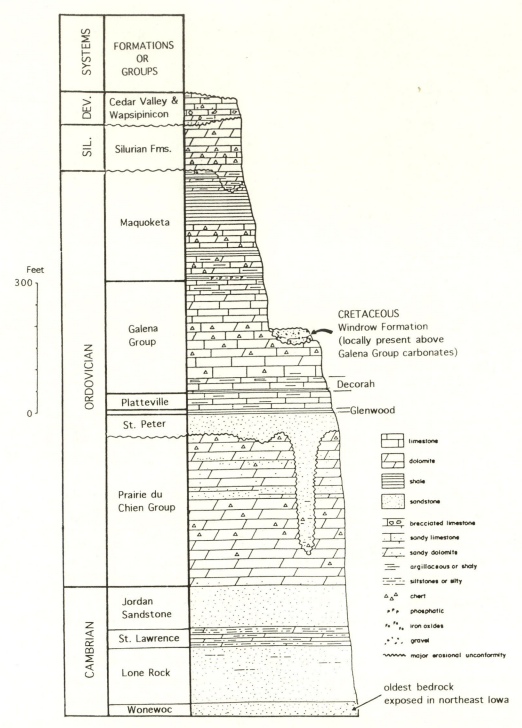

4.11 A generalized stratigraphic column for northeastern Iowa showing key Ordovician stratigraphic units. Adapted from Hoyer et al. 1986.

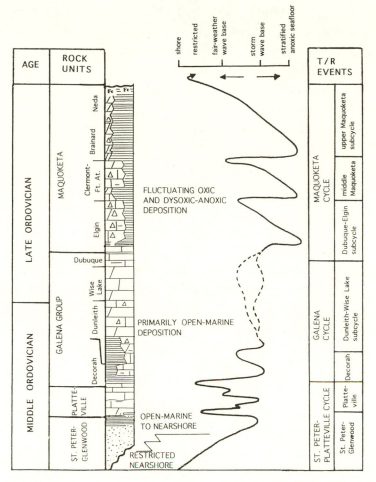

4.12 Qualitative sea-level curves for Middle and Late Ordovician seas in Iowa and northern Illinois. Oxic = normal conditions, with dissolved oxygen in bottom waters; Dysoxic = oxygen deficient; Anoxic = without oxygen. Adapted from Witzke and Kolata 1988.

PRAIRIE DU CHIEN GROUP

The Prairie du Chien Group is composed of dolomite, sandy dolomite, sandstone, and shale (fig. 4.11). Recognized as a formation in earlier literature, the Prairie du Chien is now treated as a group by the Geological Survey Bureau. The Prairie du Chien is divided into the Oneota (lower) and Shakopee (upper) formations. Rocks of the Prairie du Chien preserve a number of shallow-water indicators, such as algal structures, intraclast conglomerates, and oolites.

Prairie du Chien strata are exposed in bluffs along the Mississippi River in Allamakee County and northern Clayton County, in various roadcuts along Highway 9 in Allamakee County, and in Yellow River State Forest. Exposures occur along Highway 76 between Waukon and the Minnesota border. In addition, the Shakopee Formation is well exposed in Clayton County along the Clayton barge terminal road, where large stromatolites are prominent.

The contact between the two formations of the Prairie du Chien Group is revealed in roadcuts along Highway 76 in Allamakee County. In this area, some beds of the Oneota Formation were truncated by erosion prior to the deposition of the overlying Shakopee Formation. This erosion is attributed to a shift in the shoreline of the Prairie du Chien sea. The sea apparently retreated to the south after deposition of the carbonates of the Oneota Formation. Sandy layers in the lower Shakopee Formation (New Richmond Member) formed in a near-shore setting when the sea later advanced northward over the eroded Oneota surface.

At several locations in Allamakee County, the Prairie du Chien exhibits fracturing and mineralization, and small quantities of lead were mined from mineralized fracture zones during the 1880s. The mineral deposits did not prove to be extensive, however, and the mining ventures were short-lived, lasting only a few years.

The Prairie du Chien is still valuable in other ways. It is used for road aggregate in Allamakee County and serves as a source of groundwater in the subsurface over a large area of eastern Iowa. In Minnesota, dolomite from the Prairie du Chien is quarried for building stone.

ST. PETER FORMATION

The St. Peter Formation consists primarily of quartz sandstone, although at some localities shales are found at its base. The St. Peter interval contains a considerable quantity of shale in northwestern Iowa, reflecting the proximity of sediment sources in the Transcontinental Arch. During Middle Ordovician time, the Transcontinental Arch was exposed and shed muddy sediments into the adjacent seaway (fig. 4.3).

The Prairie du Chien–St. Peter contact is irregular (fig. 4.11). Much of this irregularity was produced by erosion of the Prairie du Chien prior to deposition of the St. Peter Formation. In northeastern Iowa, some of the unevenness of the contact may be attributed to the presence of large algal structures in the upper Prairie du Chien. The algal masses are domelike in some areas, and they formed an initial surface of irregularity on which the St. Peter was deposited.

The St. Peter Formation is well exposed in Pikes Peak State Park near McGregor in northeastern Iowa. The thickness of the formation is quite variable in the park because of the uneven nature of the contact with the underlying Prairie du Chien Group. The St. Peter is thickest where the Prairie du Chien is thinnest and thinnest where the Prairie du Chien is thickest. The composite thickness of the total interval between the base of the Prairie du Chien and the top of the St. Peter is fairly uniform, however.

A hike down the trail to Sand Cave and Pictured Rocks in Pikes Peak State Park provides a spectacular view of brightly colored sandstones of the St. Peter Formation. In this area, many of the sandstone deposits are stained by iron oxides, giving yellow, orange, and red colorations to the normally white and light gray sandstones of the formation.

The St. Peter Formation is an important source of sand for the manufacture of glass at Ottawa, Illinois, and elsewhere. In places, the St. Peter is so weakly cemented that it is mined with high-pressure fire hoses. At one time, sand from the St. Peter was mined at Clayton, Iowa. The early excavations, commencing in 1878, were open-pit operations. In 1959, underground mining was initiated.

Although no longer active as a sand mine, the underground operation at Clayton is still in use. It currently serves as an underground storage area for grain, fertilizer, coal, and other items. Ventilation and dehumidifying equipment have been added to control moisture problems, and concrete floors have been poured in about 20 acres of the former mine. Approximately 60 acres of open spaces are present overall.

In the early 1960s, the underground mine at Clayton was the designated fallout shelter for Clayton County. Boxcarloads of civil defense supplies were stored in the mine, sufficient to serve the needs of 44,000 residents for two weeks if nuclear doomsday arrived.

The St. Peter Formation serves on a limited basis as a source of groundwater. The sandstones of the St. Peter are very porous and permeable, but they are poorly cemented and cave in easily in wells. Consequently, wells are usually drilled through the St. Peter to the underlying Prairie du Chien and Jordan aquifers, and the St. Peter Formation is cased to prevent the friable sandstones from caving in.

In the subsurface of Iowa (near Redfield in Dallas County and Keota in Washington County), natural gas is stored in the porous sandstone layers of the St. Peter. Gas storage and other facets of Iowa's economic geology are discussed further in chapter 11.

GLENWOOD FORMATION

The Glenwood Formation is generally poorly exposed in Iowa. It is composed of greenish shale with minor phosphate as small nodules, grains, and fillings in fossils. Conodonts occur in the unit, but other fossils are rare. The Glenwood formed on a muddy marine shelf.

The formation thickens southeastward in Iowa's subsurface but is thin to absent in southwestern Iowa and adjacent Nebraska. The upper Glenwood interval includes thick sandstones in southeastern Iowa, Missouri, and Illinois (Starved Rock Sandstone). The sandstone interval has been referred to as the "re-Peter" by well drillers because of its similarity to the underlying St. Peter Sandstone. In the subsurface of northwestern Iowa (Plymouth County), oolitic ironstone occurs in the Glenwood interval (fig. 4.13). These deposits are marine in origin, but the iron may have been derived from Precambrian source rocks exposed in the nearby Transcontinental Arch.

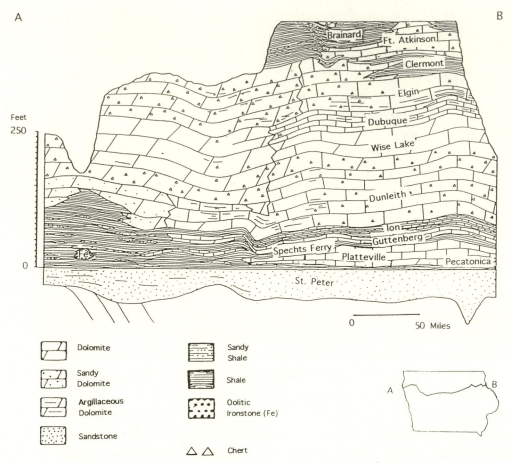

4.13 An east-west stratigraphic cross section showing Middle and Upper Ordovician strata across northern Iowa. Adapted from Witzke 1983b.

Legend:

Dolomite

Sandy Dolomite

Argillaceous Dolomite

Sandstone

Sandy Shale

Shale

Oolitic Ironstone (Fe)

△ △ Chert

PLATTEVILLE FORMATION

The Platteville Formation, named for the southwestern Wisconsin city of that name, is exposed in northeastern Iowa where it consists of two members. The lower member, the Pecatonica, is primarily dolomite. The Pecatonica contains molds of brachiopods and is sandy at the base with a prominent hardground at its top. The upper member, the McGregor, is a fossiliferous limestone with distinctive wavy bedding and abundant fossil burrows. Fossils are abundant and well preserved in this unit, including several species of brachiopods, large ostracodes, trilobites, bryozoans, gastropods, corals, echinoderm fragments, and straight nautiloid cephalopods. Some of the fossil nautiloids measure up to 10 feet in length.

The contact between the Pecatonica and McGregor members is easily recognized in surface exposures in Clayton County, but this is not generally the case in other areas. It is often difficult to distinguish between the two members in the subsurface. A small quantity of oil (approximately 400 barrels) was recovered from the subsurface near Keota in Washington County in 1963. Recovery was from porous dolomites of the Pecatonica Member and from the Starved Rock Sandstone interval.

In northwestern Iowa, the Platteville interval is quite argillaceous (fig. 4.13). These muddy sediments were probably washed into the seas from the nearby Transcontinental Arch (fig. 4.2).

DECORAH FORMATION

The Decorah Formation is one of four formations assigned to the Galena Group in Iowa. (Once recognized as a formation, the Galena is now treated as a group.) The Galena Group consists, in ascending order, of the Decorah, Dunleith, Wise Lake, and Dubuque formations.

The name Decorah Formation was first applied to shale exposures at Decorah in Winneshiek County by the versatile geologist Samuel Calvin (1904), the first permanent state director of the Iowa Geological Survey. Later, G. Marshall Kay (1929) made a comprehensive study of the Decorah Formation and divided it into three members based on lithology. Kay's investigation of the Decorah Formation and its fauna is considered to be one of the classic works in North American stratigraphy and paleontology.

Noted primarily for his contributions to stratigraphy, structural geology, and regional geology, G. Marshall Kay (1904–1975) received B.A. and M.S. degrees in geology at the University of Iowa and a Ph.D. degree from Columbia University, where he served on the faculty for over thirty years. Kay wrote *North American Geosynclines* and many other authoritative publications in the field of geology.

G. Marshall Kay was the son of George F. Kay, who was Iowa's fourth state geologist and an influential geology professor and dean at the University of Iowa. The senior Kay made important contributions to the understanding of the state's Pleistocene history.

The three members of the Decorah Formation, as proposed by G. Marshall Kay, are the Spechts Ferry (lower), Guttenberg (middle), and Ion (upper) members. These units are distinctive in composition throughout most of the Ordovician outcrop belt in northeastern Iowa.

The Spechts Ferry Member is primarily a greenish gray shale, although in the subsurface in parts of southern Iowa the unit is locally sandy. In northeastern Iowa, the Spechts Ferry contains thin layers of altered volcanic ash called bentonite. Well-preserved brachiopods and bryozoans are abundant in the Spechts Ferry in surface exposures in northeastern Iowa.

In northeastern Iowa, the Guttenberg Member consists primarily of limestone with abundant fossils and fossil fragments; two bentonite layers have been identified. Phosphatic nodules occur in the lower part of the Guttenberg in the subsurface. In the subsurface of southeastern and east-central Iowa, the Guttenberg has interbeds of brown organic shale, and some beds contain as much as 20 percent total organic carbon. The occurrence of organic-rich shales such as these suggests the presence of low-oxygen conditions on the seafloor during deposition of the Guttenberg Member. A rich fossil fauna indicates more oxygenated conditions at other times during Guttenberg deposition, however. The brown organic-rich shales of the Guttenberg occur in the outcrop belt, where they are commonly oxidized; these beds constitute the so-called oil rock of the zinc-lead-mining district of northeastern Iowa, northwestern Illinois, and southwestern Wisconsin.

The Ion Member is composed of interbedded fossiliferous shales and argillaceous limestones. Bryozoans, including the distinctive "gumdrop" bryozoan (*Prasopora* sp.), are common in the Ion at several localities in northeastern Iowa. In western Iowa, phosphatic nodules occur in the member in the subsurface. Both the Ion Member and the underlying Guttenberg Member are sandy in southwestern Iowa. The sand was apparently derived from the nearby Transcontinental Arch, which projected above the seas during Middle Ordovician time.

Witzke (1983b) established that the entire Decorah interval in the subsurface of northern and western Iowa is dominated by calcareous shales with carbonate interbeds (fig. 4.13). In these areas, the members of the Decorah Formation cannot be distinguished. Carbonate rocks in the underlying Platteville Formation also give way to shales in the subsurface of western Iowa (fig. 4.13). For this reason, the Platteville and Decorah formations in the subsurface are often logged as Decorah-Platteville undifferentiated.

DUNLEITH, WISE LAKE, AND DUBUQUE FORMATIONS

The Dunleith, Wise Lake, and Dubuque formations constitute the middle and upper parts of the Galena Group. These formations are resistant carbonates (limestone or dolomite) and form bold cliffs along the Mississippi River from Dubuque northward to Guttenberg and McGregor in northeastern Iowa. Rocks of the Galena Group also contribute significantly to the scenic topography of Winneshiek County, where they form impressive cliffs and columns along the Upper Iowa River. Geologist David Dale Owen (1840) described the "lead-bearing beds" (part of the Galena Group) near Dubuque (fig. 4.1) in the epigraph at the beginning of this chapter.

The carbonate rocks of the Galena Group are susceptible to solution by slightly acidic groundwater, and sinkholes are common in these strata in northeastern Iowa. The beautiful and well-known Coldwater Cave in Winneshiek County developed in rocks of the Galena Group, providing dramatic evidence of the work of groundwater. Commercial caves such as Wonder in Winneshiek County, Crystal Lake in Dubuque County, and Mystery and Niagara in southwestern Minnesota are in strata of the Galena Group.

The Dunleith portion of the Galena Group is highly cherty, preserves two zones of receptaculitids (green algae), and contains numerous hardgrounds (fig. 4.14). Delgado (1983) recorded sixty-seven hardgrounds in the Dunleith. Overall, hardgrounds have been reported at some 153 separate stratigraphic levels within the Dunleith, Wise Lake, and Dubuque formations. The Dunleith is well exposed in roadcuts near Guttenberg in northeastern Iowa. This formation has remarkable continuity and is recognizable throughout most of the Upper Mississippi Valley.

The Dunleith contains well-developed burrows and burrow patterns. Many of the burrows have been converted to dolomite, probably because the burrow fillings were more permeable than the surrounding rock and allowed easier access to dolomitizing fluids. Layers of well-washed fossil debris occur at various levels in the Dunleith and may represent ancient storm deposits.

A variety of marine fossils are known from the Dunleith, including brachiopods, bryozoans, ostracods, receptaculitids, sponge spicules, echinoderm remains, and gastropods. These fossils suggest that the Dunleith was laid down in a normal marine setting. The presence of large receptaculitids indicates that sunlight reached the seafloor during Dunleith deposition.

The Wise Lake Formation contains little or no chert and is less argillaceous than the underlying Dunleith Formation. The Wise Lake Formation has several hardground surfaces and preserves a zone of receptaculitids (fig. 4.14). The Dubuque Formation contains distinctive shale partings. The Wise Lake and Dubuque formations are exposed in northeastern Iowa from Dubuque to Decorah, where they form the upland bedrock surface.

Both limestone and dolomite facies are present in the three formations in surface exposures in northeastern Iowa. Dolomite facies are pervasive in the subsurface (figs. 4.13 and 4.15). Dolomite content also increases upward stratigraphically in rocks of the Galena Group. For example, rocks of the Dubuque Formation have undergone greater dolomitization than strata of the Dunleith Formation.

Although lead is no longer mined from the Galena Group in Iowa, the rocks still have economic value. Rocks of the Galena Group serve as an important source of road aggregate in northeastern Iowa, and the Galena interval serves as a minor aquifer for groundwater. In the past when stone was used as an important building material in Iowa, the Galena was frequently quarried for that purpose. Appropriately, the Julien Dubuque Monument at Dubuque was constructed from Galena stone.

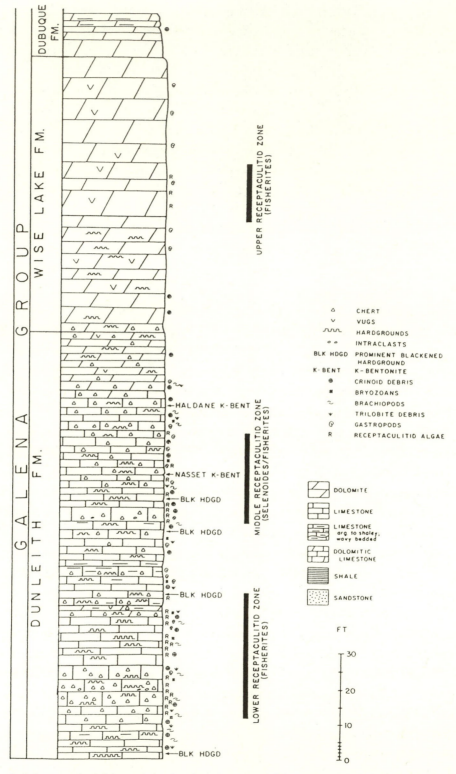

4.14 A stratigraphic section of Ordovician strata of the upper Galena Group, Guttenberg area of northeastern Iowa. Note the position of prominent hardgrounds, bentonites, and receptaculitids. Adapted from Witzke and Glenister 1987a.

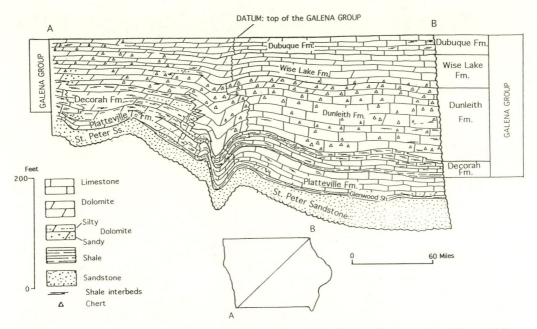

Galena Group labels: Dubuque Fm., Wise Lake Fm., Dunleith Fm., Decorah Fm., Platteville Fm., St. Peter Ss.

Legend:
- Limestone
- Dolomite
- Silty Dolomite / Sandy
- Shale
- Sandstone
- Shale interbeds
- △ Chert

Feet 200 / 0

0 ————— 60 Miles

4.15 A northeast-southwest stratigraphic cross section of the Galena Group in Iowa. The Middle Ordovician St. Peter, Glenwood, and Platteville formations are also shown. Adapted from Witzke 1983b.

The Galena Group and Lead and Zinc Mining

The Galena Group served as a host rock (fig. 4.16) for lead and zinc in the Upper Mississippi Valley Zinc-Lead District of southwestern Wisconsin, northwestern Illinois, and northeastern Iowa (fig. 4.17). The mineralization occurs where carbonate beds are intersected by vertical or nearly vertical breaks called joints. Many of the joints have been enlarged by solution activity. Most of the mining in the Dubuque area took place along vertical joints (fig. 4.18); early miners called these crevices or gash veins. The story of Julien Dubuque's Mines of Spain is an important part of Iowa's geologic history, and a brief account of that story is appropriate here.

According to Sage (1974), the British gained an advantage in their quest for North American domination by their victory over the French in the French and Indian War (Seven Years' War, 1756–1763). France, anticipating defeat, had secretly ceded New Orleans and all of the country west of the Mississippi to Spain in 1762 to prevent these lands from falling into British hands. This transfer was publicly acknowledged in the Treaty of Paris in 1763. Upon losing the war, France gave up Canada and any claim to the Ohio Valley and land east of the Mississippi, except New Orleans. Thus, in 1763, Spain was the legal owner of Iowa, while lands east of the Mississippi were controlled by England. After the American Revolution, the victorious American colonies acquired control of the area east of the Mississippi River.

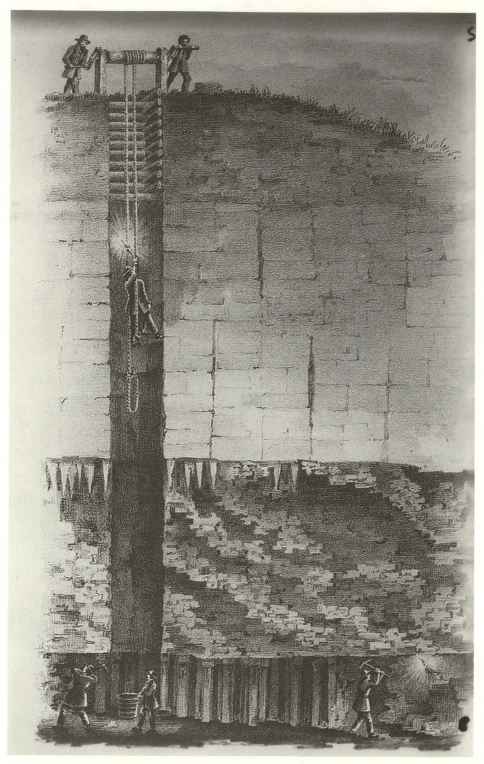

4.16 The mining of lead in the Dubuque, Iowa, area in the middle of the nineteenth century. From a report by geologist David Dale Owen. Courtesy of the Calvin Photographic Collection, Department of Geology, University of Iowa.

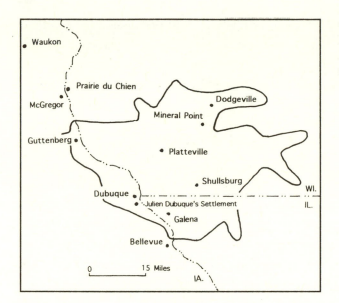

4.17 The extent of the Upper Mississippi Valley Zinc-Lead District. Adapted from Ludvigson and Dockal 1984.

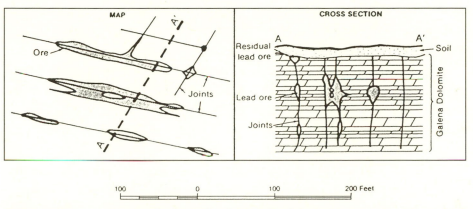

4.18 Crevice, or gash-vein, deposits of lead ore in the Galena Group. Adapted from Heyl, Agnew, Lyons, and Behre 1959 and Anderson 1983.

There were no permanent Euro-American settlements in Iowa during the Spanish regime, but the Meskwaki (Fox) Indian village of Kettle Chief was located near the mouth of Catfish Creek in what is now Dubuque. The Meskwaki possessed a productive lead mine at Catfish Creek. However, they refused to permit any Euro-Americans to work the mines until Julien Dubuque arrived on the scene.

Described as short and stocky with a dark complexion, Julien Dubuque made frequent visits to the Meskwaki village on Catfish Creek in the late 1780s. He lavished presents on the Meskwaki, became their friend, and wooed a young Meskwaki woman named Potosa. The Meskwaki adopted Dubuque as a member of their tribe and gave him the name Little Night. Potosa became Dubuque's wife.

In September 1788, the Meskwaki granted Dubuque permission to work their lead mine "as long as he shall please" and to withdraw from it without limitations. Furthermore, they advised: "He shall be free to search wherever he may think proper to do so, and to work peaceably without anyone hurting him or doing him any prejudice in his labor."

Julien Dubuque brought in ten white men to assist him with the mining operation. They served as teamsters, smelters, storekeepers, and general overseers. The actual work of mining was carried out by Native Americans — the women and elderly men. The warriors and young males considered the physical labor of mining to be beneath their dignity.

Primitive methods were used in the mining operation. No shafts were sunk, and the windlass and bucket were unknown to these early miners. Tunnels (drifts) were excavated into the hills as far as possible, and the lead ore was dragged out in buckets. The mines were worked with simple tools: shovel, hoe, pickax, and crowbar. The ore was smelted in a furnace that Julien Dubuque constructed specifically for the purpose. After smelting, the ore was reduced to bars or pigs. Dubuque took the finished product by boat to St. Louis and exchanged the lead for goods to trade with the Native Americans.

Monuments erected at the confluences of the Little Maquoketa River and the Tete des Morts Creek with the Mississippi River marked the upper and lower boundaries of Julien Dubuque's land. Dubuque later petitioned the Spanish governor in New Orleans to have the title of his land confirmed. Dubuque's plea was approved in 1796, and, as requested, his tract of land became known as the Mines of Spain. One report indicated that Dubuque's mines produced 20,000 to 40,000 pounds of lead per year. Julien Dubuque never became rich in his mining and trading ventures; he died in 1810 at the age of forty-five, deeply in debt.

In 1833, the United States government opened the land west of the Mississippi River for Euro-American settlement, which brought an influx of settlers into Iowa. Some of the new settlers vigorously pursued lead mining, and lead production increased at a fairly steady rate to a maximum production in 1846 and 1847. Production started to decline in 1857, and in 1910, the last of the large mines in Iowa closed. There are no active mines in the Dubuque area at the present time.

A limited amount of zinc ore was mined in the Dubuque area — principally zinc carbonate (smithsonite) and zinc sulfide (sphalerite). Zinc mining persisted in southwestern Wisconsin on a limited scale until 1979, when the region's last operating mine closed at Shullsburg.

How did these ores of zinc and lead form? Although some technical details are unresolved, the basic pattern is known. The ore deposits were derived from hot saline brines (80 to 200 degrees C) that were carried upward in solution. The zinc and lead ores were deposited as sulfide minerals, precipitated along fractured-and-jointed carbonate rocks.

Sulfide minerals are chemically unstable in the presence of atmospheric oxygen; they oxidize (alter in chemical state) when the local water table is lowered. In the region around Dubuque, the water table was lowered when the Mississippi River gorge was excavated by increased runoff and erosion in late glacial times. This downcutting exposed the sulfide minerals in the Ordovician bedrock to the destructive process of oxidation. Iron oxides were formed from the alteration of iron sulfides, and zinc oxides and carbonates ("drybone") were produced during the weathering of zinc sulfides. Lead sulfide (galena) weathers and oxidizes more slowly, so significant quantities of partially altered lead sulfide accumulated on the floors of joints and crevices. These concentrations of lead ore were easily separated and recovered by early miners.

Substantial quantities of zinc sulfide (sphalerite) remained in an unaltered state below the water table, and these ores were exploited later when new technologies were developed. The development of pumping systems in the late 1800s allowed mining operations to proceed below the water table.

MAQUOKETA FORMATION

The Maquoketa Formation was named for exposures in the Little Maquoketa River valley in Dubuque County, but none of the five members of the formation have a type locality there. Four of the members (Elgin, Clermont, Fort Atkinson, and Brainard) are named for localities in northeastern Fayette County or southwestern Winneshiek County. A fifth member, the Neda, is poorly exposed in Iowa; its type locality is in eastern Wisconsin.

The Elgin Member is composed primarily of shale and argillaceous dolomite. Phosphatic beds are present also. The so-called depauperate zone — an interval containing small phosphatic fossils — occurs in the Elgin Member. The depauperate zone, which has a sporadic distribution throughout the state, is interesting because of its diminutive fossils. At one time, the small fossils were attributed to impoverished living conditions on the Maquoketa seafloor, conditions that produced organisms with stunted growth. Today, the fauna is interpreted as especially adapted for a seafloor environment with recurrent oxygen stresses (fig. 4.19).

The Elgin Member is exposed at the well-known Graf locality in Dubuque County (fig. 4.20). There, a variety of fossils occur, including nautiloid cephalopods, graptolites, brachiopods, trilobites, gastropods, scaphopods, bivalves, and crinoid debris.

The Clermont Member is primarily shale. The overlying Fort Atkinson Mem-

4.19 A reconstruction of the lower Maquoketa faunal assemblage of diminutive organisms. A stand of algae is shown as a possible food source and attachment site for the invertebrates. Symbols: L = *Leptobolus* sp., an inarticulate brachiopod; Li = *Liospira* sp., a snail; N = *Nuclites* sp., a bivalve; Pr = *Praenucula* sp., a clam; S = *Septemchiton* sp., a chitin (primitive mollusk); P = *Plagioglypta* sp., a scaphopod (a type of mollusk). Adapted from Bretsky and Bermingham 1970.

ber is a cherty carbonate in Fayette and Winneshiek counties, but elsewhere it is composed of shale. The Brainard Member is characteristically represented by a thick sequence of shale (fig. 4.12). In parts of eastern Iowa, it is impossible to differentiate the Clermont, Fort Atkinson, and Brainard members because they are so similar in composition.

A distinctive assortment of fossils occurs in the "*Cornulites* zone" within the upper Brainard (fig. 4.20). The zone is named for *Cornulites*, a tapering worm-like tube, but brachiopods, bryozoans, horn corals, echinoderm debris, gastropods, bivalves, nautiloids, and trilobites also occur in the zone.

The uppermost member of the Maquoketa Formation is the Neda Member. It is found in Iowa only where the Maquoketa is thickest. Pre-Silurian erosion has removed the Neda in many areas of the state. The Neda contains red shales and, at some localities, bears ferruginous (iron-bearing) and phosphatic ooids, disks, and clasts. Horizontal and vertical burrows penetrate the Neda. This member probably represents nearshore marine deposition during a shallowing phase of the Late Ordovician sea.

The Maquoketa Formation occurs throughout the state, except where re-

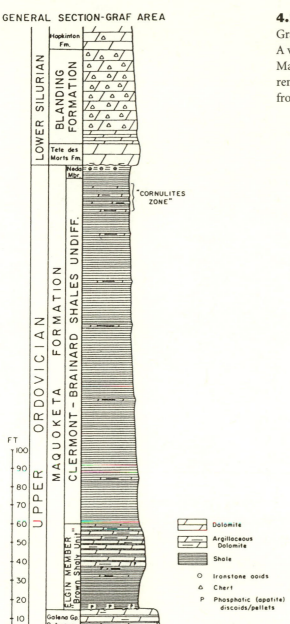

GENERAL SECTION-GRAF AREA

4.20 A stratigraphic section in the Graf area, Dubuque County, Iowa. A variety of fossils occur in the Maquoketa Formation at this renowned fossil locality. Adapted from Witzke and Glenister 1987b.

Legend:

- Dolomite
- Argillaceous Dolomite
- Shale
- O Ironstone ooids
- △ Chert
- P Phosphatic (apatite) discoids/pellets

moved by erosion in southeastern, northeastern, north-central, and northwestern Iowa. In eastern Iowa, the contact of the Maquoketa Formation and the underlying Galena Group is sharp. In western and southwestern Iowa, however, the Galena and Maquoketa lithologies are similar, and the contact between the two formations is more difficult to recognize. The Maquoketa Formation is overlain unconformably by rocks of Silurian age.

In the subsurface, the Maquoketa is characterized by lateral variations in com-

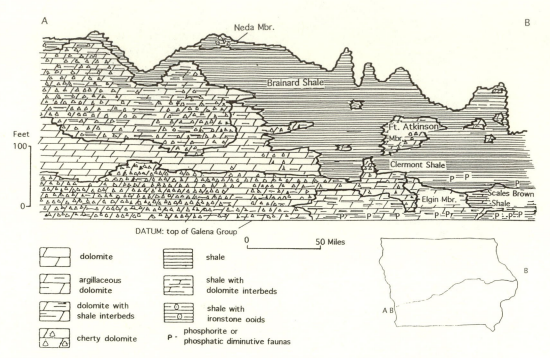

4.21 An east-west stratigraphic cross section of the Maquoketa Formation in Iowa. Shale facies dominate in eastern Iowa, whereas dolomite is more abundant in western Iowa. Adapted from Witzke 1983b.

position (fig. 4. 21). The formation becomes very dolomitic in the subsurface of western Iowa.

In eastern Iowa, the Maquoketa Formation contains a thick interval of shale and is impervious to fluids. Because of this, the formation has been used to store liquefied petroleum gas. Large caverns were excavated in the Maquoketa Formation in the subsurface under Johnson and Polk counties, and liquefied petroleum gas is stored there. Storage of this type is important because the demand for liquefied petroleum gas is seasonal. Underground storage capacity allows the pipeline industry to meet customer demands during times of peak usage.

The Fort Atkinson Member of the Maquoketa Formation provided stone for one of Iowa's earliest forts, Fort Atkinson in Winneshiek County. The fort was constructed in 1840 from stone from a quarry adjacent to the site of the fort. The quarry site today, the type locality of the Fort Atkinson Member, is part of the state preserves system. Fort Atkinson is significant from a historical point of view because it was a United States military installation built to protect Native Americans from Native Americans (the Winnebago from their enemies, the Sioux, the Sauk, and the Meskwaki).

4.22 Life on an Ordovician seafloor. Clockwise from the lower right corner: four stalked crinoids (sea lilies); part of a crinoid stem; an inarticulate brachiopod in a burrow; a bivalve (clam) in a burrow; an articulate brachiopod resting on the seafloor; a gastropod (snail), crawling; a trilobite; a conical shell of uncertain affinity — possibly a mollusk or an annelid worm; a trilobite; additional conical shells; a trilobite; a branching bryozoan; two conical shells; an attached articulate brachiopod; a nautiloid cephalopod grasping an articulate brachiopod; attached articulate brachiopods; a branching bryozoan; a high-spired snail; ten articulate brachiopods attached to the seafloor; and two fragments of crinoid stems. Adapted from McKerrow 1978.

Ordovician Life

The warm, shallow Ordovician seas were well suited for marine invertebrates (fig. 4.22), and Iowa's rock record preserves many of them as fossils. All of the major groups of marine invertebrates are known in Iowa's rock record. In addition, marine algae are common in the Prairie du Chien and Galena groups. Representative Ordovician fossils are shown in figure 4.23. Discussions of key fossil groups follow.

BRACHIOPODS

Brachiopods secrete two shells. The animal lives inside the shells, which serve as protective armor for the organism's soft parts. Although brachiopods still live today, they were much more abundant in the past. Today, bivalves (pelecypods) have largely replaced brachiopods on modern seafloors.

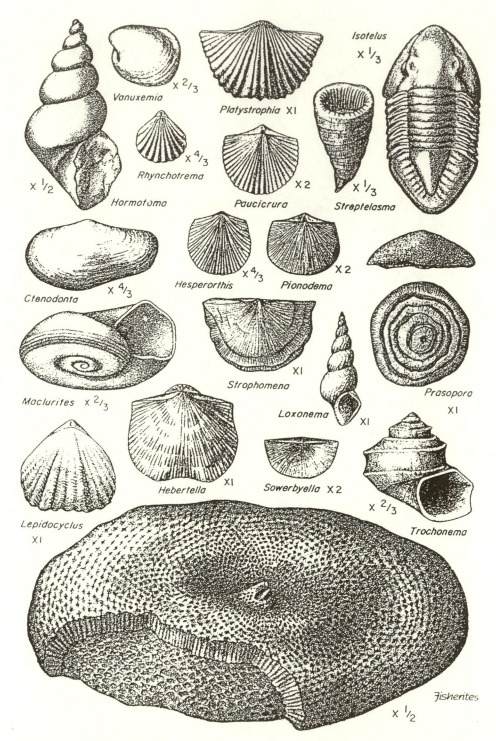

4.23 Representative Ordovician fossils. Gastropods (snails): *Hormotoma, Loxonema, Trochonema, Maclurites*; bivalves (clams): *Vanuxemia, Ctenodonta*; brachiopods: *Platystrophia, Paucicrura, Hesperorthis, Pionodema, Strophomena, Lepidocyclus, Hebertella, Sowerbyella, Rhynchotrema*; bryozoans: *Prasopora*; corals: *Streptolasma*; trilobites: *Isotelus*; green algae: *Fisherites* (previously called *Receptaculites*). From Willman et al. 1975. Courtesy of the Illinois Geological Survey.

Two major groups of brachiopods are recognized: articulate brachiopods and inarticulate brachiopods. The inarticulate brachiopods are more primitive and secrete a shell composed of calcium phosphate and chitin (a complex organic substance). These materials generally are not well preserved in the geologic record. Articulate brachiopods construct shells of calcium carbonate, which is very durable. Articulate shells are hinged and interlock with tooth-and-socket structures. Brachiopods live attached to the seafloor, at rest on the bottom of the sea in an unattached position, or within the sediment of the seafloor (fig. 4.22).

Brachiopods are well represented in the Platteville, Decorah, and Maquoketa formations of Iowa. A few representative types are shown in figure 4.23.

BRYOZOANS

Bryozoans, or bryozoa, are often called moss animals. They are colonial organisms, consisting of tiny individuals that live in skeletons of various shapes. Most skeletons are composed of calcium carbonate. Skeletons may be incrusting, branching, netlike, lacy, or hemispherical in shape. Branchlike and hemispherical varieties are common in Iowa's Ordovician rocks. Bryozoans are particularly common in the Platteville and Decorah formations. *Prasopora*, the "gumdrop" bryozoan, is abundant in the Decorah Formation of northeastern Iowa (fig. 4.23).

CORALS

The individual coral animal has a sacklike body with tentacles around its mouth. Corals secrete a calcium carbonate cup, or skeleton, in which the soft body parts reside. Coral skeletons can be either solitary or colonial. The solitary, horn-shaped types are found in the Ordovician rocks of Iowa (fig. 4.23).

RECEPTACULITIDS

Fisherites (previously known as *Receptaculites*), *Ischadites*, and *Selenoides* are common in the Galena Group of northeastern Iowa. *Fisherites* is particularly abundant at certain horizons in the Galena Group and serves as a local index fossil for these strata. Once known as the sunflower coral because of its distinctive appearance, *Fisherites* is not related to corals at all (fig. 4.23). Receptaculitids were classified with sponges at one time, but they are now grouped with algae.

MOLLUSKS

Mollusks are a diverse group of animals, most of which secrete an external shell or shells of calcium carbonate. The phylum Mollusca is divided into six subdivisions, called classes. Three of these classes are well represented in the fossil

4.24 The scaphopod *Plagioglypta iowaensis* shown in its presumed life position on the Ordovician seafloor. Adapted from Bretsky and Bermingham 1970.

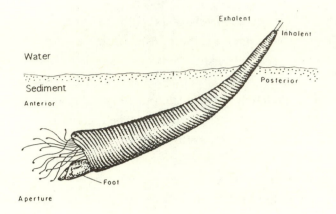

record, namely bivalves (pelecypods or clams), gastropods (snails), and cephalopods. Two additional classes, the scaphopods and chitons, are also present in the Ordovician rocks of Iowa.

Snails and clams (fig. 4.23) are found throughout the Middle and Upper Ordovician rocks of Iowa. These mollusks were represented as burrowers and crawlers on the Ordovician seafloor (fig. 4.22).

Nautiloid cephalopods (fig. 4.22) are common in the Middle Ordovician rocks of the state. Large specimens more than 10 feet in length have been found in the Platteville, Decorah, and Maquoketa formations. Calvin (1904) noted a piece of a huge nautiloid nearly 5 feet in length and 10 inches in diameter at the larger end. He estimated that the complete specimen would have measured approximately 12 feet in length.

Tens of thousands of straight-shelled nautiloids occur in the lower part of the Maquoketa Formation at Graf in Dubuque County. Some of these nautiloids exist telescoped inside each other, as many as five total, representing an unusual concentration and sorting of shells in the Ordovician seas.

Cephalopods differ from other mollusks in that they have a multichambered shell. The cephalopod shell is partitioned off by calcium carbonate plates called septa. The chambers are filled with gas, and this allows the animal to float. Cephalopods can move about by a method of jet propulsion in which they eject water from a tube near their head region and scoot away in the opposite direction.

Scaphopods have a tusk-shaped shell that is open at both ends. They burrow into the sediment with their muscular foot and draw current through their shell to obtain nutrients (fig. 4.24).

The scaphopod *Plagioglypta iowaensis* occurs in abundance in the lower Maquoketa Formation as part of what has been called the depauperate fauna (fig. 4.19). The fauna consists primarily of tiny shells of gastropods, bivalves, inarticulate brachiopods, and scaphopods. Primitive segmented mollusks called chitons are also found associated with the depauperate fauna.

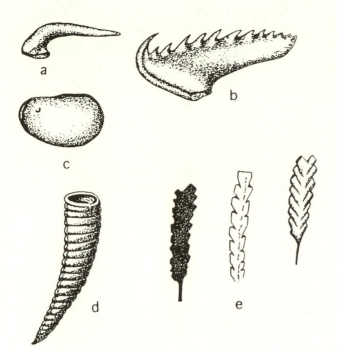

4.25 Miscellaneous Ordovician fossils: (a) conodont, phosphatic fossil representing parts of a feeding device of an early fishlike chordate (greatly magnified); (b) scolecodont, jaw parts of a marine worm (greatly magnified); (c) ostracod, a type of arthropod; (d) *Cornulites*, a conical shell of uncertain origin — possibly a mollusk or an annelid worm; and (e) graptolites from the Maquoketa Formation. Adapted from Collinson 1964 and Rose 1967.

WORMS

Scolecodonts are the fossilized jaw and mouth parts of marine annelid worms. These tiny black fossils are microscopic (fig. 4.25). Scolecodonts are common constituents in most marine sedimentary rocks.

ARTHROPODS

Arthropods are complex invertebrates that possess segmented external skeletons. Trilobites (fig. 4.23) are examples of the arthropod group. Trilobite fossils generally occur as fragments because the skeletons tend to break along the segments. Trilobite fragments are common in the Middle and Upper Ordovician strata of Iowa, but entire unbroken specimens are rare. A number of enrolled forms have been reported, and they may represent an adaptation for protecting the animal's soft underside.

Beds of the Elgin Member of the Maquoketa Formation in Fayette County yield lots of trilobite fragments and occasional complete specimens of the genus *Isotelus* (fig. 4.23). This part of the Maquoketa Formation has been referred to as the *Isotelus* beds.

Ostracods are also arthropods, although they do not have an obvious segmentation. Ostracods have a shell composed of two valves and also have segmented appendages that project from the shell. Only the shell preserves in the fossil record, however. Ostracods are typically microscopic in size, but a few Ordovician varieties are large enough to be seen with the naked eye (fig. 4.25).

Echinoderms are some of the most advanced of the invertebrates. They come in a variety of forms, but most are characterized by a general five-fold symmetry pattern. Some echinoderms, like the crinoids (sea lilies) and cystoids, are attached to the substratum with a stalk composed of buttonlike segments of calcite (fig. 4.22). Cystoid is an informal name for stalked echinoderms now assigned to the taxonomic classes Diploporita and Rhombifera.

Fragments of echinoderm stems are common in the fossil record. Some exquisite specimens of complete crinoids, cystoids, and other echinoderms have been collected from Iowa's Ordovician strata. Even rare starfish and edrioasteroids are known.

The recovery of these superb echinoderms is due in large part to the work of a group of dedicated amateurs. The fossil collections were reposited at the University of Iowa so that professional paleontologists could study them. Several new species are included in the collections. Scientific papers on these intriguing Ordovician fossils have been published by Brower and Strimple (1983) and Brower (1995).

A somewhat odd echinoderm, the carpoid *Iowacystis*, occurs in the Fort Atkinson Member of the Maquoketa Formation. The first discovery of this rare form was from a foundation slab at the Fort Atkinson Historical Site in Winneshiek County. Several other specimens have since been recovered from outcrops of the Fort Atkinson Member. The initial discovery is reported to have taken place by chance when University of Iowa paleontologist A. O. Thomas was caught in the rain while on a field trip with his students. The group sought shelter at the Fort Atkinson Historical Site, and one of the students spotted the light-colored *Iowacystis* because it stood out against the darker background of the wet rock of the fort's foundation.

Iowacystis and related carpoids are grouped together in a taxonomic division known as the iowacystids (Family Iowacystidae). They belong to a rare subdivision of echinoderms known as Homalozoa. These rare fossils, found in the Maquoketa Formation and Galena Group of Iowa and adjacent Minnesota and Illinois, were well adapted for a bottom-dwelling existence. They possessed tail-like structures rather than true stalks such as are commonly associated with crinoids and cystoids. The iowacystid carpoids apparently sculled across the bottom of the sea and browsed on organic debris that accumulated on the seafloor (fig. 4.26).

GRAPTOLITES

The name graptolite is derived from a Greek word meaning "to write," and graptolite fossils are often preserved as pencil-like markings on the bedding planes of rocks. Graptolites represent a fossil group that is assignable to the

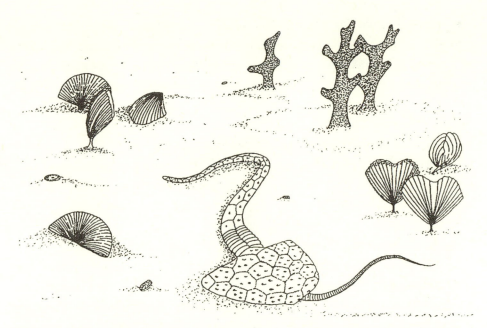

4.26 A reconstruction of an Ordovician seafloor showing an iowacystid carpoid in its inferred life position. Also shown are brachiopods and bryozoans. Based on a reconstruction by Kolata et al. 1977.

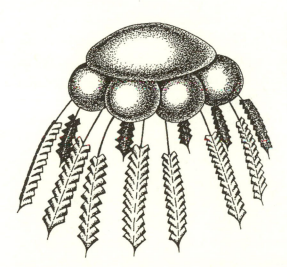

4.27 A reconstruction of an Ordovician graptolite. The individual animals of the colony lived in tiny cups arranged along slender stems. A floating variety is shown here. After Collinson 1964.

hemichordates. The hemichordates are classified just below the chordates, the group that includes the subphylum vertebrata. Although the hemichordates lack true backbones, they do have a cartilaginous rod (notochord).

Two major divisions of graptolites are recognized — the dendroids and the graptoloids. The dendroids attached to the seafloor, the graptoloids floated (fig. 4.27). Both types of graptolites are found in the Ordovician rocks of Iowa.

Graptolites are particularly abundant in the Maquoketa Formation, where they have been found preserved as carbon films, three-dimensional molds, and uncompressed colonies. The excellent preservation of these graptolites was apparently favored by low-oxygen settings on the seafloor, resulting in an absence of scavengers. Some representative graptolites from the Maquoketa Formation are shown in figure 4.25.

CONODONTS

Conodonts are tiny phosphatic fossils that look like teeth. They are mouth parts from eel-like creatures, perhaps our earliest vertebrate relatives. Conodonts are important microfossils in Iowa's Ordovician rocks, and they serve as excellent index fossils.

MISCELLANEOUS FOSSILS

Not all fossils are easy to classify; one such fossil is the conical shell called *Cornulites* (fig. 4.25). Some paleontologists consider *Cornulites* to be related to mollusks, while others believe it is more closely related to annelid worms. *Cornulites* is of interest because it occurs in abundance in the upper beds of the Maquoketa Formation and serves as a marker for that horizon (fig. 4.20).

Summary

During Early Ordovician time, ancient Iowa was located near the edge of a shallow inland sea that shifted back and forth, leaving behind a record of carbonate rocks, sandy carbonate rocks, and quartz sandstones. Algal mats and algal masses were common constituents of these seas, and their remains are now well represented in the Prairie du Chien Group of Iowa.

The sea withdrew from Iowa during the later part of Early Ordovician time, and the Prairie du Chien Group was subjected to weathering and erosion. An unconformity between the Sauk Sequence below and the overlying Tippecanoe Sequence marks this interval.

In Middle Ordovician time, the sea advanced over the irregular, eroded Prairie du Chien surface and laid down a shoreline and inner-shelf deposit of well-sorted quartz sand (the St. Peter Formation). As the Middle Ordovician sea advanced farther to the north beyond Iowa, the environment of deposition changed from a sandy seafloor to a shallow, and sometimes muddy, carbonate shelf. The marine muds and carbonate sediments that were deposited in this environment now compose the Glenwood, Platteville, Decorah, and Dunleith formations. Carbonate and mud deposition continued in Late Ordovician time when the Wise Lake, Dubuque, and Maquoketa formations were laid down. These formations contain a diverse and abundant record of Ordovician marine life. Small amounts of volcanic ash are represented by bentonite layers in the Decorah and Dunleith formations.

At times, the bottom waters of the Late Ordovician sea that covered Iowa were low or deficient in oxygen, and organic constituents were preserved in some of the shales. Although shales and carbonates are the dominant deposits of Late Ordovician time, some phosphate-rich beds also formed.

The Late Ordovician sea contained a variety of marine invertebrates, and many of these are well represented in the fossil record of the Maquoketa Formation of Iowa. For example, graptolites, cephalopods, gastropods, scaphopods, bivalves, brachiopods, and trilobites are particularly well preserved.

By the end of Ordovician time, the sea had withdrawn from Iowa, and the Upper Ordovician rocks were subjected to weathering and erosion. An unconformity separates Iowa's Ordovician and Silurian rock record.

Acknowledgments

This chapter benefited from constructive suggestions by Brian J. Witzke and the published works of David J. Delgado, James A. Dockal, Brian F. Glenister, Dennis R. Kolata, Greg A. Ludvigson, Robert M. McKay, and Brian J. Witzke. In particular, articles by Delgado, McKay, Witzke and McKay, and Witzke and Glenister increased my understanding of Iowa's surface record of Ordovician rocks. Brian Witzke's subsurface work, regional synthesis, and paleogeographic maps provide a basis for better understanding the Ordovician overall. The account of Julien Dubuque's Mines of Spain was summarized from accounts by William J. Petersen and Thomas Auge.

5 **Silurian** Dolomite and Carbonate Mounds

> It becomes very pitted on the surface, but in the mass resists the weather admirably, and tends to stand in vertical, picturesque cliffs and towers, some of which approach 100 feet in height. . . . Even over the prairies, remote from streams, particularly in the southeastern part of the county, ledges of this same horizon project through the thin drift in numberless places.
>
> Samuel Calvin (1897) describing exposures
> of Silurian dolomite in Delaware County

Silurian strata form some of Iowa's most resistant bedrock units (fig. 5.1), and a conspicuous erosional escarpment marks the edge of the Silurian System in northeastern Iowa. Rocks of Silurian age contribute significantly to the landscape and scenery in state parks and lands such as Backbone, Bellevue, Brush Creek, Echo Valley, Maquoketa Caves, Mossy Glen, Palisades-Kepler, and White Pine Hollow. Furthermore, the southeastward course of the Mississippi River from near Dubuque to Clinton may be controlled in part by the trend of the durable Silurian bedrock. Silurian rocks are noteworthy elsewhere for their resistance to erosion, too. For example, dolomites of Silurian age form the majestic waterfalls at Niagara Falls in New York.

Silurian rocks of the central and southern part of the North American craton occur primarily in structural basins that formed before Middle Devonian time. Arches and uplifts now separate these various basins, and Silurian strata have been stripped from the uplifted areas in most places. In Iowa, the thickest Silurian sections are in eastern Iowa, representing the deposits of the East-Central Iowa Basin. Silurian strata also thicken to the southwest in Iowa; they accumulated in a pre–Middle Devonian basin known as the North Kansas Basin.

Marine sedimentation was probably much more widespread across the early North American continent during Silurian time than implied by the present distribution of Silurian rocks. Figures 5.2 and 5.3 are reconstructions of the paleogeography during Silurian time, approximately 409 to 439 million years ago. The region that is now Iowa was located south of the equator in a vast inland sea. Iowa's Silurian rock record consists almost entirely of marine carbonate sediments. Input of mud and sand to the Silurian seaway was very low, except in areas next to the Taconic Mountain belt (figs. 5.2 and 5.3). Emergent lands included the Transcontinental Arch along the northwest corner of Iowa, a portion of the Canadian Shield to the northeast, and the Ozark Dome to the south (figs. 5.2 and 5.3).

5.1 Resistant Silurian dolomite of the Hopkinton Formation forms the Devil's Backbone in Backbone State Park, Delaware County, Iowa. Photo of an outing in the late 1890s by Samuel Calvin. Courtesy of the Calvin Photographic Collection, Department of Geology, University of Iowa.

5.2 The paleogeographic setting during Early Silurian time when Iowa's Mosalem, Tete des Morts, Blanding, Hopkinton, and lower Scotch Grove formations were deposited. Unshaded areas depict seas, and shaded areas represent emergent lands. Note the presence of the Taconic Mountains on the eastern seaboard and the emergent Transcontinental Arch along the northwestern corner of Iowa. To the north of Iowa, a portion of the Canadian Shield was above the sea, and to the south, the Ozark Dome was emergent. Symbols: m = evaporite crystal molds, o = carbonate ooids, s = sulfate evaporites (primarily gypsum and anhydrite), fe = oolitic ironstones, EQ = proposed paleoequator. From Witzke 1990b.

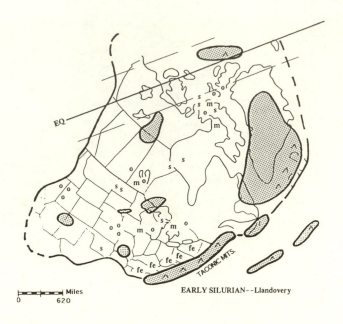

EARLY SILURIAN--Llandovery

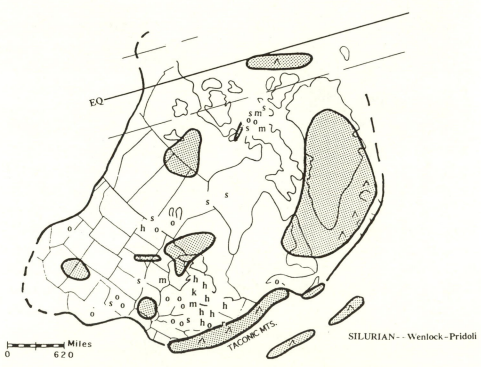

SILURIAN--Wenlock-Pridoli

5.3 The paleogeographic setting during Late Silurian time, when Iowa's upper Scotch Grove Formation and Gower strata were deposited. Symbols as in figure 5.2 with the addition of h for halite (rock salt) and k for potash evaporites. From Witzke 1990b.

Life was prolific in the Silurian seas and led locally to the development of carbonate mounds composed of masses of mud and fossil debris. Although these mounds have been called reefs, they are not at all like the reefs found in modern seas. In Michigan, Ohio, and New York, large basins formed and experienced restricted circulation and high evaporation. This produced waters with elevated salinity and led to the precipitation of evaporites (fig. 5.3) such as gypsum ($CaSO_4 \cdot 2H_2O$) and rock salt (NaCl).

Nature of the Rock Record

The Silurian System in Iowa consists primarily of a single rock type—dolomite. The Silurian dolomites of the state formed by the dolomitization of limestones, a process discussed previously in chapter 4. In dolomitization, magnesium-bearing waters react with the calcium carbonate minerals and fossils in limestones to produce the calcium-and-magnesium carbonate mineral dolomite. The name dolomite is used for both a mineral with the composition of $CaMg(CO_3)_2$ and for a rock that contains more than 50 percent of that mineral, although some geologists prefer the name dolostone for the rock.

Besides changing the composition of limestones, dolomitization also alters the original textures of rocks and fossils. Fossils that are replaced by dolomite generally preserve far less detail than their calcium carbonate precursors.

Dolomitization tends to increase the porosity of rocks. The mineral dolomite [$CaMg(CaCO_3)_2$] is denser than the mineral calcite ($CaCO_3$); if a block of calcite were dissolved and reprecipitated to form dolomite, the dolomite block would occupy about 13 percent less space than the original calcite block. In nature, when calcite is converted to dolomite, there is often a resulting increase in void space in the altered rocks. This void space (porosity) can take the form of molds, representing space formed by the solution of calcium carbonate fossils. In addition, porosity may exist between dolomite crystals (intercrystalline porosity), as small pores within the rock, and as vugs (small crystal-lined cavities within rocks).

The mineral dolomite often forms in well-shaped rhombohedral crystals, and some rocks composed of dolomite exhibit a texture that resembles granulated brown sugar. Consequently, the rock dolomite is often described as having a sucrosic (sugarlike) texture. Although Iowa's Silurian section was altered significantly by dolomitization, the primary depositional textures of the original sediments are generally discernible. Much of the initial carbonate sediment was deposited as carbonate mud and skeletal grains (fragments of brachiopod shells, pieces of echinoderms, spicules of sponges, and skeletal components of corals, bryozoans, sponges, and other fossils).

Many of Iowa's Silurian formations contain chert, a hard, fine-grained rock composed of silicon dioxide. The chert occurs as nodules within dolomite beds and as thin layers interbedded with dolomite. The chert layers and nodules

5.4 A proposed setting for the dolomitization of Iowa's Silurian strata. Note the presence of a mixing-zone environment where freshwater (groundwater) mixes with salt water (marine phreatic water) at the interface of the land and the sea during Late Silurian (Ludlovian) time. As the Silurian sea retreated, environments favorable for dolomitization formed. Adapted from Ludvigson et al. 1992.

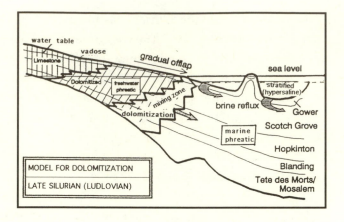

apparently formed by the silicification of carbonate rocks. Fossils, too, are commonly silicified. Silicified fossils are much better preserved than their dolomitized counterparts, confirming that silicification generally preceded the completion of dolomitization. According to Witzke (1992), the silica for silicification and chert formation was likely derived from biogenic silica produced by organisms such as siliceous sponges.

The mixing zone, discussed in chapter 4, may have provided the environmental setting that produced both dolomitization and silicification in Iowa's Silurian rocks. In the mixing zone, freshwater from the land mixes with salt water from the sea. The chemistry of this mixed fluid is intermediate between meteoric (rain) water and marine salt water. Dolomitization and silicification are both favored in the mixing zone environment. Located at the interface of the land and the sea, the mixing zone moves seaward during times of marine regressions. Such settings may explain the widespread dolomitization and silicification in Iowa's Silurian rock record (fig. 5.4).

Carbonate mounds occur in three of Iowa's Silurian formations. Although these features have been called reefs, that terminology is not altogether suitable. When the term "reef" is used in the organic or ecological sense, it suggests the presence of corals and algae that bind sediment and organisms together to produce a wave-resistant structure. The Silurian carbonate mounds are the accumulations of carbonate muds and fossils; the fossils are contained in a mud matrix. Some of the Silurian carbonate mounds do contain corals, but many do not. In fact, corals are volumetrically insignificant components of the carbonate mounds when compared with other skeletal components and carbonate muds. In addition, the ancient rugose and tabulate corals found in Silurian carbonate mounds are now extinct; they likely differed significantly from modern reef-forming corals. Many of the Silurian mounds are composed of substantial amounts of broken crinoid stems and other crinoidal fragments set in a matrix of carbonate mud. Today, stalked crinoids are rare in shelf-depth seas; they inhabit the deeper realms of modern oceans. Other Silurian carbonate mounds contain an abundance of brachiopods in a matrix of carbonate mud. Brachiopods are rare on modern marine shelves, their niches having been largely filled by molluscan bivalves.

According to Witzke (1992), the Silurian carbonate mounds of Iowa have a number of common features. (1) The central mounds are dominated by carbonate mud fabrics, and they display no evidence of an organic framework. (2) Flanking strata are present around the central mounds; these beds contain skeletal grains and other skeletal components, and they often display graded bedding. (3) Submarine cementation is apparent in all of the mounds, suggesting that they became lithified during the time of deposition. (4) Similar geometries occur in the mound facies, including isolated mounds with radial dipping flanks and mounded complexes with complex layering.

In Iowa, unconformities separate the Silurian System from the underlying Ordovician System and the overlying Devonian System. The basal beds of the Silurian were deposited on an irregular erosion surface that developed on the Maquoketa Formation during Late Ordovician time. The seas completely withdrew from Iowa during the Late Silurian, and Silurian rocks were subjected to weathering and erosion during part of Late Silurian and all of Early Devonian time.

Iowa's Silurian Rock Record and Relative Sea-Level Changes

Although they are extensively dolomitized, the Silurian strata of eastern Iowa still reveal a rich assortment of fossils. The succession of faunal assemblages provides important clues concerning changes in depositional environments and water depths during Silurian time. Markes E. Johnson (1975) contributed to a better understanding of Iowa's Silurian when he established the relative bathymetric (depth) positions of a series of benthic (bottom-dwelling) fossil communities found in the Lower Silurian strata of eastern Iowa. Relative changes in water depths can be inferred from the pattern of succession of these faunas in Iowa's Lower Silurian sequence. Johnson's initial work on the Lower Silurian strata was later extended by Witzke (1981b, 1983c, 1992) to the entire Silurian sequence in Iowa. Witzke's efforts led to the development of a relative sea-level curve for the Silurian of eastern Iowa (fig. 5.5).

The changes in sea level displayed on the curve (fig. 5.5) were apparently global in nature. Johnson and his coworkers (Johnson et al. 1991; Johnson and Lescinsky 1986; Johnson et al. 1985) found that sea-level curves based on data from different localities in North America and from localities in China and the Baltic region produced patterns that were remarkably similar. Such parallelism in patterns suggests that the major sea-level changes during Silurian time were worldwide (eustatic) in scope. What process or processes were responsible for sea-level changes of this magnitude? The advance and retreat of massive conti-

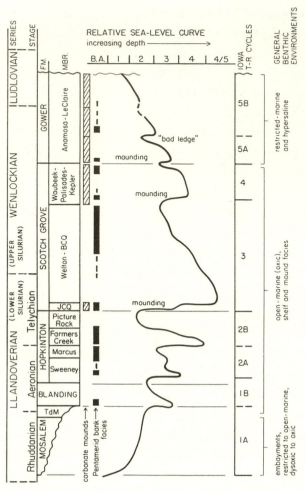

5.5 A relative sea-level curve based on faunal associations and sedimentary features found in the Silurian strata of eastern Iowa. Note the position of Iowa's principal Silurian formations in ascending order: Mosalem, Tete des Morts, Blanding, Hopkinton, Scotch Grove, and Gower. Information is provided concerning water depth and general benthic environments for each of the formations. From Witzke 1992.

nental ice sheets offers one possible explanation, and Early Silurian (Llandoverian) was a time of widespread continental glaciation on the giant southern continent of Gondwana. However, continental glaciers are unknown during Late Silurian time, so other explanations are needed to explain the sea-level fluctuations during that time interval. The relative sea-level curve in figure 5.5 provides a basis for better understanding the depositional environments of Silurian rocks.

Silurian Formations of Iowa

The Iowa Geological Survey divided the Silurian System into four formations until the late 1970s. Two of the formations, the Edgewood and the Kankakee, were assigned to the Alexandrian Series (Lower Silurian of North American usage). The other two formations, the Hopkinton and the Gower, were placed in the Niagaran Series (Middle Silurian of North American usage). Some studies suggested that the upper part of the Gower Formation might belong to the Cayugan Series (Upper Silurian of North American usage).

The name Alexandrian comes from Alexander County in southwestern Illinois, where rocks of that name were studied in the early 1900s. The name Niagaran originates in New York State, where Middle Silurian rocks are well exposed at Niagara Falls and elsewhere. The Cayugan Series was also named for exposures in New York.

James Hall (1811–1898) sought to establish the New York system of Silurian nomenclature as a standard for worldwide correlation, and he committed himself to recovering detailed fossil collections from the Silurian of New York to use for such correlations. Although Hall was not successful in establishing the New York section as the worldwide standard, he did make many contributions to geology and paleontology. His thirteen-volume series, *Paleontology of New York*, is considered one of the classics of North American paleontology. He also is credited with formulating two important concepts in geology — the geosynclinal theory and the principle of isostasy. Hall also was the first state geologist of Iowa, although he was essentially geologist in absentia in that he made few trips to Iowa and maintained his regular position with the New York Geological Survey throughout his term as Iowa's state geologist, 1855 to 1858.

The sections from which the Alexandrian, Niagaran, and Cayugan series were established have not proven to be satisfactory sections with which to correlate other Silurian beds in North America. Sections composing the Niagaran and Cayugan series of New York are not particularly fossiliferous. However, Silurian faunas are fairly well documented elsewhere in North America. A comparison of these faunas with those of the type sections of the Silurian System in the British Isles has led North American geologists and paleontologists to conclude that the succession of faunal assemblages in the two areas is quite similar.

Because of the similarity of the faunal sequences, it is possible to correlate the North American Silurian with the standard Silurian sections of the British Isles, so most North American geologists now use series terms for the Silurian based on the British sections. Thus, Llandoverian, Wenlockian, and Ludlovian are terms now used in North America instead of Alexandrian, Niagaran, and Cayugan. In British usage, the Llandoverian Series is referred to as Lower Silurian, and the Wenlockian and Ludlovian series are termed Upper Silurian. The term "Middle Silurian" is not used by British geologists. Pridolian, not represented in Iowa, is the uppermost series, named from a Czech locality.

At present, the Geological Survey Bureau recognizes six formations in the

5.6 A generalized stratigraphic column for the six dolomite formations of the Silurian of eastern Iowa. Members are also shown. Composition and content of the rocks are shown by standard symbols and the following: triangles for chert, dashes for argillaceous (clay) content, ⌒ for laminar stromatoporoids, S for stricklandiid brachiopods, P for pentamerid brachiopods, T for tabulate corals, circles for molds of crinoid stems and fragments, Ɗ for solitary (horn) corals, ≋ for laminations, and ⌒ for brachiopods. Silurian strata rest unconformably on the Ordovician Maquoketa Formation and are overlain unconformably by the Devonian Wapsipinicon Group. From Witzke 1992.

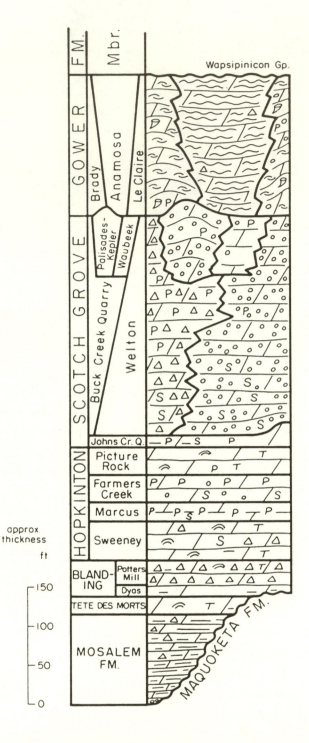

Silurian dolomite sequence of eastern Iowa (fig. 5.6). In ascending order, the Silurian column includes the Mosalem, Tete des Morts, Blanding, Hopkinton, Scotch Grove, and Gower formations. The Mosalem through lower Scotch Grove interval is assigned to the Llandoverian Series (Lower Silurian), and the middle Scotch Grove through middle Gower segment is placed in the Wenlock-

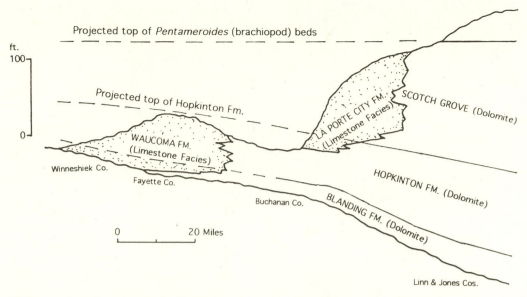

5.7 A stratigraphic cross section from Winneshiek County in northeastern Iowa to Linn and Jones counties in east-central and eastern Iowa. The stratigraphic positions of the limestones of the Waucoma and La Porte City formations are shown, as are their relationships to the dolomitized strata of the Blanding, Hopkinton, and Scotch Grove formations. Adapted from Ludvigson et al. 1992.

ian Series (Upper Silurian). The upper Gower may correlate with the Ludlovian Series (Upper Silurian), although its precise age is not known. Proposed correlations are shown in figure 5.5.

In addition to the six formations mentioned previously, the Geological Survey Bureau designated two additional formations for limestone facies within the Silurian System. The Waucoma Formation is a limestone facies that is laterally equivalent to dolomites of the Blanding and Hopkinton formations (fig. 5.7). The La Porte City Formation (fig. 5.7) is a limestone facies of the upper Hopkinton–Scotch Grove interval. Both the Waucoma and La Porte City formations (fig. 5.8) escaped dolomitization. Therefore, they provide valuable information about the original compositions and depositional textures of Silurian sediments in Iowa.

Silurian rocks outcrop in east-central and northeastern Iowa (fig. 5.9). In northeastern Iowa (Jackson, Dubuque, Clayton, Delaware, and Fayette counties) where glacial deposits are thin or absent, the Silurian bedrock forms a ridge-like topographic feature known as the Silurian Escarpment. Silurian rocks are thickest in their outcrop belt in eastern Iowa and in the subsurface of southwestern Iowa. Silurian strata, however, have been removed by erosion in northwestern and southeastern Iowa (fig. 5.9). The Plum River Fault Zone offsets Silurian bedrock in eastern Iowa and western Illinois (fig. 5.9).

Iowa's Silurian formations are discussed in the following sections in chrono-

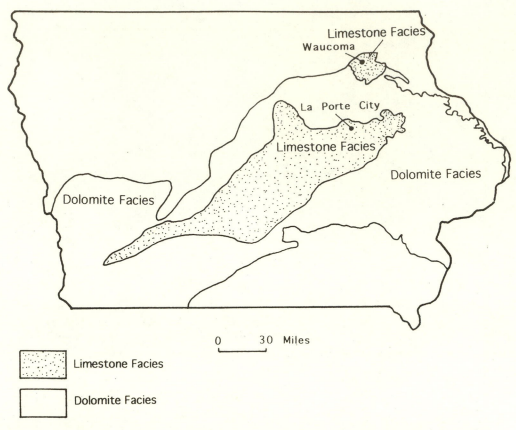

5.8 The distribution of the Silurian System in Iowa and the location of limestone facies in the Waucoma and La Porte City formations. Adapted from Ludvigson et al. 1992.

logical order from oldest to youngest. Environments of deposition of the formations are summarized in figure 5.10.

MOSALEM FORMATION

Named for exposures in Mosalem Township in Dubuque County, the Mosalem Formation is an argillaceous dolomite with chert nodules and shale partings. It has wavy bedding and in places displays a basal conglomerate containing clasts derived from the underlying Maquoketa Formation. The Mosalem is restricted in distribution and is absent where the underlying Maquoketa strata are thickest. The Mosalem filled topographic lows, probably ancient stream valleys, when the Silurian sea advanced over an irregular erosion surface developed on the Ordovician Maquoketa Formation.

Burrows are common in the formation, but otherwise fossil content is meager. Two benthic associations occur: a lower Mosalem community of inarticulate brachiopods (*Lingula*-Orbiculoid Community) and an upper Mosalem community of articulate brachiopods (*Dalmanella-Eospirigerina* Community). Both

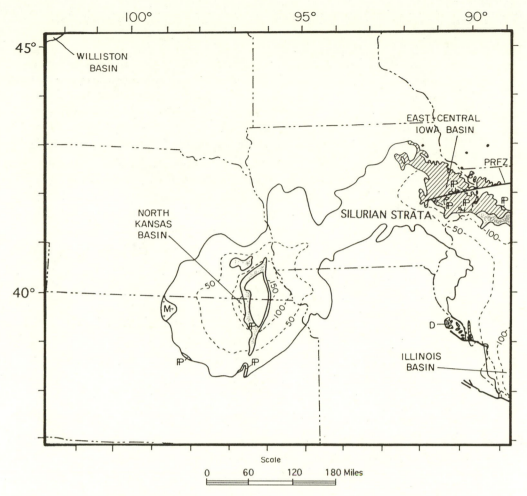

5.9 The distribution of the Silurian outcrop belt (slanted lines) in eastern and northeastern Iowa. Thickness of the total Silurian in the subsurface is shown by isopach lines; the contour interval is 50 meters (approximately 165 feet). Note that Silurian rocks are absent in northwestern and southeastern Iowa, having been removed by erosion. The Plum River Fault Zone (PRFZ) was active during mid- and late Paleozoic time and offsets Silurian strata in eastern Iowa and western Illinois. From Bunker and Witzke 1988.

floating and attached varieties of graptolites are known, and they help establish an Early Llandoverian age for the formation.

TETE DES MORTS FORMATION

The Tete des Morts Formation derives its name from its type locality within the drainage of the Tete des Morts River in Dubuque County. It is present throughout most of the Silurian outcrop belt of eastern Iowa, northwestern Illinois, and southwestern Wisconsin. The formation is difficult to distinguish in the subsurface, however. The Tete des Morts rests unconformably on the under-

5.10 Iowa's Silurian formations and their depositional environments. After Witzke 1992.

Formation	Depositional Environments
Gower Formation	Restricted marine embayment with hypersaline bottom waters during deposition of the Anamosa Member; carbonate mound facies (Brady and Le Claire members) were built on the seafloor and projected above the hypersaline waters into waters of normal salinity; the remains of crinoids, brachiopods, and corals occur in a matrix of carbonate mud within the carbonate mound facies
Scotch Grove Formation	Open marine shelf with normal marine salinity and well-oxygenated water; carbonate mound facies with abundant carbonate mud and crinoidal debris
Hopkinton Formation	Open marine shelf with normal salinity and well-oxygenated water; varied marine fauna; isolated carbonate mounds
Blanding Formation	Open marine shelf with normal salinity and well-oxygenated water
Tete des Morts Formation	Open marine shelf with normal salinity and well-oxygenated water
Mosalem Formation	Embayments of the sea; restricted marine to open marine shelf; low oxygen to normal amounts of oxygen; muddy conditions with clay derived from erosion of the shales of the Ordovician Maquoketa Formation

lying Mosalem in some sections, but at other localities it lies unconformably on the Ordovician Maquoketa Formation.

The Tete des Morts dolomites are generally very fine to medium crystalline. They occur as massive beds with pitted vertical surfaces. Recognized as forming resistant cliffs, the Tete des Morts produces bold rock faces above the more subdued slopes developed on the underlying Mosalem and Maquoketa formations. The fossils of the Tete des Morts include lamellar stromatoporoids (sponges), horn corals, and tabulate corals of the honeycomb variety. Scattered brachiopods occur also.

BLANDING FORMATION

The Blanding Formation is named for Blanding, a village in northwestern Illinois, across the Mississippi River from Bellevue, Iowa. The Blanding is recognized throughout northwestern Illinois and in northeastern and east-central Iowa. Erosional outliers occur in southwestern Wisconsin.

The Blanding is very cherty and contains both nodules and distinct beds of chert. Chert commonly represents 20 to 30 percent by volume of the formation. In the subsurface of western Iowa, the Blanding contains less chert, and it is difficult to distinguish from adjacent Silurian formations. The Blanding rests conformably on the underlying Tete des Morts Formation throughout most of the Silurian outcrop belt in eastern Iowa, but it oversteps the erosional edge of the Mosalem–Tete des Morts in the subsurface and lies unconformably on the Ordovician Maquoketa Formation (fig. 5.6). The lower Blanding and the underlying Tete des Morts and Mosalem formations are visible at Bellevue State Park in Jackson County.

A variety of fossils occur in the formation, with the upper part being more fossiliferous than the lower part. Fossils include bryozoans, horn corals, tabulate corals of the honeycomb type, trilobites, lamellar stromatoporoids, sponge spicules, crinoid debris, and several species of brachiopods. Brachiopod faunas help date the Blanding Formation as mid-Llandoverian, and fossil communities found in the formation have been used to interpret the water depth in which it formed (fig. 5.5).

HOPKINTON FORMATION

The Hopkinton Formation is named for exposures along the Maquoketa River near the town of Hopkinton in Delaware County. It is recognized throughout most of eastern Iowa and in the subsurface of central and southwestern Iowa. The Hopkinton is conspicuous in Backbone State Park, where it composes a narrow ridge from which the park gets its name (fig. 5.1). Sinkholes and other solution features occur in the Hopkinton in eastern Iowa, including caves at Maquoketa Caves State Park in Jackson County. The natural bridge at Maquoketa Caves State Park, however, is in the overlying Scotch Grove Formation.

Dolomites with very fine to coarsely crystalline textures comprise most of the formation. Nodular cherts occur also. The Hopkinton dolomites are quarried in eastern Iowa for concrete aggregate, agricultural lime, and road stone. Historically, the formation furnished raw material for the manufacture of quicklime, which was used in mortar. The Hopkinton Formation contains a variety of fossils, including brachiopods, corals, crinoids, bryozoans, nautiloids, gastropods, stromatoporoids, and trilobites. The formation is divided into four members (fig. 5.6), in ascending order: Sweeney, Marcus, Farmers Creek, and Picture Rock. A brief description of each member follows.

The Sweeney Member is named for its type locality near Mississippi Palisades State Park in Illinois, across the river from Sabula, Iowa. The Sweeney is well known from surface exposures in northwestern Illinois and eastern Iowa. It also occurs in the subsurface of central and southwestern Iowa. A tabulate coral-lamellar stromatoporoid community is present in the member. Other fossils include lacy bryozoans, stricklandiid brachiopods, and pentamerid brachiopods.

The Marcus Member is also named for a locality in northwestern Illinois near Mississippi Palisades State Park, where Silurian dolomites form bold bluffs. Marcus strata are recognized throughout much of the Silurian outcrop belt in eastern Iowa. Pentamerid brachiopods of the genus *Pentamerus* (fig. 5.11) are plentiful in the Marcus Member. Some of the brachiopods occur in life posi-

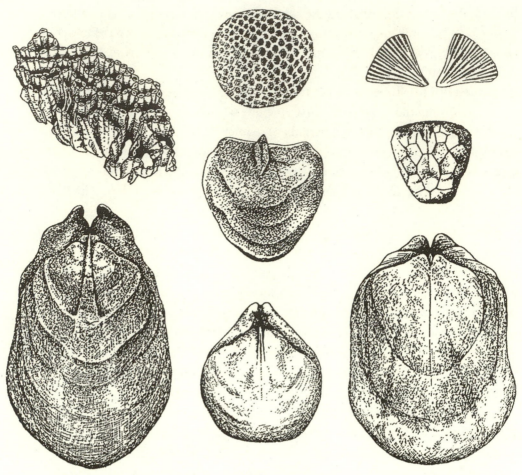

5.11 An assortment of key Silurian fossils from eastern Iowa. Top row, left to right: *Halysites labyrinthicus* (chain coral), *Cyclocrinites dactioloides* (green alga), *Petalocrinus mirabilis* (arm fans of a crinoid). Middle row: *Stricklandia laevis* (interior of a brachiopod shell), *Eucalyptocrinites ornatus* (calyx of a crinoid). Bottom row: *Pentamerus oblongus* (a common pentamerid brachiopod), *Harpidium* (*Isovella*) *maquoketa* (a pentamerid brachiopod), *Pentameroides subrectus* (a pentamerid brachiopod). From Witzke 1992.

tions, but most are found as scattered shells. Tabulate corals and lamellar stromatoporoids are present, interbedded with pentamerid-rich layers.

Named for its type locality in Farmers Creek Township in Jackson County, the Farmers Creek Member is recognized in much of the Silurian outcrop belt in eastern Iowa and northwestern Illinois. Caves occur in the member, and irregular solution surfaces are common in outcrops, giving the unit a distinctive appearance. In the subsurface, the Farmers Creek Member forms the principal water-yielding interval within the Silurian aquifer.

The Farmers Creek Member preserves a variety of fossils, including the green alga *Cyclocrinites* (fig. 5.11). Witzke and Johnson (1992) recognized the following brachiopod faunal associations (fig. 5.11), in ascending order: *Stricklandia laevis*

Community, *Harpidium maquoketa–Stricklandia laevis* Community, and *Harpidium maquoketa* Community. In addition to the brachiopod faunas, the member also yields bryozoans, crinoids, horn corals, tabulate corals, mollusks, and trilobites.

The uppermost member of the Hopkinton Formation is named for Picture Rock County Park in Jones County. There, the Picture Rock Member is recognized as a cliff former. The member occurs in the Silurian outcrop belt of eastern Iowa and northwestern Illinois and in Iowa's subsurface. It contains lamellar stromatoporoids, tabulate corals, large pentamerid brachiopods (*Pentamerus oblongus*), crinoidal debris, and conodonts.

SCOTCH GROVE FORMATION

The Scotch Grove Formation derives its name from the type locality in Scotch Grove Township, Jones County. It is an interval of dolomite and cherty dolomite between the underlying Hopkinton Formation and the overlying Gower Formation (fig. 5.6). Witzke (1985) proposed the Scotch Grove as a formational name and recognized five lithofacies: (1) flat-lying, cherty, sparsely fossiliferous dolomite; (2) flat-lying, noncherty, sparingly fossiliferous dolomite; (3) flat-lying, porous, fossiliferous dolomite; (4) mounded crinoidal dolomite; and (5) mounded sparsely fossiliferous dolomite. The five facies were designated as members (fig. 5.6). Only the mounded facies, the Johns Creek Quarry and Palisades-Kepler members, are discussed here.

The Johns Creek Quarry Member is named for a quarry adjacent to Johns Creek in Dubuque County. The member includes carbonate mounds of dense, coral-bearing dolomite with adjacent crinoid-rich beds. The mounds, sometimes called bioherms (life mounds), reach 40 to 50 feet in thickness and range between 50 to 500 feet in lateral dimension. The central mound areas consist of very fine grained dolomite (composed of particles of mud size) with large colonial rugose corals, tabulate corals, and crinoid debris. The mounds are flanked by dipping (inclined) beds with inclinations up to 30 degrees. The inclined beds, containing crinoid debris and displaying graded bedding, bury the mound facies.

Complex crinoidal mound and skeletal bank facies in the upper Scotch Grove Formation are now placed in the Palisades-Kepler Member. Named for the type locality in Palisades-Kepler State Park in Linn County, these strata were incorrectly assigned to the Gower Formation until the early 1980s. The Palisades-Kepler Member occurs at several localities in eastern Iowa. It is characterized by isolated mounds or coalesced mound complexes; adjacent flat-lying beds are known also.

Isolated mounds range from 200 to 3,000 feet across and extend 30 to 200 feet in vertical dimension. Dips within mounds vary between 0 and 50 degrees. The principal rock type of the member is crinoidal dolomite. The crinoidal material is seen as segments of crinoid stems, parts of plates, and broken debris; preservation is in the form of molds and dolomitized grains. Corals and stromatoporoids occur in the mound facies, but they do not produce an organic framework. Some mounds are highly fractured and brecciated, and much of this alteration was probably contemporaneous with deposition. Post-Silurian erosion leveled the crests of most of the mounds in the Palisades-Kepler Member.

GOWER FORMATION

The Gower Formation is named for quarry exposures in Gower Township, Cedar County. As originally defined, the formation included the distinctive laminated dolomites known as Anamosa stone and laterally adjacent mound facies composed of carbonate mud and skeletal components. As currently recognized, the Gower Formation (fig. 5.6) includes a dominant facies of flat-lying laminated dolomite (Anamosa Member), a brachiopod-rich mound facies with moderate to steeply inclined beds (Brady Member), and mounded and flat-lying facies of fossiliferous dolomites of varied character (Le Claire Member). The Gower Formation is truncated by erosion and overlain unconformably by Middle Devonian strata. Its maximum thickness is in eastern Iowa, where thicknesses of 100 to 150 feet are known.

The Anamosa Member, named for exposures at Stone City, west of Anamosa, is a distinctive rock unit with prominent laminations. Both wavy and planar laminations occur. At the microscopic level, the laminae display slight variations in crystal size. Individual laminae are laterally continuous and can be traced throughout a given quarry. Although the wavy laminations resemble stromatolites, the planar varieties are varvelike in appearance. Do these planar laminations represent cycles of seasonal deposition? Although such an explanation is speculative, it is plausible.

Fossils are rare in the Anamosa Member, and many typical marine fossils are conspicuously absent. For example, echinoderms, bryozoans, and trilobites are unreported. Burrows are generally absent; otherwise, the distinctive laminations so characteristic of the member would have been destroyed. Molds of gypsum crystals have been reported in the member, suggesting that it was deposited in waters of higher than normal salinity. Perhaps the sea was too briny to support a normal marine fauna when the Anamosa beds formed. The member accumulated under quiet subtidal conditions, below wave base. See figure 5.12 for a model depicting the environmental setting during the time of deposition.

The Brady Member is named for exposures in the Brady Quarry in Cedar County, and it is recognized at localities in Cedar, Jackson, Johnson, Jones, and Linn counties. The Brady is a discontinuous mound facies surrounded by laminated dolomites of the flat-lying Anamosa Member. Thicknesses of Brady strata are two to five times greater than those of their Anamosa counterparts.

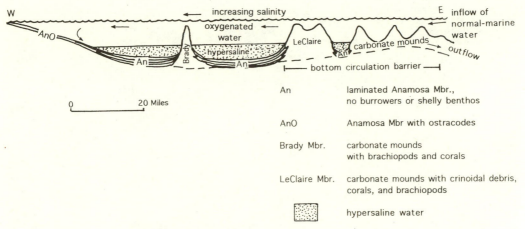

W — increasing salinity — E inflow of normal-marine water

oxygenated water ←

AnO

Brady

hypersaline

LeClaire

carbonate mounds

An

outflow

An An An

|— bottom circulation barrier —|

0 ___ 20 Miles

An	laminated Anamosa Mbr., no burrowers or shelly benthos
AnO	Anamosa Mbr with ostracodes
Brady Mbr.	carbonate mounds with brachiopods and corals
LeClaire Mbr.	carbonate mounds with crinoidal debris, corals, and brachiopods
	hypersaline water

5.12 A model to explain the deposition of the Gower Formation. As shown here, much of the Anamosa Member formed in hypersaline bottom waters that supported little or no life. The Brady and Le Claire members represent carbonate mounds that were built by a prolific production of carbonate mud and skeletal fossils. Shelly faunas thrived only where mounds projected into the zone of oxygenated waters of normal salinity. Overall, the depositional setting was a salinity-stratified embayment of the Silurian seaway that was inhospitable for bottom-dwelling life. Adapted from Witzke 1992.

The mounds of the Brady Member formed topographic highs on the Silurian seafloor.

The Brady Member is primarily fine-grained fossiliferous dolomite, representing an accumulation of carbonate muds and brachiopod shells. Although the Brady facies is dominated by brachiopods, tabulate and rugose corals are present in some beds, and domal stromatolites are known.

Strata in the Brady Member are moderately to steeply inclined. The exact thicknesses of the Brady mounds are unclear because they are truncated beneath Devonian or Quaternary deposits. However, exposures in the type area preserve thicknesses of 100 feet or more. Lateral dimensions of mounds range between 300 and 5,000 feet. Dips within most Brady mounds reflect initial depositional slopes, but increases in inclination by later compaction events are also known. Beds dip radially away from central mound areas, with dips commonly between 0 and 45 degrees. Although dips in the 0-to-45-degree range are the norm for the member, steeply dipping to overturned beds also occur. Witzke (1992) interprets these structurally complex beds as resulting from postdepositional compaction and slumping of flanking sediments around rigid central mounds that had previously undergone cementation.

The Le Claire Member is named for the town of Le Claire in Scott County. It is composed of flat-lying and mounded dolomites that grade laterally into laminated beds of the Anamosa Member. The Le Claire is currently recognized in Clinton and Scott counties in Iowa and in three nearby counties in Illinois. The member contains a variety of dolomites, representing deposits of carbonate muds

5.13 Building stone from the Anamosa beds, Weber Stone Company quarries, Stone City, Iowa. Note the fine laminations characteristic of the Anamosa Member. Photo by the author.

and skeletal fossils. Crinoid debris, brachiopods, and corals are common. Bedding in the Le Claire varies from horizontal to steeply inclined. Overturned beds and slump structures occur in at least one locality. The fauna of the Le Claire is varied and includes brachiopods, corals, crinoids, trilobites, bryozoans, nautiloids, and stromatoporoids.

Economic and Historical Significance of the Anamosa Facies

The uniform bedding and texture of the Anamosa Facies make it an ideal building stone (fig. 5.13). Exposures of the Anamosa beds along the Wapsipinicon River west of Anamosa in Jones County caught the attention of early Euro-American settlers. The stone, attractive in appearance and covered by relatively thin overburden, turned out to be easily quarried (fig. 5.14). It was widely used for windowsills in the construction of houses. The United States Army utilized stone from the Anamosa and Stone City area in the early 1840s, several years before Iowa became a state in 1846. Many of the early buildings at the Rock Island Arsenal in Illinois were constructed of stone from Stone City quarries.

5.14 Bealer's Quarry, about 1900. During the last half of the 1800s, building stone was quarried at several sites near Anamosa and Stone City, Iowa. Courtesy of the Calvin Photographic Collection, Department of Geology, University of Iowa.

The quarry industry at Stone City once employed more than a thousand men and served as one of eastern Iowa's most important industries during the last half of the nineteenth century. In the years 1859 to 1895, over 150,000 railroad car shipments of stone originated from the Stone City area. The stone was shipped to all six of Iowa's bordering states for use in the construction of railroad bridges, bridge piers, foundations, and buildings. Three large buildings in downtown Minneapolis were constructed from Anamosa stone.

Most of the buildings of the Iowa Men's Reformatory at Anamosa are made from stone quarried from the local Silurian bedrock by convict labor. At Cornell College in Mt. Vernon, the first college buildings were erected in 1852 from stone that had been transported across the prairies by team and wagon from the Stone City quarries approximately 15 miles to the north. More recently, the Herbert Hoover Presidential Library at West Branch has been constructed from stone quarried at Stone City.

The quarry industry at Stone City continues, but on a limited scale. Only one major quarry operates currently. Many of the early stone buildings of Stone City were torn down, and the stone blocks were transported to Cedar Rapids for use as building stone. A few stone houses, a stone store, and a stone church still re-

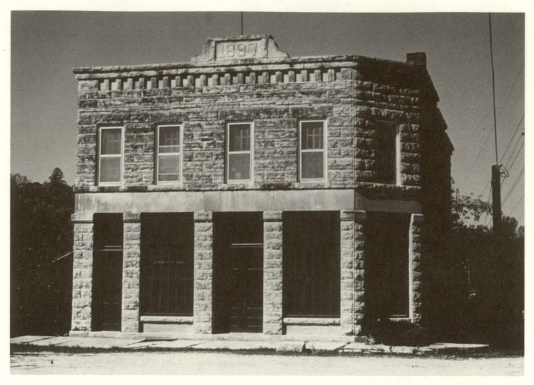

5.15 The General Store, Stone City, Iowa, was constructed in 1897 from Anamosa building stone from local quarries. Photo by the author.

main in Stone City as examples of the type of construction that was once so common in eastern Iowa (fig. 5.15).

Grant Wood, the well-known artist, selected the Stone City area as the site for a summer artist colony in the 1930s. One of Wood's paintings, *Stone City*, portrays the area as seen from the hills overlooking the town.

Silurian-Devonian Unconformity

The Ordovician and Silurian rocks of Iowa were warped, uplifted, and eroded during Late Silurian and Early Devonian time. Middle Devonian rocks now lie unconformably on either Silurian or Ordovician strata. The Middle Devonian rests on Silurian rocks throughout much of eastern Iowa, but in northeastern Iowa (Howard and Winneshiek counties), the Devonian lies on Ordovician rocks. The paleotopography produced by this interval of erosion during Late Silurian through Early Devonian time influenced the deposition of Middle Devonian strata in Iowa. This topic will be explored further in chapter 6.

Many of the Gower carbonate mounds were truncated by erosion during Late Silurian–Early Devonian time. Figure 5.16 shows Middle Devonian beds unconformably overlying beds of a truncated carbonate mound in the Lime City

5.16 The steeply inclined beds of a carbonate mound in the Silurian of Iowa as revealed in one of the Lime City quarries in southern Cedar County, Iowa. Samuel Calvin, who did early work on Iowa's Silurian, is shown in this picture, taken about 1900. Courtesy of the Calvin Photographic Collection, Department of Geology, University of Iowa.

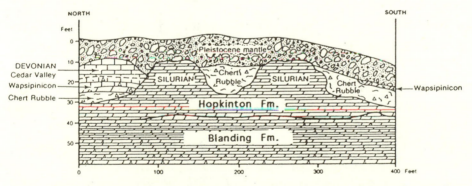

5.17 The Silurian-Devonian unconformity in a quarry exposure in Bremer County, Iowa. Devonian strata of the Wapsipinicon and Cedar Valley groups overlie Silurian strata at this locality. Adapted from Dorheim and Koch 1962.

Quarry in Cedar County. In northeastern Iowa, where the Gower Formation is absent, Middle Devonian rocks generally lie unconformably on the Hopkinton Formation. A spectacular view of an unconformity between the Silurian and Devonian systems was revealed in 1962 by quarry operations in Bremer County (fig. 5.17). There is considerable relief on top of the Silurian, and pockets of clay and chert rubble, the products of Late Silurian and Early Devonian weathering, overlie the Silurian dolomites.

Silurian Life

What was the nature of life in the Silurian seas of Iowa? For the most part, the fossil record of the Silurian is similar to that of the Ordovician, except that the Silurian fossils are often more poorly preserved because of the extensive dolomitization that affected most of the Silurian rock record. All of the major invertebrate groups that were present in Iowa's Ordovician strata are also known from the Silurian of the state. Nearly all of the Silurian species were different from those of the Ordovician, however. Representative Silurian fossil groups include brachiopods, crinoids, stromatoporoids, mollusks, worms, conodonts, arthropods, corals, bryozoans, and graptolites. Fossil algae occur in both Ordovician and Silurian strata, too.

A remarkable array of Silurian fossils was discovered by William Hickerson Jr. in 1996. Hickerson recovered at least forty varieties of soft-bodied organisms in a limestone quarry in Clinton County. This unusual Silurian fauna includes worms, primitive fish, shrimplike creatures, and other rare forms. A report on the specimens is currently being prepared for publication.

A reconstruction of life on a Silurian seafloor is shown in figure 5.18, and representative Silurian fossils are illustrated in figure 5.19. A brief discussion of selected fossil groups follows.

COLONIAL CORALS

Silurian life in Iowa shows an increase in colonial forms. Colonial corals occur in all of Iowa's Silurian formations, with the exception of the Mosalem Formation. The chain coral (*Halysites*) and the honeycomb coral (*Favosites*) are examples of two colonial tabulate corals that are common in the Silurian of Iowa (fig. 5.19). Tabulate corals are characterized by an absence or poor development of septa and by the presence of horizontal partitions called tabulae.

An individual, soft-bodied coral animal (polyp) lived in each of the openings of the chainlike and honeycomb-like colonies (fig. 5.19). The tabulae in these corals were added to partition off the chamber in which the corals resided. The living coral animal was situated on the uppermost partition and could extend its mouth and arms upward to extract nutrients from the sea. The individual coral polyps moved upward as new tabulae were added, and the lower levels of the colonial framework were abandoned. In a sense, the coral colonies were like miniature high-rise apartment complexes with only the penthouse levels occupied.

Tabulate corals are sufficiently abundant in some of the Silurian formations to constitute laterally continuous beds of coralline rock. These beds represent portions of the Silurian seafloor where coral growth flourished and extended laterally for some distance as coral gardens. Tabulate corals also occur in some carbonate mound facies of the Hopkinton, Scotch Grove, and Gower formations.

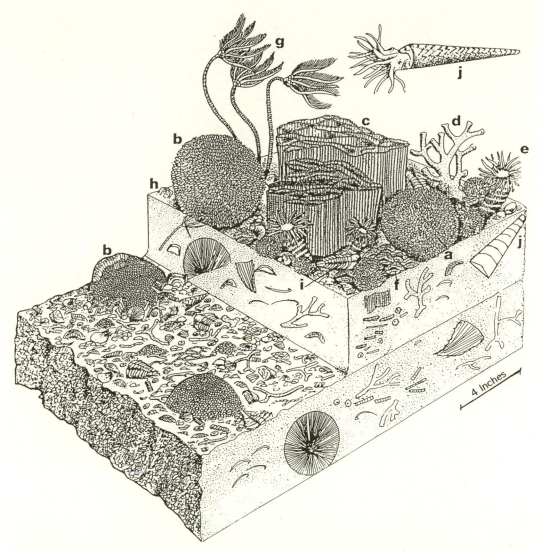

5.18 Life on a Silurian seafloor: (a) tabulate coral, (b) the tabulate coral *Favosites*, (c) the tabulate chain coral *Halysites*, (d) bryozoans, (e) horn corals, (f) spiriferid brachiopods, (g) crinoids, (h) strophomenid brachiopod, (i) trilobite, and (j) nautiloid cephalopod. From McKerrow 1978.

The corals, however, do not produce a framework structure; the mounds are essentially the accumulation of carbonate mud and fossil debris. Colonial rugose corals are also locally important, especially *Arachnophyllum*.

SOLITARY CORALS

Solitary corals are common in the Hopkinton and Scotch Grove formations but are known from all Silurian formations in Iowa. One such coral deserves special mention. The slipper coral (square coral), *Goniophyllum pyramidale*, occurs in the Scotch Grove Formation near Scotch Grove in Jones County. This oc-

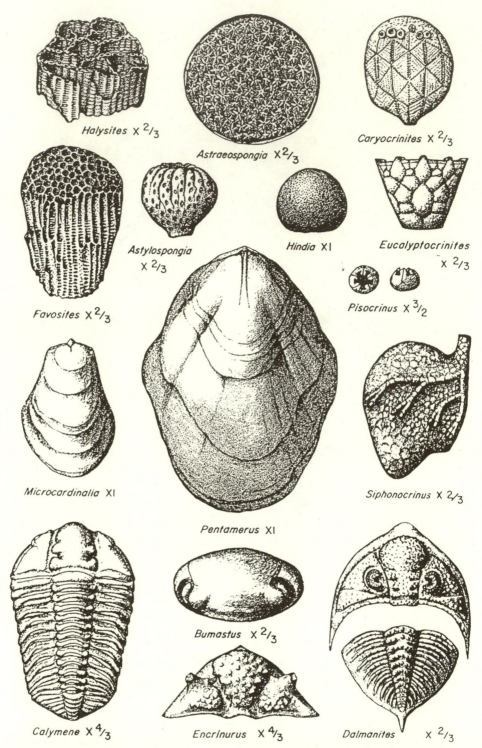

5.19 Representative Silurian fossils. Corals: *Halysites, Favosites*; sponges: *Astraeospongia, Hindia, Astylospongia*; cystoids: *Caryocrinites*; crinoids: *Eucalyptocrinites, Pisocrinus*; brachiopods: *Pentamerus, Microcardinalia* (= *Stricklandia*); trilobites: *Calymene, Bumastus, Dalmanites, Encrinurus*. From Willman et al. 1975. Courtesy of the Illinois Geological Survey.

currence is important because the only other abundant occurrence of the slipper coral is on the Baltic island of Gotland.

STROMATOPOROIDS

Stromatoporoids are extinct sponges, although they were once classified with corals. The stromatoporoids, or stroms as they are commonly called, constructed skeletons of various shapes and sizes. Their calcium carbonate skeletons occur in diverse shapes, including subspherical, branching, and matlike varieties. Stroms are preserved in dolomite strata in the Tete des Morts, Blanding, Hopkinton, Scotch Grove, and Gower formations. They also occur in the Waucoma and La Porte City limestone facies, where thin sections allow the stroms to be taxonomically classified.

ALGAE

Fossil algae or algal structures are associated with Silurian rocks of the Mosalem, Hopkinton, Scotch Grove, and Gower formations. Algae may also be responsible for some of the irregular laminations found in the Anamosa Member of the Gower Formation. A globose alga, *Cyclocrinites dactioloides* (fig. 5.11), occurs in the Hopkinton Formation of Iowa and is well known in Silurian paleocommunities elsewhere. Markes Johnson (1988) observed that these algae are the size of large marbles when found associated with the stricklandiid brachiopod community (estimated water depth of 200–300 feet) in the lower part of the Farmers Creek Member of the Hopkinton Formation, whereas this same algal species reaches golf-ball size in the upper part of the member, where it occurs in association with the pentamerid brachiopod community (estimated water depth of 100–200 feet). The apparent explanation for the size differences in the algae is that the shallower seafloor setting of the pentamerid association provided more sunlight, allowing the algae to grow to a larger size.

ECHINODERMS

Echinoderms, particularly the remains of crinoids, are common in the Hopkinton, Scotch Grove, and Gower formations. Crinoid debris, consisting of stem segments and plates, is an important constituent of the carbonate mounds of the Scotch Grove and Gower formations.

BRACHIOPODS

Brachiopods are found throughout the Silurian of Iowa, but one occurrence is particularly noteworthy. The brachiopod *Pentamerus* (fig. 5.20) is found by the thousands in the Hopkinton Formation. Some specimens are preserved in life positions. Specimens of *Pentamerus* are preserved almost exclusively as internal molds. The original shell material was destroyed during dolomitization, and

5.20 The pentamerid brachiopod community: (a) *Pentamerus* in life position, (b) chain corals, (c) horn corals, (d and f) other brachiopods, and (e) bryozoans. From McKerrow 1978.

what is preserved is the sediment that filled the shells. This filling of sediment (an internal mold) superficially resembles the form of a deer's foot or pig's foot, and it is not uncommon for an uninformed collector to discover a fossil "pig's foot" in the Hopkinton Formation in eastern Iowa.

Ancient benthic communities have been recognized in Iowa's Silurian rock record based on key species of brachiopods. These communities and other faunal associations are discussed in the following section.

Silurian Communities

A community is a group of organisms living together and linked by their effects on one another and their environment. Several benthic communities have been reconstructed from the Silurian System of the British Isles. Different lines of evidence indicate that the communities are depth associated. For example, when plotted on paleogeographic maps, the communities occur as bands parallel to the ancient shoreline. Presumably, the faunas the farthest from the shoreline lived in deeper waters than those closest to the shoreline. Of particular significance are examples of submarine volcanic flows that buried deeper-water communities. These flows prepared the way for subsequent colonization of shallower-water communities by building up the seafloor and decreasing the water depth.

Similar Silurian benthic communities have been recognized in New York and Nevada. These communities, as well as those of the British Isles, were situated where the slope of the seafloor was sufficient to support several different communities contemporaneously.

Markes Johnson (1975) recognized four major benthic communities in the Silurian rock record of eastern Iowa, and he estimated the water depths for each of the communities. The four communities and their approximate water depths are: (1) an inarticulate brachiopod (*Lingula*) and algal stromatolite community (0–30 feet); (2) a coral-stromatoporoid community consisting of abundant tabulate corals (*Favosites*, *Halysites*, and others), lamellar stromatoporoids, horn corals, gastropods, and articulate brachiopods (30–100 feet); (3) a pentamerid brachiopod community characterized by an abundance of brachiopods of the genus *Pentamerus* and some tabulate corals (100–200 feet); and (4) a stricklandiid brachiopod community containing articulate brachiopods such as *Stricklandia* or *Costistricklandia* (200–300 feet).

With few exceptions, the Iowa communities are similar in faunal content to those of the Silurian continental margin areas of New York, Nevada, and the British Isles. However, the depositional setting in which the Iowa communities formed was a platformlike sea bottom, several hundred miles from the nearest shelf margin. The depth of water was quite uniform over this seafloor for long distances, and the bottom topography was relatively flat. Because of the uniformity of water depth and submarine topography, it is likely that only one major benthic community existed on broad areas of the seafloor at any given time. The replacement of one benthic community by another in the Silurian rock column of Iowa was probably caused by changes in water depths in the seas brought about by widespread fluctuations in sea level. Several cycles of deepening and shallowing are suggested by the distribution of faunal communities in Iowa's Silurian rocks (figs. 5.5 and 5.21). The Silurian seas that covered the state were probably only a few feet deep during the shallowest of conditions, but water depths of 200 to 300 feet were likely common during times of maximum rise in sea level. The variations in sea level may have been caused, in part, by cycles of glaciation and deglaciation in the Southern Hemisphere continents.

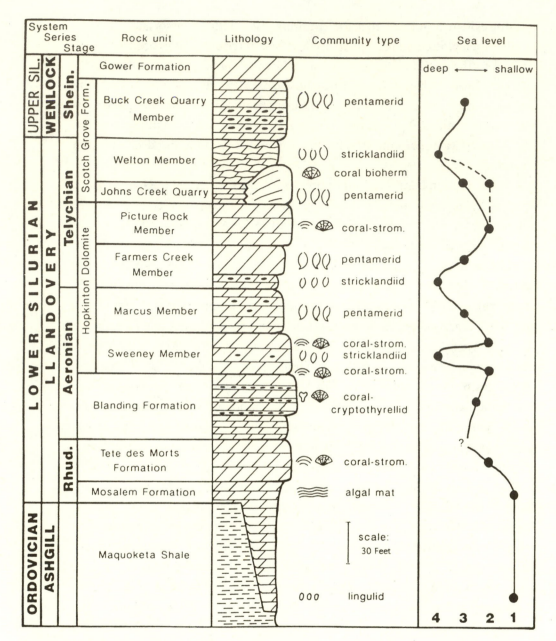

5.21 Silurian fossil communities of Iowa and their approximate water depth: community 1, the lingulid brachiopod–algal stromatolite community (inferred water depth of 0–30 feet); community 2, the coral-stromatoporoid community (inferred water depth of 30–100 feet); community 3, the pentamerid brachiopod community (inferred water depth of 100–200 feet); and community 4, the stricklandiid brachiopod community (inferred water depth of 200–300 feet). The numbers shown on the lower right corner refer to the four communities. Terminology for communities and estimations of water depths are based on Johnson 1988.

Summary

Although dolomite comprises nearly all of Iowa's rock record, that record is not as monotonous and uneventful as was once thought. Six dolomite formations (Mosalem, Tete des Morts, Blanding, Hopkinton, Scotch Grove, and Gower) and two limestone facies (Waucoma and La Porte City) comprise the Silurian of Iowa. Benthic fossils in Silurian strata are useful in estimating the water depths of the ancient Silurian seas and indicate that Iowa's Silurian record formed on a shallow seafloor with fluctuating water depths, ranging from a few feet deep during the shallowest conditions to 200 to 300 feet deep during the deepest deposition. Five T-R cycles are recorded in Iowa's Silurian rocks. The bulk of Iowa's Silurian record (upper Mosalem through Scotch Grove formations) reflects deposition on an open marine shelf with well-oxygenated waters and normal marine salinities. The lower Mosalem Formation preserves some evidence for restricted marine settings with low levels of oxygen. The Anamosa Member of the Gower Formation was deposited in a restricted marine embayment with hypersaline bottom waters.

A variety of fossils occur in Iowa's Silurian strata, including algae, brachiopods, corals, crinoids and crinoid debris, gastropods, graptolites, nautiloids, stromatoporoids, and trilobites. Crinoidal debris, brachiopods, corals, and stromatoporoids are important constituents of carbonate mounds in the Scotch Grove and Gower formations. These ancient mounds are composed primarily of carbonate mud, fossils, and skeletal debris; they are unlike modern coral reefs that represent organically bound, wave-resistant structures.

The Silurian rocks of Iowa have economic value. They serve as a bedrock aquifer in eastern Iowa and as a source of agricultural lime, road aggregate, and aggregate for concrete. The Anamosa Member of the Gower Formation furnished building stone during the early years in Iowa and is still quarried for that purpose near Stone City.

Acknowledgments

I acknowledge the constructive suggestions of Brian J. Witzke. Brian completed both a Master's thesis and a Ph.D. dissertation on the Silurian of Iowa, and he has been actively investigating Silurian paleontology and stratigraphy in Iowa for more than twenty years. I relied heavily on Brian's research in preparing this chapter. In particular, I found his 1992 guidebook to be very useful. In addition, I made extensive use of the published works of Markes E. Johnson.

6 **Devonian** A Variety of Marine Deposits

For a long time the seafloor was almost barren; later it teemed with seaweeds, corals, and other organisms. Diminutive clams plowed through the muds; snails crept over it; sea lilies moved to and fro above it. There were hordes of brachiopods, or lamp-shells, some lying free on the bottom, though others fastened themselves to it or to other shells. Distant relatives of the nautilus swam or crawled about; sea urchins devoured scumlike plants, and animals called bryozoa built colonies that looked like corals. Fish swam lazily above them, some breathing with lungs instead of gills.

Fenton and Fenton (1958) describing conditions that existed during deposition of the Devonian Lime Creek Formation

The seas withdrew from the craton during the Late Silurian, approximately 415 million years ago, and did not return until the Middle Devonian, approximately 385 million years ago. Considerable warping affected the interior of the continent before its resubmergence in the Middle Devonian. This warping affected the Precambrian, Cambrian, Ordovician, and Silurian systems, and rocks of these systems were subjected to prolonged weathering and erosion during part of the Late Silurian and all of the Early Devonian. The unconformity between Iowa's Silurian and Devonian systems marks the boundary between the Tippecanoe and Kaskaskia sequences proposed by L. L. Sloss (1963) (fig. 1.8). Approximately 30 million years of rock record is missing from Iowa's stratigraphic column at this major continent-wide unconformity.

Although Iowa has no record of Lower Devonian strata, the state possesses a rich and varied assortment of Middle and Upper Devonian rocks and fossils. The Devonian System is one of the state's most fossiliferous units (fig. 6.1).

In northeastern Iowa, Devonian rocks unconformably overlie an eroded surface of Ordovician and Silurian rocks. Upper Devonian strata onlap onto Precambrian bedrock next to the Sioux Ridge in extreme northwestern Iowa. Devonian rocks of the central midcontinent region crop out in eastern Iowa and southern Minnesota; they are bounded to the north by the Transcontinental Arch, to the west by the Cambridge Arch and Central Kansas Uplift, and to the south by the Chautauqua Arch–Ozark Uplift–Sangamon Arch (fig. 6.2). The East-Central Iowa and North Kansas basins (fig. 5.9) were primarily Silurian features, but they did accumulate a record of sedimentary rocks during the initial phases of Middle Devonian deposition. The total Devonian isopach (thickness)

6.1 A reconstruction of life in a Devonian sea, such as the one that covered Iowa during Late Devonian time when the Lime Creek Formation was deposited. The remnants of marine creatures are abundant in Iowa's fossil record. Drawing by Lori McNamee; based on information in Fenton and Fenton 1958.

map shows that Iowa's thickest Devonian sequence is in central Iowa (fig. 6.2). This region, termed the Iowa Basin by Bunker et al. (1988), was an area of shallow marine, tidal-flat, and evaporite deposition during much of the Middle and Late Devonian. The Devonian seaway deepened across southeastern Iowa into the Illinois Basin, where deeper-water sedimentation prevailed during much of

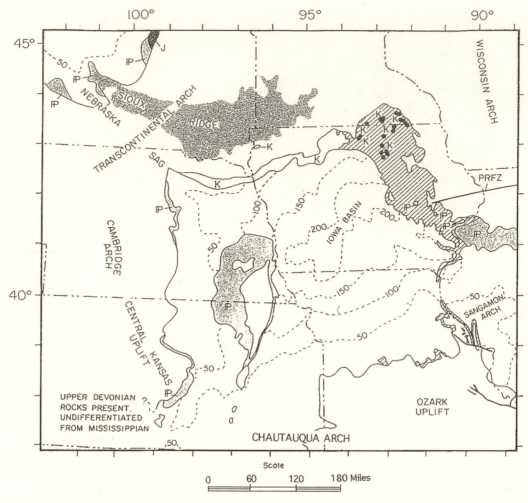

6.2 The distribution of the Devonian outcrop belt (slanted lines) in eastern Iowa, southern Minnesota, and eastern Missouri. Thickness of the total Devonian is shown by isopach lines; the contour interval is 50 meters (approximately 165 feet). Note the presence of erosional outliers of Pennsylvanian (℘) and Cretaceous (K) on Devonian bedrock in eastern Iowa. The Plum River Fault Zone (PRFZ) is shown in eastern Iowa and adjacent Illinois. From Bunker and Witzke 1988.

Devonian time. Witzke et al. (1988) describe the stratigraphy and deposition of Iowa's Middle Devonian–early Late Devonian rocks.

Regional Setting

During the Middle and Late Devonian, the seas advanced and retreated over Iowa several times. Mud washed into the Devonian sea from several sources, including a mountainous region on the eastern seaboard (Acadian Mountains). Ancient Iowa was a shallow marine shelf (Midcontinent Carbonate Shelf), where water depths ranged from shoreline and intertidal settings to offshore depths of

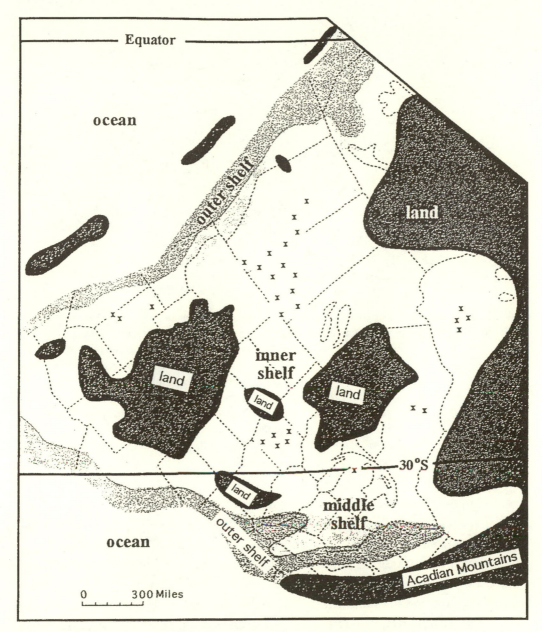

6.3 A paleogeographic map of Middle Devonian time (approximately 375 million years ago) when Iowa was located in the Southern Hemisphere in a warm, tropical seaway. Evaporites (shown by the x symbols) precipitated in areas of high evaporation where the seas reached above-average salinity. From Witzke and Bunker 1994.

approximately 100 to 200 feet (figs. 6.3 and 6.4). Iowa's Devonian rock record formed in shallow, tropical seas at low latitudes in the southern hemisphere.

The seaways expanded and contracted across the interior of the North American continent several times during the Devonian, leaving behind a cyclic record of sedimentary deposits. Witzke et al. (1988) estimate the duration of a single

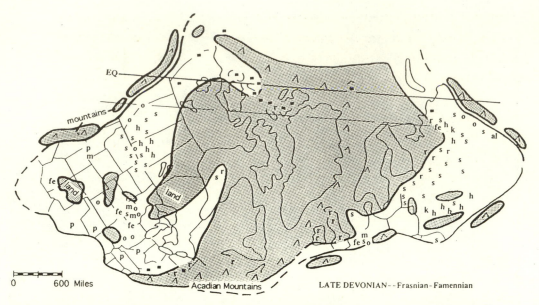

6.4 The interpreted paleogeography for Late Devonian time for the ancient Euramerica region, approximately 370 million years ago. The Acadian Mountains provided clastic sediments for the Devonian record of the eastern United States. The interior of North America was occupied by a shallow sea during this time. The Transcontinental Arch projected above the sea at the northwestern tip of ancient Iowa, and the Ozark Dome was emergent to the south. Symbols: m = evaporite crystal molds, o = carbonate ooids (oolites) and coated grains, p = phosphorites or phosphatic sediments, s = sulfate evaporites (primarily gypsum and anhydrite), h = halite (rock salt), fe = oolitic ironstones, r = redbeds, k = potash salts, al = bauxite (an aluminum-rich deposit), rectangles = coal deposits, EQ = proposed paleoequator. Shaded areas depict land, and unshaded areas indicate ancient seas. From Witzke 1990b.

large-scale cycle of transgression and regression to be 1 to 3 million years. Some episodes of shallowing in eastern Iowa left the seafloor exposed to the atmosphere and subject to intervals of weathering and desiccation. During times of maximum transgression, the Devonian seaway covered most of Iowa and portions of adjacent states, Michigan, several eastern states, the Dakotas, western Canada, and the southwestern United States (fig. 6.4). Evaporites (gypsum and anhydrite) precipitated from seawater in areas of high evaporation and increased salinity.

The to-and-fro shifting of the Devonian seas over Iowa was probably the result of worldwide fluctuations in sea level. What produces sea-level changes of this magnitude? One promising explanation relates global changes in sea level during Devonian time to processes such as the growth and decay of ocean-ridge systems, episodes of submarine volcanism, and the changing configuration of ocean basins. In addition, the advance and retreat of continental glaciers may have played a role in sea-level fluctuations toward the end of the Devonian Period (during the Famennian Age).

Middle and Upper Devonian rocks are well represented in Iowa and contain a variety of index or zone fossils, allowing correlation with the Devonian sections of Europe. In Europe, the Middle Devonian is divided into Eifelian (lower) and Givetian (upper) divisions. The Upper Devonian in Europe is split into Frasnian (lower) and Famennian (upper) divisions. Similar terminology is now used in the classification of the Devonian of Iowa.

The Devonian formations of Iowa and their inferred environments of deposition are shown in figure 6.5. The recognition of T-R cycles is very important in understanding the stratigraphic framework of Iowa's Middle and Upper Devonian rocks. A qualitative sea-level curve for late Middle through early Late Devonian is illustrated in figure 6.6. Several rock types and environments of deposition repeat within the Devonian rock column. For example, open marine deposits consisting of biostromes (beds of corals and/or stromatoporoids), mudflat facies, and evaporites occur at several stratigraphic intervals. This repetition was caused by the cyclic shifting (advance and retreat) of the Devonian seas over the interior of the continent. Seven T-R cycles mark the Devonian record of Iowa that is shown in figure 6.6.

Lower Devonian

No Lower Devonian strata are currently recognized in Iowa, although in the past, a subsurface unit characterized by gray chert and limestone was assigned to the Lower Devonian. These beds (the La Porte City Formation) are now placed in the Silurian System by geologists at the Geological Survey Bureau. The La Porte City Formation is a limestone facies, equivalent to dolomites of the Silurian Scotch Grove and upper Hopkinton formations.

Middle Devonian

The Middle Devonian rock record in Iowa consists primarily of limestones and dolomites, although shales and evaporites are also present at some localities. The rocks that comprise Iowa's Middle Devonian record were formed in shallow marine environments representing subtidal, intertidal, and supratidal settings. Middle Devonian time in Iowa saw the deposition of carbonates, shales, and evaporites of the Wapsipinicon Group. The characteristics of these rocks suggest that the Wapsipinicon Group (fig. 6.5) formed primarily in a shallow, restricted sea. Some carbonates probably accumulated initially as carbonate muds on supratidal and tidal mudflats. Later, these deposits were converted to dolomite. Portions of the Wapsipinicon seas were restricted in circulation and briny in composition; evaporites precipitated in these areas and on coastal mudflats.

Few fossils occur in the Wapsipinicon Group, suggesting that the early Middle Devonian seas were generally inhospitable for marine life. Perhaps the seas were

6.5 Iowa's Devonian formations and their depositional environments. After Witzke et al. 1988; Bunker 1988; and Bunker et al. 1988.

Series/Group	Formation	Composition	Depositional Environments
Upper Devonian "Yellow Spring" Group	Maple Mill Formation	Shale and siltstone	Shallow marine shelf in northern Iowa grading into a deeper basinal setting to the south and southeast; low oxygen levels on the seafloor
	Aplington Formation	Dolomite	Shallow marine shelf grading laterally into a muddy shelf
	Sheffield Formation	Shale	Muddy marine shelf; low oxygen levels on the seafloor
	Sweetland Creek Formation	Shale	A lateral equivalent of the Lime Creek Fm.; deposited in a deeper setting; little or no oxygen present in bottom waters; bottom waters devoid of life
	Lime Creek Formation	Shale, calcareous shale, and limestone	Deposited during a T-R cycle; shallowing upward pattern; grades from a shallow shelf and shelf margin in northern Iowa into a deeper setting in eastern Iowa, where the Lime Creek interval is represented within the Sweetland Creek Fm.; evaporites in the subsurface of central Iowa; dominated by carbonates in the subsurface of northwestern Iowa
Upper Devonian "Upper" Cedar Valley Group	Shell Rock Formation	Fossiliferous carbonates (limestone and dolomite) with some shale in the type area of northern Iowa; includes laminated, brecciated, intraclastic, and evaporitic facies in the subsurface; well-developed stromatoporoid biostromes	Primarily an open marine shelf; deposited during a T-R cycle; contains some mudflat and evaporite deposits
	Lithograph City Formation	Limestone, shale, and dolomite; including very fine grained (lithographic)	Deposited during a T-R cycle; displays shallowing upward characteristics;

Series/Group	Formation	Composition	Depositional Environments
	Lithograph City Formation (*cont.*)	limestone in northern Iowa; evaporites in the subsurface of central Iowa; stromatoporoid biostromes in northern Iowa; tidal-channel facies in Johnson County with calcarenites (sand-size fragments of limestone and fossils); local lag deposits of fish bones and fish parts in Johnson County	includes tidal-channel deposits locally in eastern Iowa; hypersaline conditions in central Iowa where evaporites formed
Middle Devonian "Lower" Cedar Valley Group	Coralville Formation	A variety of carbonate (limestone and dolomite) facies, including fossiliferous units and coral and stromatoporoid biostromes; the fossiliferous units are overlain by laminated, brecciated, and evaporitic facies; thick deposits of evaporites (gypsum and anhydrite) in the subsurface of central Iowa; shaley intervals in northern and central Iowa	Deposited during a T-R cycle; displays a shallowing upward pattern
	Little Cedar Formation	Various limestone and dolomite facies; often sandy at the base; some shale; abundant corals, brachiopods, echinoderms, stromatoporoids, and other marine invertebrates; biostromes of corals and stromatoporoids; evaporites in the subsurface	Deposited during a T-R cycle; primarily open marine shelf conditions but contains shallow mudflat facies near the top and evaporites in the subsurface; environments in southeastern Iowa represent a more open marine and deeper setting than those of northern and western Iowa
Middle Devonian Wapsipinicon Group	Pinicon Ridge Formation	Generally unfossiliferous carbonates (dolomites and limestones) but burrows, stromatolites, ostracodes, and fish fossils are known; variably shaley, laminated, or brecciated; locally sandy; gypsum and anhydrite occur in the subsurface of central and southern Iowa; contains solution-collapse breccias	Marine seaway with restricted circulation; shallow marine mudflats; evaporite deposition in a shallow sea or supratidal setting; limited marine life; later development of solution-collapse breccias

(*continued*)

Series/Group	Formation	Composition	Depositional Environments
	Spillville-Otis formations	Dolomite and limestone; bird's-eye structure and mud cracks; contains marine invertebrates such as brachiopods, mollusks, and trilobites	Open marine shelf and coastal mudflats; Spillville Fm. deposited in a more open marine setting than the Otis Fm.; the two formations were separated by a northeastern-trending pre-Devonian Silurian escarpment
	Bertram Formation	Unfossiliferous dolomite, laminated in part; sandy; sandy shale in basal portion; bird's-eye structure; restricted distribution	Coastal mudflats
Lower Devonian	No rock record	No rock record	Erosion and weathering

too saline to support a typical marine fauna in Iowa during this time. The widespread occurrence of evaporites in the Wapsipinicon Group indicates that the seas were often briny. Evaporite rocks (gypsum and anhydrite) are abundant in both the Wapsipinicon Group and in the overlying Cedar Valley Group. In order for evaporites to precipitate from seawater, the salinity of the water needs to increase significantly. In marine environments, the salinity increases when water evaporates from restricted seas with limited circulation. Evaporites also form on arid tidal flats.

In theory, gypsum ($CaSO_4 \cdot 2H_2O$) starts to precipitate after evaporation reduces a body of normal seawater to about a third of its original volume. If evaporation continues, the remaining water becomes a concentrated salt brine. The mineral halite (NaCl) begins to precipitate when the body of seawater reaches approximately a tenth of its original volume. In order for an appreciable amount of gypsum to form, the salinity of the body of water must remain at the saturation point for gypsum for a long time. This requires the simultaneous addition of seawater and removal of concentrated brines.

Figure 6.7 illustrates two models for the formation of evaporites in shallow marine and coastal settings. Similar environments exist today in the Persian Gulf, where marine evaporites form in restricted lagoons and on arid supratidal flats. Marine waters wash onto supratidal flats during times of storms. Evaporation increases the salinity of these waters, which accumulate in the pores of the sediments. Precipitation of evaporite minerals in a supratidal setting is chiefly from saline pore waters in the sediments.

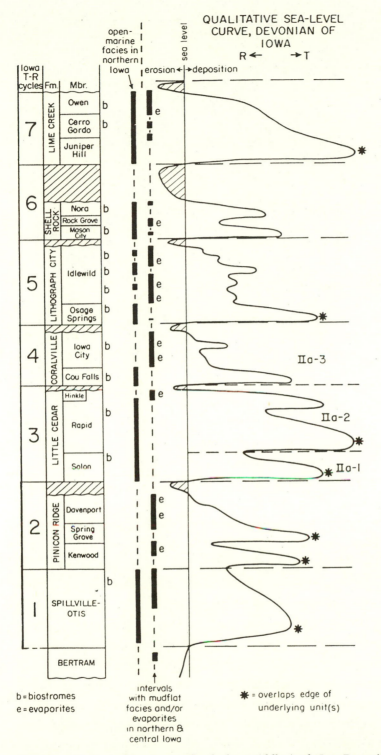

6.6 A qualitative sea-level curve for the late Middle–early Late Devonian showing seven transgression-regression (T-R) cycles. Note that several facies and rock types repeat in Iowa's Devonian record. Adapted from Johnson, Klapper, and Sandberg 1985; Witzke et al. 1988; and Bunker 1988.

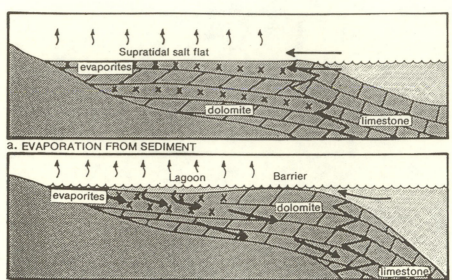

a. EVAPORATION FROM SEDIMENT

b. EVAPORATION FROM STANDING WATER

6.7 Two models to explain the origin of evaporites in shallow marine environments: (a) evaporites from very saline waters on a supratidal (above tide) flat; (b) evaporites by precipitation of saline waters in a restricted lagoon. Evaporites shown by the x symbols. After Dott and Batten 1971.

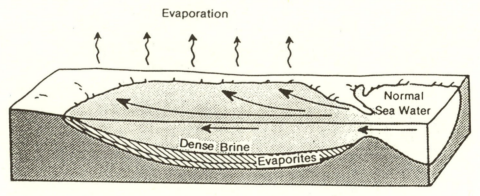

6.8 A model for an evaporite basin. After Dott and Batten 1971.

An additional model to explain evaporite formation is illustrated in figure 6.8. In this model, high rates of evaporation in a restricted basin produce brines of sufficient concentrations to cause the precipitation of gypsum or halite.

Evaporites, such as those of the Persian Gulf, are often associated with dolomites. Dense, magnesium-rich brines form during the evaporation process and later seep down through limestone beds or sediments that lie below the surface. As the brines move through the limestone, they react with the calcium carbonate minerals of the limestone and convert them to dolomite by a process of dolomitization (fig. 6.7). Some dolomite beds of the Wapsipinicon Group may have formed by this evaporative-reflux process.

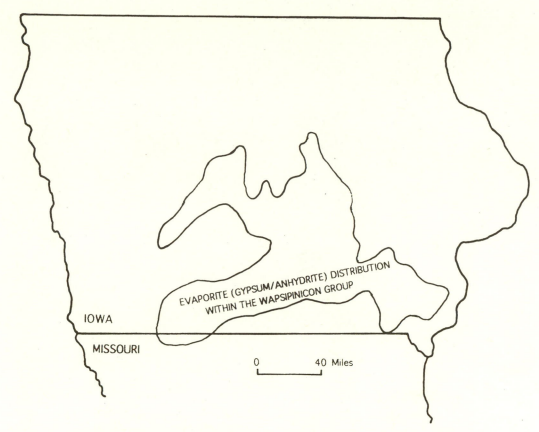

IOWA

MISSOURI

EVAPORITE (GYPSUM/ANHYDRITE) DISTRIBUTION WITHIN THE WAPSIPINICON GROUP

0   40 Miles

6.9 The distribution of evaporites within the Wapsipinicon Group. Adapted from Witzke et al. 1988.

Anhydrite ($CaSO_4$) is associated with gypsum in the Middle Devonian evaporites of Iowa. Gypsum, the hydrated form of calcium sulphate, precipitates under most conditions of evaporation. Gypsum, however, converts to anhydrite in the subsurface at burial depths of 100 to 200 feet. Both evaporite minerals are highly soluble, and neither is common at the surface in areas with humid temperate climates. In Iowa, the Devonian evaporites are known only from the subsurface (figs. 6.9 and 6.10). Because gypsum and anhydrite are highly soluble, they dissolve in groundwater. This process of solution produces caverns and voids in the rock sequence that can result in the collapse and brecciation of overlying rock units.

Carbonate breccias occur in the Wapsipinicon and Cedar Valley groups (fig. 6.5). These rocks contain angular chunks of limestone and dolomite in a carbonate matrix. Many such breccias probably were formed by the solution of evaporite beds and the resulting collapse and brecciation of overlying carbonate beds.

Whereas restricted marine conditions were the norm during deposition of the Wapsipinicon beds in Iowa, more normal marine conditions prevailed during

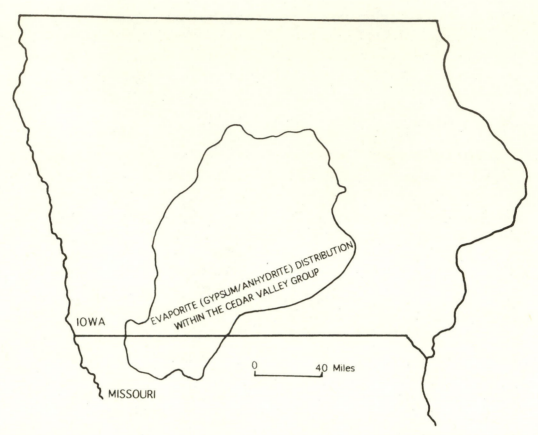

6.10 The distribution of evaporites within the Cedar Valley Group. Adapted from Witzke et al. 1988.

the deposition of the Cedar Valley Group. Deposition of the Cedar Valley beds began with the advance of a sea that spread over exposed and eroded Wapsipinicon strata. This transgression incorporated sand from several sources and rock fragments from the Wapsipinicon strata into the basal deposits of the Cedar Valley Group. As the transgression continued, a variety of fossiliferous carbonate sediments formed. These deposits comprise the Little Cedar Formation (fig. 6.5) and contain a diverse assemblage of marine organisms, reflecting deposition in warm, shallow seas with normal marine circulation. Brachiopods, colonial and solitary corals, echinoderms, stromatoporoids, and bryozoans were the dominant marine organisms of the early Cedar Valley seas. Locally in eastern Iowa, solitary and colonial corals were sufficiently abundant to form fossil-rich beds (biostromes) and a few small mounded deposits (bioherms).

The upper beds of the Little Cedar Formation reflect deposition in progressively shallower waters and represent an environmental sequence ranging from shallow subtidal, through intertidal, to supratidal conditions. These beds are less fossiliferous than are the beds of the lower and middle parts of the formation. A similar pattern of deposition occurs in the overlying Coralville Formation (fig. 6.5), which also represents a shallowing upward sequence produced by a T-R cycle of the Devonian sea (fig. 6.6). The Little Cedar and Coralville formations of the Cedar Valley Group belong to the Middle Devonian, whereas the overlying Lithograph City and Shell Rock formations are placed in the Upper Devonian.

Upper Devonian

Cyclic deposition continued during the Late Devonian with deposition of the Lithograph City and Shell Rock formations (fig. 6.6). Each of the formations records a T-R cycle of the Devonian sea, although there are cyclic variations within both formations. Overall, the Lithograph City and Shell Rock formations display shallowing upward patterns, and biostromes composed of stromatoporoids (fig. 6.11) are common in both formations.

Mudflat facies characterize the upper portions of all four of the formations in the Cedar Valley Group, and they are particularly well developed in the Lithograph City beds (fig. 6.6). Some beds contain mud cracks (fig. 6.12) and bird's-eye structures (fig. 6.13), both of which form by desiccation and are typically associated with high intertidal or supratidal settings. In such environments, fine-grained carbonate sediments (lime muds) are exposed to the atmosphere. As the sediments dry on the exposed tidal and supratidal flats, they shrink and form mud cracks (fig. 6.12).

There are two varieties of bird's-eye structures. Originally, the term "bird's-eye" was given to tiny eye-shaped blebs of sparry calcite found in fine-grained carbonate rocks. As presently recognized, however, bird's-eye structures can exist as either planar voids or as isolated bubblelike vugs. Such structures are common in modern supratidal environments. The isolated vugs apparently form from gas bubbles trapped in fine-grained supratidal muds, whereas the planar structures represent shrinkage features produced by the desiccation of exposed sediments. The vugs of bird's-eye structures from modern environments are generally filled with either calcite or anhydrite. However, the bird's-eye structures in Iowa's Devonian record are composed almost exclusively of sparry calcite.

Tidal channel deposits occur in the Cedar Valley Group in Benton and Johnson counties. In Johnson County, the tidal channel deposits (State Quarry Limestone) formed contemporaneously with deposits of the Lithograph City Formation. The tidal channel facies consists primarily of cross-bedded clastic limestones (calcarenites). Similar tidal channels exist in present-day settings along the southern shore of the Persian Gulf.

Dolomitization has altered the Cedar Valley Group extensively in northern Iowa, making environmental interpretations difficult. Other lithologic changes

6.11 A biostrome within the Lithograph City Formation composed of branching and subspherical stromatoporoids, Black Hawk County, Iowa. Photo by the author.

in the Cedar Valley sequence are known from the subsurface, where gypsum and anhydrite are present at several levels (figs. 6.5 and 6.10).

Conditions in the Devonian seas changed after deposition of the carbonates of the Shell Rock Formation of the Cedar Valley Group, and widespread muddy conditions prevailed during the remainder of Late Devonian time. Mud washed into the seas from a variety of sources (fig. 6.3). Some of these muds comprise the shales of the Lime Creek Formation. Locally in north-central Iowa, carbonates were deposited along with mud to form argillaceous limestones. Upper Devonian rocks of the Lime Creek Formation are exceptionally fossiliferous at localities in Floyd and Cerro Gordo counties, and these strata provide an important record of the marine life of the time (fig. 6.1). All of the major invertebrate groups occur in the Lime Creek fauna, and fossil preservation is superb. Elsewhere, the Lime Creek Formation and its equivalents are sparingly fossiliferous, although a few brachiopods and a variety of microfossils are sometimes present.

Throughout the Devonian seas in Iowa and elsewhere, important groups of brachiopods, colonial corals, mollusks, bryozoans, echinoderms, and stromatoporoids died out in Late Devonian (late Frasnian) time. This was a time of global

6.12 Mud cracks in fine-grained limestone of the Lithograph City Formation, Benton County, Iowa. Photo by the author.

6.13 Bird's-eye structure in the Lithograph City Formation, Black Hawk County, Iowa. Photo by the author.

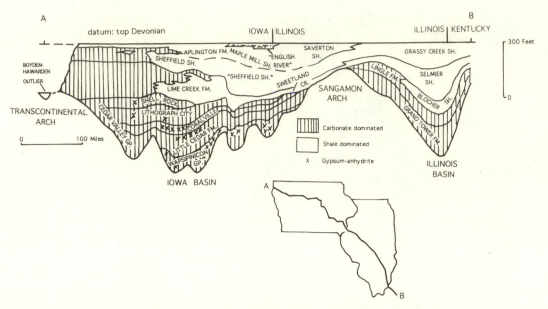

6.14 A generalized stratigraphic cross section of the Middle and Upper Devonian strata of the Iowa and Illinois basins and major tectonic features. Note the presence of the "English River" Siltstone in southeastern Iowa and a thick sequence of carbonates in the subsurface of northwestern Iowa. Location of evaporites (gypsum and anhydrite) shown by the x symbols. Shale is common in the Lime Creek, Sheffield, and Maple Mill formations. The Boyden-Hawarden outlier contains a nonmarine fauna of Late Devonian age. Adapted from Witzke et al. 1988.

mass extinctions. According to Day (1995), of the seventy-one species of brachiopods found in Iowa's Lime Creek Formation and its equivalents, only five survived into the early Famennian (youngest stage of the Devonian). The late Frasnian extinctions were probably not caused by a single calamity, but rapid rises in sea levels were apparently contributing factors to the demise of bottom-dwelling invertebrates. Rising sea levels greatly reduced the area of shallow-shelf habitats all over the globe and produced widespread deep-water settings with oxygen-deficient bottom waters. According to Johnson et al. (1985), a rapid sea-level rise marks the Frasnian-Famennian boundary within the Upper Devonian. In addition, two similar rises occurred just before the boundary.

The uppermost Devonian formations in north-central Iowa (Sheffield, Aplington, and Maple Mill) consist primarily of shales, argillaceous dolomites, and siltstones. These strata, once assigned to the Mississippian System, contain diagnostic Late Devonian conodont faunas. Besides containing conodonts, the strata in this interval also bear plant spores, other assorted microfossils, and brachiopods. For the most part, the shales of the Sheffield and Maple Mill formations formed on a muddy seafloor with low levels of oxygen in the bottom waters.

A discussion of Iowa's Devonian formations follows. These formations are exposed in east-central and north-central Iowa (fig. 6.2). Considerable information is also available on the Devonian rocks in Iowa's subsurface (fig. 6.14).

Devonian Formations of Iowa

The Devonian System in Iowa includes thirteen major formations. The Bertram, Otis, Spillville, Pinicon Ridge, Little Cedar, and Coralville formations comprise the Middle Devonian (fig. 6.5). The Lithograph City, Shell Rock, Lime Creek, Sweetland Creek, Sheffield, Aplington, and Maple Mill formations represent Iowa's Upper Devonian record. Currently, the Devonian formations of Iowa are assigned to the Wapsipinicon, Cedar Valley, and Yellow Spring groups (fig. 6.5). However, the name Yellow Spring Group will likely be changed to the New Albany Shale Group in the near future.

The Devonian strata of the state are of considerable economic value, a topic that will be treated further in chapter 11. Devonian rocks are quarried extensively for road and concrete aggregate in eastern Iowa, and gypsum is mined from Devonian strata from the subsurface in southeastern Iowa. Portland cement is produced from Devonian carbonates at Mason City and at Buffalo, near Davenport. The Devonian carbonates along with the underlying Silurian dolomites constitute the Silurian-Devonian aquifer and serve as important sources of high-quality groundwater across eastern and north-central Iowa.

Each of the major formations of the Devonian System of Iowa is discussed in the following sections. Formations are described in ascending order from oldest to youngest.

BERTRAM FORMATION

The Bertram Formation, the basal formation of the Wapsipinicon Group, is geographically restricted. It occupies a small, asymmetrical basin in east-central Iowa, next to the Plum River Fault Zone. Named for a locality in Linn County and originally assigned to the Silurian System, the Bertram is primarily an unfossiliferous dolomite, in part laminated and sandy. Brecciation and fracturing are common in the unit. The presence of bird's-eye structures suggests episodes of exposure to the atmosphere and desiccation; possible gypsum pseudomorphs imply intervals of aridity, high evaporation, and increased salinity.

OTIS AND SPILLVILLE FORMATIONS

The Otis Formation, named for the former Otis railroad station in southeastern Linn County, overlies either the Bertram Formation or eroded Silurian strata (fig. 6.15). It reaches a maximum thickness of approximately 50 feet where it overlies the Bertram Formation next to the Plum River Fault Zone in eastern Iowa. The Otis Formation usually consists of a lower dolomite interval (Coggon Member) and an upper limestone and dolomite segment (Cedar Rapids Member). The formation is entirely limestone in the Davenport area of eastern Iowa.

The Coggon Member is not particularly fossiliferous. Scattered molds of

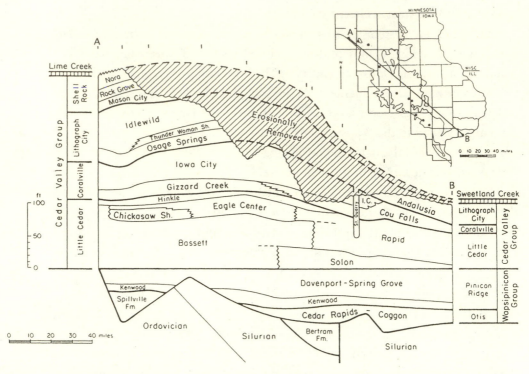

6.15 A generalized stratigraphic cross section from north-central Iowa to the Iowa-Illinois border in east-central Iowa showing inferred stratigraphic relationships of the formations and members of the Wapsipinicon and Cedar Valley groups. A post-Silurian escarpment separates the Spillville Formation from the Otis Formation (Cedar Rapids and Coggon members). From Witzke et al. 1988.

brachiopods (*Emanuella* sp.) occur, along with less common gastropods, trilobites, conodonts, and rostroconchs (an extinct class of mollusk). The Cedar Rapids Member contains a few brachiopods, gastropods, and bryozoans. Spirorbid worms have been noted also. The varied lithologies of the Cedar Rapids Member include laminated carbonates with bird's-eye structures and mud cracks.

The Spillville Formation is the basal Devonian formation in northeastern Iowa and southeastern Minnesota. Named by Klapper and Barrick (1983) for quarry exposures at Spillville in Winneshiek County, these strata were assigned previously to the Cedar Valley Group. The Spillville Formation is primarily fossiliferous dolomite. It contains brachiopods, solitary and colonial corals, trilobites, tentaculites, crinoid debris, gastropods, bivalves, rostroconchs, nautiloids, conodonts, and other marine fossils. The basal part of the formation contains shale and argillaceous dolomite. The Spillville is a lateral equivalent of the Otis Formation; the two formations originated in different environmental settings, separated at the time of deposition by a northeast-trending, pre-Silurian escarpment (fig. 6.15).

PINICON RIDGE FORMATION

The Pinicon Ridge Formation of the Wapsipinicon Group is an unfossiliferous assortment of rocks lying between the underlying Otis and Spillville formations and the overlying strata of the Cedar Valley Group. Named for the type region at Pinicon Ridge State Park in Linn County, the formation includes, in ascending order, the Kenwood, Spring Grove, and Davenport members. The Pinicon Ridge Formation overlaps the edges of the Spillville and Otis formations and in places rests on the eroded edges of Ordovician and Silurian rocks (fig. 6.15).

The Pinicon Ridge Formation consists of unfossiliferous carbonate rocks, shales, evaporites, and breccias. The lower portion of the formation, the Kenwood Member, is primarily an unfossiliferous argillaceous dolomite and dolomitic limestone, containing scattered to abundant silt. Gypsum and anhydrite occur in the subsurface of southeastern Iowa. The stratigraphic assignment of these evaporites is unclear at present; they may correlate with part of the Kenwood interval.

The Spring Grove Member is comprised primarily of dolomite, typically laminated and with a strong petroliferous odor. Breccias occur at the base of the unit in several localities. Anhydrite is present in the subsurface in southern Iowa. Although the Spring Grove Member is generally unfossiliferous, it has yielded ostracodes, burrows, stromatolites, and fish remains in the Quad Cities area.

The Davenport Member consists mostly of very fine grained limestone, dolomitic limestone, and dolomite. The member is extensively brecciated across most of eastern Iowa. Breccia clasts were derived from limestones within the member and from the overlying Solon Member of the Little Cedar Formation. The breccias probably formed by the solution of evaporite beds (anhydrite or gypsum), followed by the collapse of overlying layers of limestone. Later, the angular clasts of limestone were cemented together to form a breccia. The solution and collapse processes were operating, at least in part, during deposition of the open marine limestones of the lower Little Cedar Formation. Clasts of limestones from the Little Cedar beds are present in the breccias of the Davenport Member.

LITTLE CEDAR FORMATION

The Little Cedar Formation is the basal formation of the Cedar Valley Group. It rests on strata of the Wapsipinicon Group or on older rocks where the Wapsipinicon Group is absent. The Little Cedar Formation underlies the Coralville Formation. The type locality of the Little Cedar Formation is at Chickasaw Park Quarry, by the Little Cedar River in southwestern Chickasaw County.

The formation was deposited during a large-scale T-R cycle of the Devonian sea (fig. 6.6).

In northern Iowa, the Little Cedar Formation consists, in ascending order, of the Bassett, Chickasaw Shale, Eagle Center, and Hinkle members (fig. 6.15). Two members, the Solon and the Rapid, constitute the formation in eastern and southeastern Iowa (fig. 6.15).

The Solon Member is primarily limestone, with scattered shaley or carbonaceous partings. A shale interval is normally present at the base and grades into a sandstone facies in northern Missouri and western Illinois. To the north in Iowa, the limestones of the Solon Member give way to argillaceous dolomites of the lower Bassett Member (fig. 6.15).

The Solon Member possesses an abundant and diverse fauna, including colonial and solitary rugose corals, tabulate corals, stromatoporoids, a wide assortment of brachiopods, and other marine invertebrates. The *independensis* zone of Merrill Stainbrook (1941) spans the lower half of the member. It is named for a characteristic brachiopod, *Independatrypa independensis*. The profunda beds of Stainbrook occur in the upper half of the Solon Member and are named for the colonial rugose coral *Hexagonaria profunda*. The profunda beds also contain other rugose corals, tabulate corals, and stromatoporoids. Locally, the upper Solon interval is biostromal, with abundant corals and/or stromatoporoids. To the north, near its transition with the Bassett Member, the Solon Member exhibits packed accumulations of brachiopod shells, representing an ancient shell bank.

The Rapid Member is primarily fine-grained argillaceous limestone with shaley partings and lenses of coarser clastic limestone. Three divisions characterize the member: (1) a lower fossiliferous interval of fine-grained limestone with shaley partings (*bellula* zone of Stainbrook [1941]); (2) a middle segment of fossiliferous and unfossiliferous limestones (*Pentamerella* beds of Stainbrook), capped by widespread coralline biostromes; and (3) an upper unit of fine-grained limestones (calcilutites) interbedded with lenses of echinoderm-bearing clastic limestones (calcarenites). The upper unit is locally glauconitic and commonly cherty in the lower part. It represents the *waterlooensis* zone of Stainbrook and is named for the brachiopod *Desquamatia waterlooensis*.

The Rapid Member contains a diverse marine fauna, including conodonts, brachiopods, bryozoans, echinoderm debris, crinoids, blastoids, cystoids, echinoids, edrioasteroids (a rare group of echinoderms), starfish, and corals. The middle Rapid interval displays two conspicuous biostromes. The member is present across east-central and southeastern Iowa, western Illinois, and northeastern Missouri. It is replaced to the north in Iowa by strata of the middle and upper Bassett, Chickasaw Shale, Eagle Center, and Hinkle members (fig. 6.15).

CORALVILLE FORMATION

The type section of the Coralville Formation is at Conklin Quarry, near Coralville in Johnson County. Originally recognized as a member within the Cedar Valley Limestone, the Coralville is now accorded formational status. The Coralville Formation includes a lower fossiliferous interval with an abundant marine fauna (Cou Falls or Gizzard Creek members) and an upper carbonate unit with restricted marine faunas (Iowa City Member). As shown in figure 6.15, the Gizzard Creek Member occurs in north-central Iowa, whereas the Cou Falls Member is present in the east-central part of the state. The upper member (Iowa City Member) is dominated by carbonates with laminated and brecciated textures; evaporites occur in the subsurface. The Coralville Formation formed during a single T-R depositional cycle of the Devonian sea (fig. 6.6). Coralville strata overlie the Little Cedar Formation (fig. 6.15).

The Cou Falls Member in the Johnson County area consists of two faunal intervals known as the *Cranaena* zone (lower) and the *Idiostroma* beds (upper). The *Cranaena* zone includes coralline biostromes with the colonial rugose coral *Hexagonaria*, massive stromatoporoids, favositid (honeycomb) corals, and solitary rugose corals. Brachiopods (including *Cranaena*, the name-bearer for the interval) are common in some beds of the zone. The *Idiostroma* beds are also biostromal; they are named for the branching stromatoporoid *Idiostroma*. These biostromes contain massive stromatoporoids, colonial rugose corals such as *Hexagonaria*, solitary rugose corals, and tabulate corals of the honeycomb variety. In addition, a sparse conodont fauna occurs in the Cou Falls Member.

The Gizzard Creek Member is primarily dolomite with some dolomitic limestone. It is argillaceous in part and has calcite-filled vugs. Locally, intraclasts are present. The member contains scattered to abundant fossil molds. Faunas of the Gizzard Creek Member consist of only a few species. Brachiopods, conodonts, crinoid debris, gastropods, bryozoans, branching stromatoporoids, favositid corals, and other corals have been reported.

The Iowa City Member is variable in composition and displays significant lateral variations. Bird's-eye structures and mud cracks record intervals of desiccation in Johnson County and areas to the north. Conodonts have not been recovered from the member, but other marine fossils are known, including corals, stromatoporoids, algal oncolites and stromatolites, calcareous algae, foraminifers, ostracodes, and gastropods. A biostromal interval contains abundant branching stromatoporoids (*Amphipora* bed of Kettenbrink [1972a, b]). Evaporites (gypsum and anhydrite) occur within the member in the subsurface of central Iowa, and these deposits represent the thickest occurrence of evaporites within the Cedar Valley Group.

As with the previously described formations of the Cedar Valley Group, the Lithograph City Formation formed during a T-R cycle of the Devonian sea (fig. 6.6). The Lithograph City Formation was proposed by Bunker et al. (1986) for an interval of Upper Devonian rocks lying disconformably between the underlying Coralville Formation and the overlying Shell Rock Formation or Sweetland Creek Shale. Previously, these rocks were miscorrelated with the Coralville strata of east-central Iowa. The type locality for the Lithograph City Formation is an area of abandoned quarries next to the former town of Lithograph City in northern Floyd County. High-quality stone for lithographic engravings was quarried there in the early 1900s.

In northern Iowa, the Lithograph City Formation consists, in ascending order, of the Osage Springs, Thunder Woman Shale, and Idlewild members. Two distinct facies of the formation are recognized outside of the northern Iowa outcrop belt. The State Quarry Member, a tidal channel facies, occurs in Johnson County in eastern Iowa; the Andalusia Member is represented in southeastern Iowa and adjacent areas of Missouri and Illinois (fig. 6.15).

The Osage Springs Member is a fossiliferous dolomite and dolomitic limestone, with a strong resemblance to the Gizzard Creek Member of the Coralville Formation. However, the Osage Springs can be distinguished by its higher stratigraphic position and by differences in faunal content. The member contains a distinctive conodont fauna, along with brachiopods, stromatoporoids, echinoderm debris, bryozoans, gastropods, corals, and burrows. In some localities, stromatoporoid biostromes occur.

The Thunder Woman Shale Member is primarily gray dolomitic shale and silty shale. In northernmost Iowa and adjacent Minnesota, the shale is replaced by carbonates. The Thunder Woman Shale is relatively unfossiliferous, but a few fish parts and conodont fragments occur in the unit in the subsurface of north-central Iowa. Burrows have been reported from surface exposures.

The Idlewild Member is varied in composition, consisting of: (1) laminated and pelletal, very fine grained (lithographic and sublithographic) limestones and their dolomitized equivalents, in part with mud cracks, bird's-eye structures, or evaporite molds; (2) nonlaminated dolomites and limestones, in part with mud cracks, bird's-eye structures, or evaporite molds; (3) calcareous shales and fossiliferous dolomites; and (4) limestones, with scattered to abundant brachiopods and/or stromatoporoids and locally with stromatoporoid biostromes. In the subsurface of central Iowa, the Idlewild Member contains gypsum and anhydrite, mostly in the lower part of the unit. In southeastern Iowa, the Idlewild lithologies are replaced by fossiliferous carbonates of the middle and upper Andalusia Member.

The Idlewild Member contains ostracodes, stromatolites, gastropods, echinoderm debris, favositid corals, bryozoans, brachiopods, and a restricted conodont fauna. Stromatoporoids are often abundant and locally form biostromes (fig. 6.11).

The State Quarry Member is confined to Johnson County, where it occupies broad channels up to a mile in width. The channels are incised into the underlying Coralville and Little Cedar formations, with the deepest incision to the level of the middle Rapid Member. The State Quarry beds are primarily fossiliferous calcarenites and calcilutites, the remnants of ancient tidal channels. Crossbedding and lag deposits of fish bones are known at some localities.

Conodonts occur in the State Quarry beds, along with echinoderm debris, brachiopods, and stromatoporoids. In addition, various fossils have been reported: corals, gastropods, nautiloids, spirorbid worm tubes, ostracodes, trilobites, calcareous algae, foraminifera, and fish debris.

The Andalusia Member is composed of argillaceous and fossiliferous dolomitic limestone, limestone, and dolomite; fossiliferous calcareous shales are present in the lower part of the unit. Coral and stromatoporoid biostromes occur in the upper third of the member. Hardgrounds with encrusted auloporid tabulate corals are present in the lower and upper parts. The Andalusia Member is the sole representation of the Lithograph City Formation southeast of Johnson County in eastern Iowa. Andalusia strata are replaced by beds of the Osage Springs and Idlewild members to the west-northwest in Iowa (fig. 6.15).

In addition to corals and stromatoporoids, the Andalusia Member contains conodonts, brachiopods, echinoderm debris, bryozoans, bivalves, gastropods, rostroconchs, nautiloids, and fish debris. Biostromes vary in content but include solitary corals, favositid tabulate corals, or massive stromatoporoids.

The Lithograph City Formation is quarried extensively for road and concrete aggregate. At Mason City, the Lithograph City Formation supplies high-quality raw material for the manufacture of portland cement. In the early 1900s, stone from the Lithograph City Formation was used in lithography. Lithography is the process whereby text and drawings can be reproduced from smooth slabs of dense, fine-grained (lithographic) limestone. The lithographic stones are inked with a preparation of wax, soap, lampblack, and water. Lithography was a common printing technique prior to being replaced by metal engraving after World War I.

At the turn of the twentieth century, the town of Lithograph City was founded near the Floyd-Mitchell county line for the purpose of quarrying lithographic stone. Stone from the quarries at Lithograph City was judged to be of excellent quality and received several awards at the 1904 Louisiana Purchase Exposition in St. Louis. The quarry industry at Lithograph City had a short history, however.

The quarries operated for only a brief period of time, and lithography was soon replaced by metal-engraving processes in the printing industry. In 1918, the Devonian Products Company took over the quarry operations at Lithograph City and began to produce crushed rock and related products; the town's name was changed to Devonia. Eventually, the company closed and the town lost population. By the late 1930s, many of the town's streets were plowed and converted to cropland. Today, there is little physical evidence for the presence of Lithograph City, but the name lives on in Iowa's rock column.

SHELL ROCK FORMATION

The Shell Rock Formation is represented by fossiliferous carbonates, with some shale in the type area in northern Iowa. In its western outcrop area and in the subsurface, the formation includes laminated, bird's-eye-bearing, brecciated, and intraclastic facies. C. H. Belanski (1927) named the formation (Shellrock stage) from exposures near Nora Springs in northern Iowa. He subdivided the formation into three "substages" (members); in ascending order, these are the Mason City, Rock Grove, and Nora members. The formation was deposited in a T-R cycle of the Devonian sea (fig. 6.6).

The Mason City Member consists primarily of fossiliferous limestones and dolomitic limestones. Hardground and discontinuity surfaces are present, and locally near Rockford, they preserve encrusted edrioasteroids and auloporid corals along with attached cystoids (fig. 6.16). The member becomes more dolomitic and less fossiliferous west of the type area.

The Rock Grove Member is composed of limestone and dolomite, with some shale in the type area of northern Iowa. To the west and southwest, mud cracks and bird's-eye structures occur. Minor evaporites mark the middle Shell Rock Formation in the subsurface of central Iowa and probably correlate with the Rock Grove interval.

In its type area, the Nora Member is characterized by prominent stromatoporoid biostromes, separated by a shaley unit. To the west and southwest, the member becomes more dolomitic and contains fewer stromatoporoids.

All three members of the Shell Rock Formation bear fossils. The Mason City Member yields articulated crinoids, rhombiferan echinoderms (cystoids), edrioasteroids, and disarticulated echinoids. Biostromal beds in the Mason City and Nora members are dominated by stromatoporoids. Brachiopods and echinoderm debris occur in all three members. A variety of mollusks have been reported from the formation, including bivalves, gastropods, nautiloids, and scaphopods. Conodont faunas help establish correlations with Upper Devonian strata elsewhere. The Shell Rock Formation is overlain by the Lime Creek Formation.

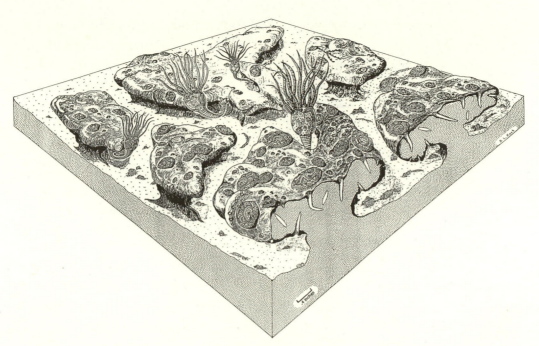

6.16 A reconstruction of a shallow seafloor in north-central Iowa during deposition of the Late Devonian Shell Rock Formation. A discontinuity surface in the Mason City Member displays a remarkably well preserved fauna of cystoids, edrioasteroids, corals, and bryozoans. The organisms were preserved in their growth positions. From Koch and Strimple 1968.

LIME CREEK FORMATION AND EQUIVALENTS

The Lime Creek Formation in Floyd and Cerro Gordo counties is noted for its superbly preserved and spectacularly abundant fossils. Dozens of choice specimens can be collected in a few minutes. The fossils of the Lime Creek Formation are well known to both professional paleontologists and amateur collectors. During field investigations for Iowa's first geological survey, Edward Hungerford and J. D. Whitney collected spiriferid brachiopods from the fossiliferous Lime Creek beds west of Rockford. These species, new to science, were named and illustrated by Hall and Whitney (1858) in their report on the geological survey of Iowa.

Nearly two hundred species of marine invertebrates and microfossils have been described and illustrated from the Lime Creek Formation. Many of the fossil groups in the formation have not been studied in detail, however. In addition, some of the earlier work is in need of restudy and taxonomic revision.

In its type area of northern Iowa, the Lime Creek Formation displays a shallowing upward succession of shales and carbonate rocks that formed during a single T-R cycle during Late Devonian (Frasnian) time (fig. 6.6). The depositional setting of the Lime Creek Formation is illustrated in figure 6.17. The Lime Creek Formation of northern Iowa represents the deposits of a carbonate shelf

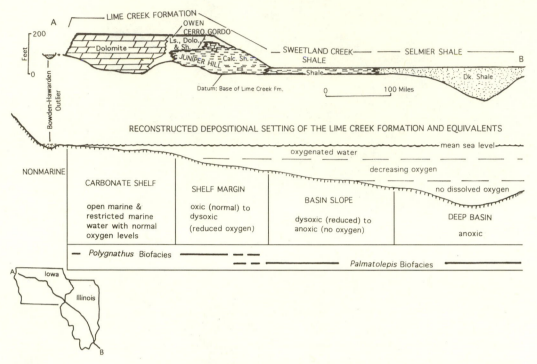

6.17 A stratigraphic cross section of the Lime Creek Formation and equivalent strata (above) and reconstructed depositional setting (below). Note that the Lime Creek Formation of the type area in Floyd and Cerro Gordo counties was deposited in a shelf-margin setting, whereas the laterally adjacent Sweetland Creek Shale of eastern Iowa formed in a deeper (basin slope) environment. Also shown are the distribution of shallow-water conodont faunas (*Polygnathus* Biofacies) and deeper-water, offshore faunas (*Palmatolepis* Biofacies). Adapted from Day 1995.

and shelf margin; lateral equivalents in eastern Iowa and Illinois reflect deposition in deeper water. These deeper-water facies include the Selmier Shale of Illinois and the Sweetland Creek Shale of eastern Iowa and west-central Illinois.

The Lime Creek Formation consists, in ascending order, of the Juniper Hill, Cerro Gordo, and Owen members. The Juniper Hill Member is primarily gray calcareous shale and mudstone. A basal unit of phosphatic fossils (fish plates and conodonts) occurs at the Hackberry Grove locality in Cerro Gordo County. Calcareous shelly fossils are rarely seen in surface exposures of the member, but a moderately diverse brachiopod fauna has been noted in subsurface sections in Cerro Gordo County. A sparse fauna from surface exposures of the Juniper Hill Member includes conodonts, lingulid and other brachiopods, sponges, and carbonized plant remains.

The Cerro Gordo Member is exceptionally fossiliferous in surface exposures near Rockford, Iowa. There are three well-known exposures of the member in the area between Charles City and Mason City in northern Iowa. The best known of the three is the former Rockford Brick and Tile quarry, now the Floyd County Fossil and Prairie Area. Bird Hill State Preserve exposes fossiliferous beds of the Cerro Gordo Member along an east-west county road in eastern Cerro Gordo County. The stratigraphic section known as Hackberry Grove is located in the Clay Banks Natural Area, maintained by the Cerro Gordo County Conservation Board. All three members of the Lime Creek Formation are present at the Clay Banks preserve, the designated type section of the Cerro Gordo Member.

The Cerro Gordo Member consists of fossiliferous calcareous shales, nodular argillaceous limestones, and bedded argillaceous limestones. Fossils weather free from the rock matrix and are easy to collect. At least thirty-eight species of brachiopods occur in the Cerro Gordo Member, and many fine specimens can be collected in Cerro Gordo and Floyd counties. Other common fossils found in the Cerro Gordo beds include bivalves, gastropods, solitary and colonial corals, and bryozoans. Both nautiloid and ammonoid cephalopods occur, but they are not particularly abundant.

The Owen Member consists of limestones, dolomitic limestones, dolomites, and shales. Fossils from the Owen include brachiopods, mollusks, corals, and stromatoporoids. The base of the Owen is marked by a prominent stromatoporoid biostrome in eastern Cerro Gordo County. Faunal diversity is lower in the Owen Member than in the underlying Cerro Gordo Member. The Owen strata record shallow-water deposition during a regressive phase of the Late Devonian sea.

Lateral equivalents of the Lime Creek Formation in the Amana area in Iowa County have been called the Amana Beds. These beds contain a diagnostic Late Devonian fauna composed of conodonts, brachiopods, and ammonoids. The name Independence Shale is applied to lateral equivalents of Lime Creek strata in several localities in eastern Iowa, including sites in Buchanan County near Independence. The Independence Shale occupies caverns, sinkholes, and other solution features that formed on Silurian and Devonian carbonates prior to the Late Devonian (Lime Creek) transgression across the central midcontinent. Stratigraphic leaks of the Lime Creek Formation and its equivalents (Independence Shale) filled many karst features and in this manner came to rest on, within, and below older strata. This led to considerable stratigraphic confusion, as will be discussed later in this chapter.

The Lime Creek Formation is equivalent to the Sweetland Creek Shale of Muscatine County (fig. 6.17). The Sweetland Creek Shale contains a prolific conodont fauna, including species adapted for a deep-water, offshore setting. In contrast, the conodonts of the type Lime Creek area in northern Iowa are indicative of a shallow-water shelf environment (fig. 6.17). The Sweetland Creek Formation was deposited in a deep-water, basin-slope setting, where the bottom waters contained little or no dissolved oxygen and the seafloor was inhospitable for bottom-dwellers. The fauna of the Sweetland Creek Formation consists of swimming or floating creatures such as conodonts, whose remains fell to the seafloor and were preserved as fossils.

SHEFFIELD, APLINGTON, AND MAPLE MILL FORMATIONS

As originally defined, the Yellow Spring Group included the Sheffield, Aplington, Maple Mill, and English River formations. The Lime Creek Formation has since been assigned to the group, and the English River Siltstone is now recognized as a member within the Maple Mill Formation. The Yellow Spring Group is not completely exposed at any one locality. Its name is derived from Yellow Spring Township in Des Moines County in southeastern Iowa, where subsurface samples from a well help define the group. A well in Webster County in north-central Iowa provides additional information to characterize the variation within the Yellow Spring sequence. However, uncertainties remain in our understanding of this part of the rock column, and correlations of strata of the Yellow Spring Group of southeastern Iowa with the Yellow Spring Group of northern Iowa are not completely resolved. Figure 6.5 summarizes the composition and depositional environments of the group.

The Sheffield Formation is named for the town of Sheffield in Franklin County, where the formation consists of gray shale with minor amounts of dolomite. Shale was once quarried at Sheffield for the manufacture of brick and tile. In southeastern Iowa, the Sheffield interval includes dark gray and black shales. Conodonts, spores, and a few additional microfossils occur in the formation.

The Aplington Formation is poorly exposed in north-central Iowa, and the type section in a quarry at Aplington in Butler County is no longer completely accessible. The characteristic composition of the Aplington is argillaceous dolomite with discontinuous beds of chert. In southeastern and south-central Iowa, the dolomite grades laterally into shale. Brachiopods and conodonts occur in the formation in north-central Iowa.

The Maple Mill Formation is named for exposures at Maple Mill on the English River in Washington County. The formation consists of light gray to greenish gray shales; it is calcareous in places and contains beds of siltstone. Fossils found in the Maple Mill Formation include conodonts, fish remains, and large spores.

The name English River Siltstone was first used for exposures of dolomitic siltstone that occur along the English River in Washington County. The unit is essentially restricted in distribution to southeastern Iowa. The English River is now considered to be a member of the Maple Mill Formation.

Devonian Life

Marble Rock in Floyd County, Shell Rock in Butler County, and Coralville in Johnson County are all located in the Devonian outcrop belt of Iowa, and they derive their names from some aspect of the local Devonian bedrock. Shell Rock and Coralville suggest something of the fossiliferous nature of Iowa's Devonian, one of the state's most fossiliferous rock systems.

Iowa's Devonian fossils have attracted the attention of many investigators over the years. Louis Agassiz, the distinguished Harvard naturalist, visited the Iowa City area in the 1860s and noted the significance of the ancient coral banks found in the local rock record. C. H. Belanski amassed noteworthy collections of Late Devonian fossils from the Cerro Gordo and Floyd county areas in the 1920s. Carroll Lane Fenton, coauthor of the widely known *The Fossil Book*, did some of his early paleontological work on the Late Devonian fossils of Iowa. One of the most devoted students of Devonian fossils was Merrill A. Stainbrook.

Merrill Addison Stainbrook (1897–1956) maintained an active interest in Devonian paleontology throughout his lifetime. Born and raised on a farm near Brandon in southwestern Buchanan County, Stainbrook lost the sight in his left eye in a childhood accident. This partial loss of vision did not keep Stainbrook from participating in outdoor activities. He became interested in rocks and fossils and collected extensively from the nearby exposures of Devonian strata.

Stainbrook's principal teacher at the University of Iowa was A. O. Thomas, an enthusiastic and dedicated student of Iowa's Devonian faunas. Stainbrook completed his Ph.D. degree at the University of Iowa in 1927 and from 1927 to 1948 taught at Texas Tech University at Lubbock. Throughout most of his years in Texas, Stainbrook returned to Iowa in the summer months to pursue his investigations of Devonian paleontology. During part of this time, he served as a consulting geologist on Devonian formations for the Iowa Geological Survey.

Stainbrook is best known for his work on the paleontology of the Devonian Cedar Valley Group and the Independence Shale. Stainbrook believed that the Upper Devonian Independence Shale fossils came from beds that were stratigraphically below the Cedar Valley Group and therefore that the Cedar Valley strata were younger. Most geologists recognized that the fauna from the Independence Shale was similar to the Upper Devonian Lime Creek Formation fauna, which is clearly younger in age than the Cedar Valley fauna. These geologists explained the stratigraphic position of the Independence Shale as the result of stratigraphic leak, the filling in of Independence Shale (Lime Creek Formation) in a cavernous terrain on and within rocks of the Cedar Valley Group. Stainbrook was not convinced, however. Through an immense amount of field-work and a study of numerous drill cuttings, Stainbrook found many localities where he was convinced that the Independence Shale was in place below the Cedar Valley Group. Stainbrook vigorously supported this view to the time of his death.

One of Stainbrook's opponents in the Independence Shale controversy was G. Arthur Cooper of the Smithsonian Institution, an acknowledged expert on Devonian brachiopods. Although Stainbrook and Cooper strenuously debated the age of the Independence Shale, they were on friendly terms. Copper commented: "I regarded Merrill Stainbrook very highly as a person and as a stratigrapher and paleontologist. Our disagreements were indeed of a very friendly nature, even though we were widely apart in our opinions. I regard his paleontological work very highly. His generic work on the brachiopods is very good. I think Merrill Stainbrook was one of the outstanding collectors in this country. I collected with him extensively in the Devonian of Michigan and Iowa, and believe that his one eye was quicker at seeing fossils that the two eyes of our best collectors."

Merrill Stainbrook is honored today by the presence of the Stainbrook Geological Preserve, a state preserve near Lake Macbride in Johnson County. Appropriately, the reserve contains fossil-bearing exposures of Stainbrook's beloved Devonian strata. The great flood of 1993 introduced Devonian fossils to the masses. Merrill Stainbrook would have been pleased.

Among the effects of the summer floods in Iowa in 1993 was the overflow of the emergency spillway at Coralville Lake. The overflow lasted twenty-eight days, and floodwaters excavated at least 15 feet of surficial sediments to expose a spectacular expanse of fossiliferous Devonian bedrock (the Rapid Member of the Little Cedar Formation). During the six months following the flooding, more than 250,000 visitors came to view this remarkable sample of a tropical seafloor and its ancient life. The broad horizontal rock exposures, scoured by the flood waters, provided the public with a pathway into the past.

Nationwide newspaper accounts heralded the flood-exposed fossils. The ABC nightly news, CNN, and Paul Harvey broadcast the story. Visitors flocked to the site. Numerous school groups from eastern Iowa took advantage of the learning opportunity afforded by a visit, and the locality soon became an outdoor laboratory for many. Following a contest by the Corp of Engineers, the Coralville Lake spillway locality was officially named Devonian Fossil Gorge.

A ten-year-old Girl Scout gained a better understanding of fossils and geologic time as a result of her trip to Devonian Fossil Gorge. She was quoted in *USA TODAY*: "It's weird. You sit on these rocks and think they're older than my parents — or even my grandparents."

Yes, the abundant fossils at Devonian Fossil Gorge are old, older still than our grandparents' grandparents. Approximately 375 million years old, these fossils are nearly two hundred times older than the earliest representatives of our genus, *Homo*.

A bountiful array of fossils is visible in the rock exposures of Devonian Fossil Gorge. Brachiopods are abundant, and some rocks are crowded with their shells. Crinoids (sea lilies) are plentiful, although most occur as pieces of plates and fragments of stems. However, some complete crinoids are visible, preserved with stem, head, and arms intact. Both solitary and colonial corals are numerous, as are bryozoans. In addition, a careful observer can spot sponges, trilobites, and primitive fish.

The Devonian seas of Iowa supported a great variety of marine life, and all of the major groups of marine invertebrates are represented in the rock record of the state. Fish were abundant and diverse, too. Although the Devonian Period has been called the Age of the Fish, the clear leaders in terms of abundance in Iowa's rock record are brachiopods, corals, and stromatoporoids. A lifelike reconstruction of a Devonian seafloor can be seen at Iowa Hall on the campus of the University of Iowa. Figures 6.18 and 6.19 illustrate the marine life typical of Devonian time. Brief discussions of the more important Devonian fossil groups follow. Representative fossils are shown in figures 6.20 and 6.21.

BRACHIOPODS

Brachiopod shells make up an important part of the Devonian fossil record of Iowa. The remains of both articulate and inarticulate brachiopods are known. Brachiopods are particularly abundant in the rocks of the Cedar Valley Group and the Lime Creek Formation. The spiriferid group of brachiopods is characterized by a spiral-shaped internal structure that served as a support for the lophophore (feeding organ). Many spiriferids have a wide hinge line with pointed, winglike margins (figs. 6.21 and 6.22). Brachiopod shells in the Cedar Valley Group and Lime Creek Formation served as hosts for other organisms (fig. 6.22). Some of the more common Devonian brachiopods are illustrated in figures 6.20 and 6.21.

BRYOZOANS

Bryozoans, the moss animals, are fairly common in Iowa's Devonian record. These colonial fossils occur as stemlike, branching, lacy, and encrusting varieties. Although the calcium carbonate skeletons of bryozoans appear no more complex than those of corals, the soft anatomies of the two are significantly different. Bryozoans have a more complex digestive system than do corals. Fossil bryozoans are common in strata of the Cedar Valley Group and the Lime Creek Formation.

6.18 A reconstruction of a Middle-Late Devonian seafloor similar to the setting where Iowa's Cedar Valley Group was deposited. Life included: (a) stalked crinoids; (b) solitary rugose corals; (c) lacy bryozoans; (d) trilobites; (e) spiriferid brachiopods, some with encrusting spirorbid worms; (f–h) other articulate brachiopods; and (i) tabulate corals. From McKerrow 1978.

CORALS

Modern corals are impressive builders; they have constructed over 700,000 square miles of reef rock. In addition, today's reef systems contain a wide array of marine life, housing perhaps a quarter of all marine species. Modern types of corals appeared on earth during the Mesozoic. The ancient rugose and tabulate

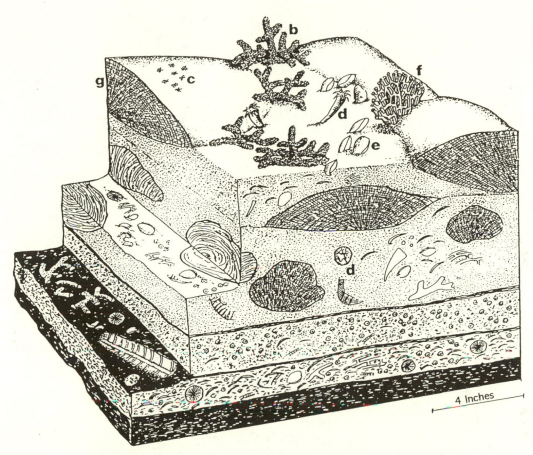

6.19 A reconstruction of a Middle-Late Devonian seafloor similar to the setting where Iowa's Cedar Valley Group was deposited. Life included: (a) nautiloid cephalopods, (b–d and f) corals, (e) articulate brachiopods, and (g) stromatoporoids. Stromatoporoids (ancient spongelike creatures) and corals contributed to the formation of biostromes (fossil-rich beds) like the one shown here. From McKerrow 1978.

corals of Iowa's Devonian are classified in the same phylum as modern corals. However, they were incapable of building prominent wave-resistant structures like those constructed by modern corals and algae. Nevertheless, the rugose and tabulate corals of Iowa contributed to the formation of impressive, laterally continuous beds of ancient life (biostromes).

The soft-bodied animals that built and inhabited the ancient coral skeletons were probably similar to their living relatives in modern reefs. These tiny animals

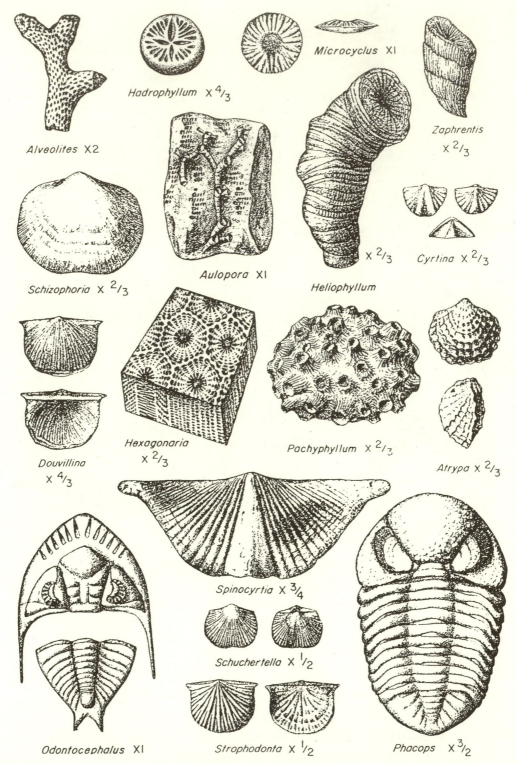

6.20 Representative Devonian fossils. Corals: *Alveolites, Hadrophyllum, Microcyclus, Zaphrentis, Heliophyllum, Aulopora, Hexagonaria, Pachyphyllum*; brachiopods: *Douvillina, Cyrtina, Atrypa, Spinocyrtia, Schuchertella, Strophodonta, Schizophoria*; trilobites: *Odontocephalus, Phacops*. From Willman et al. 1975. Courtesy of the Illinois Geological Survey.

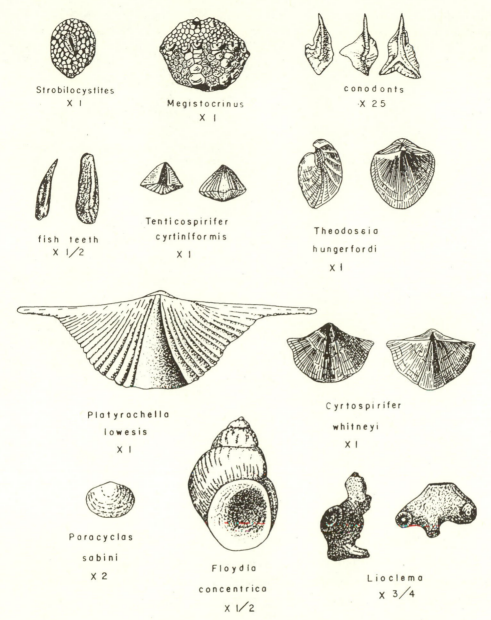

6.21 Assorted Devonian fossils from Iowa. Cystoids: *Strobilocystites*; crinoids: *Megisto-crinus*; conodonts; fish teeth; brachiopods: *Tenticospirifer*, *Platyrachella*, *Theodossia*, *Cyrto-spirifer*; gastropods: *Floydia*; bivalves: *Paracyclas*; bryozoans: *Lioclema*. From Anderson 1983.

possess specialized cells, but they have no organs. Their bodies are like sacks or tubes, with a mouth ringed by tentacles.

The Devonian rocks of Iowa contain an abundance of well-preserved corals. Tabulate corals and both solitary and colonial rugose corals are common. Rugose corals have septa (radial partitions that extend inward from the margin of

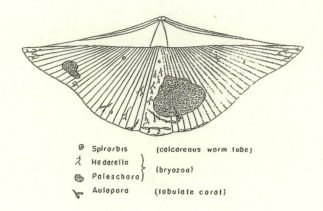

6.22 A spiriferid brachiopod shell that served as a host for *Spirorbis* (a calcareous worm tube), *Aulopora* (a tabulate coral), and two varieties of encrusting bryozoans. After Ager 1961.

the coral skeleton). Septa are absent or poorly developed in tabulate corals. Horn-shaped corals are a common variety of solitary rugose coral (fig. 6.20). Skeletal structures are often preserved in exquisite detail in fossil rugose corals, probably because the skeletons are composed of durable calcite ($CaCO_3$).

Corals are important constituents of biostromes in the Cedar Valley Group of Iowa. The Little Cedar and Coralville formations contain well-developed biostromes in east-central Iowa. Tabulate corals, colonial and solitary rugose corals, and stromatoporoids are the key components of the biostromes. The biostromes represent the occurrences of extensive and prolific life on the Devonian seafloors. One of the biostromes can be traced more than 40 miles along the outcrop belt, has a width of at least 20 miles, and ranges in thickness from 1 to 4 feet. Figure 6.19 illustrates a biostromal accumulation of corals, stromatoporoids, and other invertebrates.

STROMATOPOROIDS

As mentioned previously, the exact zoological affinity of the stromatoporoids was uncertain for many years. Today, paleontologists assign them to the phylum Porifera (sponges). In the past, the stromatoporoids, or stroms, were classified with corals and coralline organisms. Stromatoporoids are extinct now, but some types clearly resemble a group of modern sponges called sclerosponges. Iowa's Silurian and Devonian rocks reveal a wide assortment of stromatoporoids. They occur in many forms, including subspherical, branching, and laminar matlike varieties (fig. 6.23). Encrusting forms are also known. Stromatoporoids formed biostromes individually (fig. 6.11) and in combination with corals (fig. 6.19). All four of the formations of the Cedar Valley Group in Iowa contain prominent stromatoporoid biostromes. Mass extinctions during Late Devonian time (Frasnian-Famennian boundary) took a heavy toll on the stromatoporoids, and stroms are not significant in Iowa's rock record after Late Devonian (Frasnian) time.

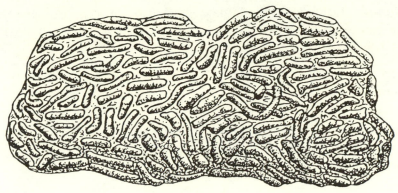

Branching stromatoporoid X 1

Subspherical stromatoporoid X 3/4

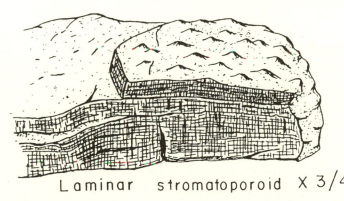

Laminar stromatoporoid X 3/4

6.23 Three types of fossil stromatoporoids. Branching, subspherical, and laminar varieties are shown. From Anderson 1983.

ECHINODERMS

Stalked echinoderms (cystoids) and disklike edrioasteroids occur preserved in their growth positions in the Shell Rock Formation near Nora Springs in northern Iowa (fig. 6.16). Cystoids had smaller arms than crinoids, and they possessed specialized porous plates in their calyx (main body region). Well-preserved cystoids also occur in the Little Cedar Formation (fig. 6.21). Iowa's Devonian cystoids are some of the youngest known geologically; worldwide, cystoids are absent in rocks younger than Devonian.

Whole crinoids (fig. 6.18) are fairly common in the Little Cedar Formation, and their remains can be seen at Devonian Fossil Gorge. More than a dozen colonies of crinoids have been recovered from Iowa's Devonian. The Devonian discoveries are not as impressive as Iowa's Mississippian crinoid colonies, but they do provide useful information about the ecology of ancient crinoids.

MOLLUSKS

The phylum Mollusca is well represented in the Devonian rocks of Iowa by gastropods (snails), bivalves (clams), and cephalopods. In addition, rostroconchs and scaphopods are known. Cephalopods are squidlike organisms; many of the ancient varieties had chambered shells. Both nautiloid and ammonoid types of cephalopods occur in Iowa's Devonian. They differ in several ways, including the contrast in their sutures. The suture is the line formed by the intersection of the shell partition (septum) with the inner surface of the shell wall. The sutures of ammonoids are more complex than the sutures of nautiloids. Ammonoids evolved rapidly, and many species of ammonoids have short geologic-time ranges, making them excellent index fossils. They are readily identifiable because of their characteristic suture patterns. The presence of ammonoids in the Lime Creek Formation establishes quite precisely the geologic age of the formation.

Both ammonoids and nautiloids were probably accomplished swimmers, like the modern pearly *Nautilus*. The gas-filled shells of cephalopods were widely distributed after death, contributing to their utility as index fossils. A straight nautiloid swims above a Devonian seafloor in figure 6.19, whereas a coiled ammonoid rests on the bottom of the sea in figure 6.1.

WORMS

The remains of fossil worms are known from scolecodonts (jaw and mouth parts), burrows, and calcareous worm tubes. *Spirorbis* is the genus name for a common calcareous worm tube; it is abundant in the Devonian and elsewhere in Iowa's rock record. These small, calcareous tubes look like snail-shaped configurations and are commonly found affixed to brachiopods (fig. 6.22) and other marine invertebrates. Similar worm tubes occur in modern marine settings.

ARTHROPODS

Trilobites (figs. 6.18 and 6.20) occur in the Devonian rocks of Iowa, although they are not particularly abundant. Exquisite specimens have been recovered from the Little Cedar Formation, however. Ostracodes (small, bean-shaped arthropods that are largely microscopic) are fairly common in microfossil residues of certain of Iowa's Upper Devonian formations.

SPONGES

Spicules, tiny internal body supports for sponges, occur in microfossil residues from the Lime Creek Formation. More complete sponges are also known from the Little Cedar Formation. Stromatoporoids are now classified with the sponges, and they are one of the most common of Iowa's Devonian fossils.

SPORES

Spores, reproductive bodies from primitive plants, are common in Iowa's Devonian section. They occur in the Lime Creek and Maple Mill formations and throughout the Cedar Valley Group. Many of the spores were probably from terrestrial sources and were later transported to a marine setting. Curtis Klug (1992) has documented the ranges of spores in the Little Cedar, Coralville, and Lithograph City formations.

CONODONTS

Conodonts, tiny phosphatic microfossils from early chordates, are abundant in many of Iowa's Devonian formations. They are excellent index fossils and allow for precise correlation of Iowa's strata with Devonian rocks elsewhere. The individual conodonts (fig. 6.21) were part of a feeding apparatus for an early fish-like animal. The conodont animal is now extinct, but fossil impressions suggest that it resembled an eel or hagfish (fig. 6.24).

Fish

Fossil fish are fairly common in the Devonian rock record of Iowa and include the remains of placoderms, sharks, and bony fish (figs. 6.1 and 6.25). Our knowledge of fossil fish comes primarily from teeth, dental plates, and isolated bones. Fish remains are usually dark in color (black, brown, brownish black, or dark gray), and they stand out against the light-colored beds that enclose them. Although complete specimens of fish are rare, articulated skulls and whole placo-

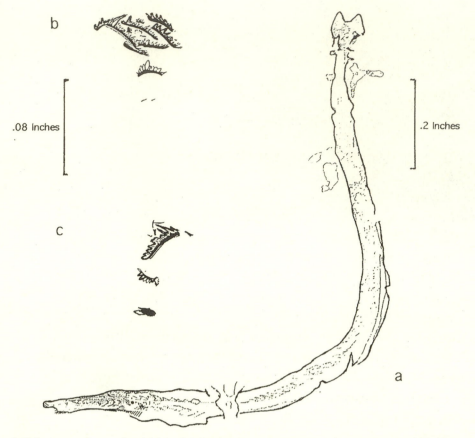

b

.08 Inches

.2 Inches

c

a

6.24 A conodont animal (a) with bilobed head. The individual toothlike conodont elements occur within assemblages (b and c). They were part of a feeding apparatus for an eel-like chordate. The specimen shown here, approximately 1.5 inches long, is from Lower Carboniferous strata of Scotland. Adapted from Briggs et al. 1983.

derms are known from Devonian strata in the Quad Cities and Waterloo areas. Although many of the Devonian fish were small, there were some exceptions. An armored fish from the Devonian of Ohio had an estimated length of over 25 feet.

Among the most common fish remains in Iowa are bones and toothlike plates of placoderms. The placoderms were a primitive fish with an internal framework of cartilage instead of bone. Some placoderms were partly enclosed by bony plates, giving this group of fish an armored look. These bony plates, composed of calcium phosphate, are often preserved in the fossil record. A common group of placoderms, the ptyctodontids, had flat, ovoid dental plates, apparently for crushing shells and other prey. Ptyctodont plates occur in the Little Cedar, Coralville, and Lithograph City formations in the Coralville Lake area of eastern Iowa.

Another subdivision of the placoderms includes the arthrodires, whose heads were armored with thick, bony plates. Some arthrodires had bony spines. Plates of arthrodires occur in the Rapid Member of the Little Cedar Formation at

6.25 A Devonian seafloor with two placoderm fish in the foreground and a shark in the background. After Collinson 1964.

Devonian Fossil Gorge near Coralville Lake. According to Witzke and Bunker (1994), some of these arthrodires reached lengths of 6 to 10 feet. They were the largest animals in Iowa's Devonian seas.

In the Coralville Lake area, the remains of early sharks occur in the Rapid Member of the Little Cedar Formation and in the State Quarry Member of the Lithograph City Formation. Teeth and spines constitute the record of this group. Sharks have a cartilaginous internal framework rather than a true bony skeleton, and the cartilaginous material is not preserved in the fossil record.

Several types of bony fish are known from Devonian rocks in the Coralville Lake area. For example, the Rapid and State Quarry members reveal teeth, jaws, and spines, and tidal-channel facies in the State Quarry Member contain tooth plates of lobe-finned fish. The Devonian Maple Mill Shale of southeastern Iowa also contains beds with concentrations of fish remains. Additional fish fossils occur elsewhere in Iowa's Devonian record, including a variety of disarticulated fish remains from a core near Hawarden in northwestern Iowa. The Upper Devonian rocks and fossils from Hawarden appear to be of nonmarine origin. They may represent an ancient lake or marsh environment.

Volume 18 of the *Iowa Geological Survey Annual Report* by Charles R. Eastman (1907) is devoted almost exclusively to a discussion of Devonian fishes of Iowa. Much of the volume is based on material from the State Quarry beds near North Liberty in Johnson County.

Summary

Iowa's Devonian record began with deposition of the Wapsipinicon Group. The Wapsipinicon Group (Middle Devonian) consists of the Bertram, Otis, Spillville, and Pinicon Ridge formations. The Bertram was deposited in a very restricted setting, dominated by coastal mudflats. The Otis-Spillville interval formed in a T-R cycle of the Devonian sea. Two T-R cycles are represented in the

Pinicon Ridge Formation. Common rocks of the Wapsipinicon Group include dolomites, shales, evaporites, and solution breccias. The evaporites have commercial importance in southeastern Iowa.

Sedimentation continued during the rest of Middle Devonian time and into Lake Devonian with deposition of the Cedar Valley Group. The Cedar Valley Group (Middle and Upper Devonian) includes, in ascending order, the Little Cedar, Coralville, Lithograph City, and Shell Rock formations. Each formation records a T-R cycle of the sea. The upper beds of each formation reflect deposition in progressively shallower water. Carbonates (limestones and dolomites) are the principal rock types in the Cedar Valley Group. Evaporites occur in the subsurface. The formations of the Cedar Valley Group contain a wide array of well-preserved fossils, including brachiopods, bryozoans, corals, conodonts, stromatoporoids, and other marine invertebrates. A variety of fish are known, too. Corals and/or stromatoporoids form laterally extensive biostromes. The flood of 1993 exposed a wide expanse of fossiliferous beds of the Cedar Valley Group (Little Cedar Formation) below the emergency spillway at Coralville Lake. Now known as Devonian Fossil Gorge, the locality attracts thousands of visitors and has introduced the general public to the life of an ancient seafloor.

The Lime Creek Formation (Upper Devonian) formed during still another T-R cycle of the Devonian sea in Iowa. The Lime Creek Formation is primarily shale and calcareous shale with minor argillaceous limestone. In Floyd and Cerro Gordo counties, the formation is exceptionally fossiliferous. Brachiopods, mollusks, bryozoans, corals, and other marine invertebrates weather free from the rock matrix of the Lime Creek Formation, making fossil collecting easy. The Lime Creek Formation in northern Iowa accumulated in a shallow shelf and shelf-margin setting; lateral equivalents in eastern Iowa (Sweetland Creek Shale) formed in deeper water in a basin-slope environment. The Independence Shale is also a Lime Creek equivalent. It occupies caverns, sinkholes, and other solutional features that developed on carbonates of the Cedar Valley Group.

In Late Devonian time, after deposition of the Lime Creek Formation, a global mass extinction took its toll on marine invertebrates. Particularly hard hit were brachiopods, colonial corals, mollusks, bryozoans, echinoderms, and stromatoporoids. Stromatoporoids are insignificant in Iowa's rock column after Late Devonian (Frasnian) time.

The youngest Devonian deposits in Iowa are part of the upper Yellow Spring Group. These Upper Devonian strata (Sheffield, Aplington, and Maple Mill formations) consist primarily of shales with minor siltstones and dolomites. They represent the sediments of a muddy marine shelf with bottom waters that contained little or no oxygen.

Acknowledgments

Our knowledge of the Devonian rock record grew significantly during the past decade, leading to new insights and better understandings. In preparing this chapter, I relied heavily on the contributions of Bill J. Bunker, Brian J. Witzke, Jed E. Day, Fred Rogers, Gilbert Klapper, James Barrick, Curtis R. Klug, and Orrin W. Plocher. Their research provides the basis for many of the revisions of Iowa's Devonian, and references to their publications are at the end of the text. Bill Bunker reviewed this chapter and provided helpful suggestions.

Crinoidal life at the time of deposition must have been prolific beyond all
conception. The organisms made up almost wholly of hard, thick plates
composed of carbonate of lime, furnished an abundance of material for the
accumulation of extensive beds. An idea of the immense quantity of crinoidal
hard parts making up some layers may be obtained when it is remembered
that the average crinoid is composed of from thirty to fifty thousand plates
of all sizes, and that the remains of half a dozen individuals would occupy
the space of about a pint.

Charles Keyes (1919) describing the origin of crinoidal limestones
from the Mississippian System of Iowa

Seas inundated the North American heartland several times during Mississippian time, approximately 360 to 320 million years ago. These were the last widespread carbonate-producing seas to invade Iowa. The Mississippian seas teemed
with life, including prolific occurrences of stalked crinoids (fig. 7.1).

Although limestone was the predominant deposit during Mississippian time,
other rocks formed as well. Dolomites and cherts are associated with many of
Iowa's Mississippian formations, and evaporites (gypsum and anhydrite) occur
in the subsurface of the state. Siliciclastic rocks (shales and quartz siltstones and
sandstones) indicate that terrigenous sediments washed into the Mississippian
seas from source lands to the east of Iowa and from emergent areas within the
craton. Figure 7.2 depicts a general paleogeographic setting for ancestral North
America during Mississippian time.

The rock record deposited during the Mississippian Period is known as the
Mississippian System. The Mississippian System derives its name from strata exposed in the Mississippi River valley between Burlington, Iowa, and the St. Louis
area of Missouri. Therefore, the Mississippian rocks of southeastern Iowa are
of historic significance to the geological profession, and these exposures serve
as important reference sections for the standard rock column of North America. The Mississippian System, roughly equivalent to the Lower Carboniferous
of Europe, is subdivided into four series (Kinderhookian, Osagean, Meramecian, and Chesterian), whose names are derived from localities in Missouri and
Illinois.

7.1 A reconstruction of life on a Mississippian seafloor, similar to the setting during deposition of the Burlington Formation of Iowa. Note the abundance of stalked crinoids (sea lilies). Courtesy of the Field Museum of Natural History, Negative number GEO80871, Chicago.

Regional Mississippian Sedimentation and Sequence Stratigraphy

The origins of Iowa's Mississippian rocks are best understood in reference to environmental changes produced by fluctuations in sea level. It has been recognized in recent years that variations in regional sedimentation patterns are often a result of worldwide changes in sea level. Sequence stratigraphy is the new subdiscipline of geology that provides a framework for understanding these changes in patterns of sedimentation. The concepts of unconformity-bounded packages

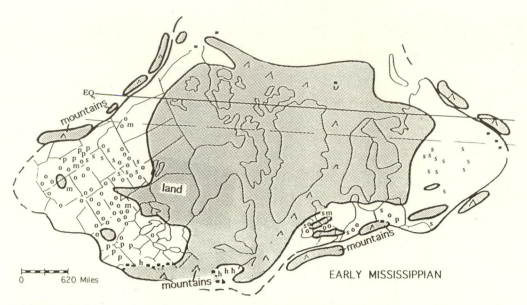

EARLY MISSISSIPPIAN

7.2 The interpreted paleogeography for Early Mississippian time for the Euramerican region. Shaded areas indicate land and unshaded areas depict seas. Symbols: o = carbonate ooids (oolites) and coated grains, m = evaporite crystal molds, EQ = proposed paleoequator. Located south of the equator, ancient Iowa was covered by warm, subtropical seas. From Witzke 1990b.

of strata (sequences), T-R cycles, and the influence of water depth on sedimentary facies were introduced in previous chapters. These ideas are expanded further in this chapter with respect to patterns of sedimentary deposits in the Mississippian rock record of Iowa. Before moving to that discussion, a review of Dunham's (1962) classification of carbonate rocks is in order.

Dunham's classification scheme is based on the depositional textures of limestones and the relative abundance of lime mud compared to carbonate grains such as oolites, pellets, and fossil fragments. Dunham's classification (fig. 7.3) also considers whether grains are bound together by mud or held by chemically precipitated cements. The classification separates rocks that were not bound together at the time of deposition into two categories: those that contain lime mud (silt- and clay-size particles of calcium carbonate) and those that are free of lime mud. Rocks that contain lime mud have textures that are either mud supported, as in lime mudstones and wackestones, or grain supported, as in packstones and grainstones. In packstones, the carbonate grains are in contact and provide a structural framework for the rock, but mud is present between the grains. Grainstones also have grains that are in contact, but the space between the grains is filled primarily with chemically precipitated cement. In general, mud-supported textures in carbonate rocks are associated with quiet-water settings such as carbonate shelves below wave base or protected shoreline areas. Grainstones usually reflect origins within the zone of wave agitation, where lime muds are winnowed

Depositional texture recognizable					Depositional texture not recognizable
Original components not bound together during deposition				Original components were bound together	
Contains mud (clay and fine silt-size carbonate)		Lacks mud and is grain supported			
Mud-supported		Grain-supported			
Less than 10% grains	More than 10% grains				
Mudstone	Wackestone	Packstone	Grainstone	Boundstone	Crystalline

LIME MUD MATRIX

SPARRY CALCITE CEMENT

7.3 Dunham's classification of carbonate rocks. Adapted from Dunham 1962; Scholle 1978; and Tucker and Wright 1990.

by currents. In addition, Dunham's classification recognizes that some carbonate rocks are the products of in-place growth by organisms such as colonial corals or algae.

The identification of depth-related carbonate facies in the Lower Carboniferous (Mississippian) of England led to the regional recognition of major T-R cycles in the Carboniferous seas. The British geologist W. H. C. Ramsbottom (1973) described the following facies patterns, from shallowest (onshore) to deepest (offshore): carbonate mudstones with some stromatolites or desiccation features (inferred to be supratidal-to-intertidal deposits), oolitic limestones, bioclastic limestones, argillaceous limestones with skeletal debris, and offshore shales composed of siliciclastic muds. Geologists in North America recognized similar onshore-to-offshore facies belts in Mississippian carbonate rocks in Montana.

Building on these studies, Witzke et al. (1990) established the following progression of onshore-offshore carbonate facies for the Mississippian sequence of southeastern Iowa: (1) nonmarine carbonates of lacustrine (lake) origin, associated with fluvial (stream) and paludal (marsh) settings; (2) pelletal and very fine grained lime mudstones, in part with stromatolites, bird's-eye structures, other desiccation features, or evaporites (associated with supratidal or sabkha settings); (3) lime mudstones and wackestones with pellets, oolitic in part, faunas of low diversity (back-shoal facies); (4) oolitic grainstones representing shoal deposits; (5) packstone-grainstone carbonates with skeletal fossils, abraded in part (interpreted as deposits that formed near fair-weather wave base); (6) discontinuous skeletal packstones-grainstones, rich in crinoidal debris, interbedded with skeletal wackestones-mudstones, faunas of moderate diversity (above storm wave base); and (7) carbonates with mud-supported textures (mudstones and wackestones), benthic faunas that may show a decrease in diversity (below storm wave base). An eighth facies belt occurs in some T-R cycles in central and southern Illinois. This facies belt, unrepresented in Iowa, is characterized by offshore shales deposited in a deep-water setting.

Oolitic limestones, noted in carbonate facies 3 and 4, attest to current action and wave agitation. Lime precipitates on the surface of grains as they roll and shift around on warm, shallow seafloors. Typically, the final product is an oolite, a sand-size body of laminated calcium carbonate (fig. 7.4). Oolites form today on the shallow, wave-agitated seafloors of the Bahama Banks, and their ancient counterparts are common in Iowa's Mississippian record (fig. 7.5).

The Mississippian sequence of southeastern Iowa records ten major T-R cycles (fig. 7.5). Siliciclastic sediments (quartz sandstones, quartz siltstones, and shales) are associated with some of the depositional cycles. The progradation of these siliciclastic sediments appears to coincide with the regressive phase of certain sedimentary cycles (fig. 7.5). Woodson and Bunker (1989) interpreted T-R cycles for the Mississippian column of north-central Iowa; their proposed cycles closely parallel those established for southeastern Iowa. In both areas, the depositional cycles begin with thin basal transgressive units and culminate with thicker regressive deposits.

Mississippian strata constitute the bedrock in a diagonal belt 20 to 40 miles wide from Lee and Des Moines counties in the southeastern corner of the state northwestward to southern Cerro Gordo and Hancock counties in north-central Iowa (fig. 7.6). Limestones of the Mississippian System are quarried throughout the outcrop belt for road aggregate. Collectively, the Mississippian formations serve as a major bedrock aquifer underlying about two-thirds of Iowa. The Mississippian aquifer is a particularly good source of groundwater in the north-central part of the state, where large yields of good-quality water have been recovered from wells intersecting large crevices in the Mississippian carbonates. Less favorable conditions are found in south-central Iowa, where crevices are scarce and water yields are low. In addition, water from the Mississippian aquifer in south-central Iowa is of poor quality. The poor quality of the water is probably due, in part, to the presence of Mississippian evaporites in the subsurface. Evaporites are quite soluble and contribute undesirable sulfate ions to the groundwater.

Correlations of Mississippian formations between southeastern and north-central Iowa are not completely resolved. For this reason, the stratigraphic successions of the two regions are discussed separately.

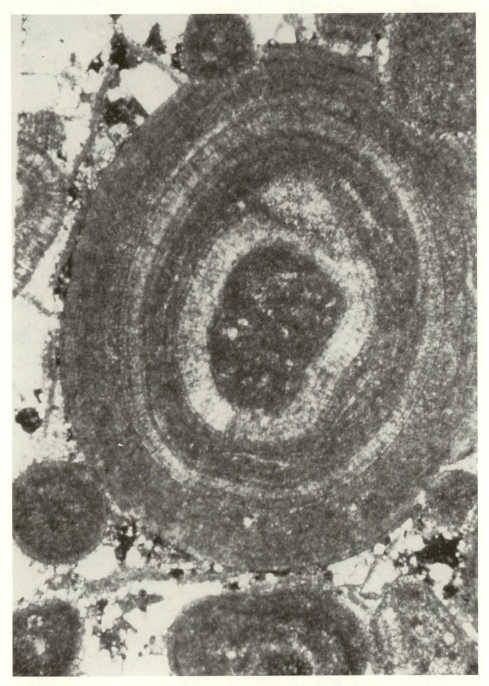

7.4 Oolitic limestone from the Starrs Cave Formation, Burlington, Iowa. Individual oolite grains range in diameter from .02 to .04 inch and indicate times of agitation on the Mississippian seafloors of Iowa. Photo by the author.

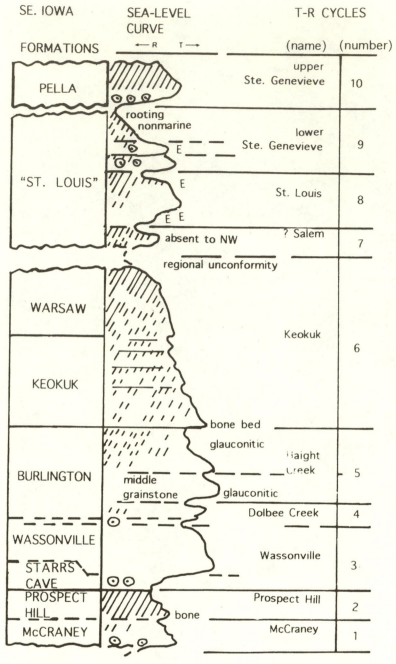

SE. IOWA FORMATIONS	SEA-LEVEL CURVE ←R T→	T-R CYCLES (name)	(number)
PELLA		upper Ste. Genevieve	10
"ST. LOUIS"	rooting nonmarine / E	lower Ste. Genevieve	9
	E	St. Louis	8
	E E / absent to NW	? Salem	7
	regional unconformity		
WARSAW		Keokuk	6
KEOKUK			
	bone bed / glauconitic		
BURLINGTON	middle grainstone	Haight Creek	5
	glauconitic		
WASSONVILLE		Dolbee Creek	4
STARRS CAVE		Wassonville	3
PROSPECT HILL		Prospect Hill	2
McCRANEY	bone	McCraney	1

▨ Siliciclastic Influx

E   Evaporite (gypsum or anhydrite)

◉   Oolite

7.5 A qualitative sea-level curve interpreted for the Mississippian record of southeastern Iowa. T = transgressive, or deepening, trends; R = regressive, or shallowing, trends. Ten T-R cycles are named and numbered at the right. From Witzke et al. 1990.

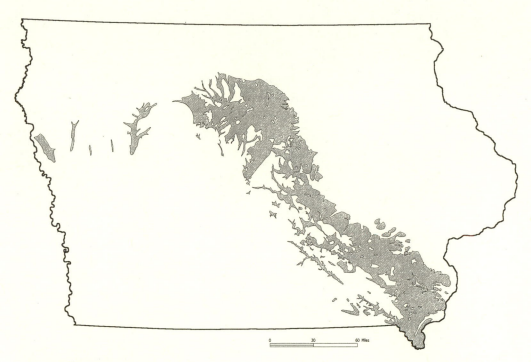

7.6 The distribution of Mississippian bedrock in Iowa. Information provided by Geological Survey Bureau, Iowa City, 1998.

Mississippian Formations of Southeastern Iowa

Geologists at the Geological Survey Bureau assign the Mississippian rock record of southeastern Iowa to ten formations (fig. 7.7). Important reference sections for the Mississippian System of North America occur along the Mississippi River valley in southeastern Iowa (figs. 7.8 and 7.9), western Illinois, and eastern Missouri. Significant exposures of the Kinderhookian, Osagean, and Meramecian series are represented in Iowa. In addition, the Chesterian Series is represented in the state by the Pella (Ste. Genevieve) Formation.

The Kinderhookian Series (Lower Mississippian), as originally defined, included some units that are now assigned to the Upper Devonian in Iowa. The Kinderhookian Series as presently recognized in southeastern Iowa includes, in ascending order, the McCraney, Prospect Hill, Starrs Cave, and Wassonville formations (fig. 7.7). The Burlington, Keokuk, and Warsaw formations provide the state with an excellent record of the Osagean Series (fig. 7.7). The type sections of the Burlington and Keokuk formations are located in southeastern Iowa, near the cities for which they are named. The Warsaw Formation derives its name from exposures near Warsaw, Illinois, across the river from Keokuk. The Salem (Spergen) and "St. Louis" formations of the Meramecian Series (fig. 7.7) are represented in Iowa by lateral facies of formations that are thicker and more uniform in

7.7 The Mississippian formations of southeastern Iowa and their depositional environments. After Witzke et al. 1990.

Series	Formation	Composition	Depositional Environments
Chesterian (Upper Mississippian)	Pella	Calcareous shale, marls, and a variety of limestones; very fossiliferous overall	Deposited in T-R cycle 10 on a shallow marine shelf; upper part of the formation records a regression of the sea and an influx of land-derived sediment
Meramecian (Middle Mississippian)	"St. Louis"	Dolomite, dolomitic limestone, various limestone facies, carbonate breccias, gypsum and anhydrite in the subsurface, quartz sandstone and other siliciclastics	Records T-R cycles 8 and 9; primarily the deposits of a marine shelf setting with occasional influxes of terrigenous sediments; evaporite deposits indicate some restricted marine settings with highly saline waters; abundant carbonate breccias suggest solution-collapse processes associated with later dissolution of evaporites; upper part of the formation records a variety of nonmarine environments and preserves an exceptional fauna of amphibians and fish
Meramecian (Middle Mississippian)	Salem (Spergen)	Various limestone facies; recognized only in extreme southeastern Iowa	Deposits of a marine shelf, possibly recording T-R cycle 7; better known in Illinois and Indiana
Osagean (Middle Mississippian)	Warsaw	Dolomite and shale	Deposited on a marine shelf during a time of regression; a shallowing-upward sequence within the upper part of T-R cycle 6
Osagean (Middle Mississippian)	Keokuk	Various fossiliferous limestones, chert, dolomite, and shale	Deposited on a marine shelf during T-R cycle 6, primarily in a middle shelf setting; upward increase in siliciclastic clay implies an influx of terrigenous mud from land sources
Osagean (Middle Mississippian)	Burlington	Crinoidal limestone; in part, cherty, dolomitic, or glauconitic	Deposited on a marine shelf during T-R cycles 4 and 5; reflects deposition at various shelf depths, both above and below storm wave base

Series	Formation	Composition	Depositional Environments
Kinderhookian (Lower Mississippian)	Wassonville	Cherty dolomite, dolomitic limestone, and fossiliferous limestone	Deposited on a marine shelf during T-R cycle 3
Kinderhookian (Lower Mississippian)	Starrs Cave	Oolitic limestone and fossil-iferous limestone	Deposited in T-R cycle 3; associated with shoaling environments above fair-weather wave base
Kinderhookian (Lower Mississippian)	Prospect Hill	Dolomitic quartz siltstone with interbedded shales; often with concentrations of fish bones and teeth at the base	Deposited in T-R cycle 2 in a marine setting; quartz silt indicates an influx of land-derived sediments
Kinderhookian (Lower Mississippian)	McCraney	Fine-grained limestone and coarse-grained dolomite	Deposited in a shallow, restricted subtidal setting during the first transgression of the Mississippian seas in Iowa

composition farther south in the Mississippi River valley. (The Geological Survey Bureau currently uses quotation marks with the term "St. Louis" to emphasize that the formation terminology used in Iowa only partly correlates with the type St. Louis of Missouri.) The Pella Formation (fig. 7.7) correlates with the Ste. Genevieve Formation of Missouri. The Ste. Genevieve and Pella formations are assigned to the Chesterian Series (Upper Mississippian). Brief descriptions of the Mississippian formations of southeastern Iowa follow.

MCCRANEY FORMATION

The McCraney Formation is composed primarily of very fine grained light-colored limestone and coarse-grained brown dolomite. The alternation of the light- and dark-colored rocks gives the formation a distinctive banded appearance in outcrops in Des Moines County in southeastern Iowa. Although the formation is sparingly fossiliferous, it has yielded chonetid brachiopods, other brachiopods, bivalves, gastropods, corals, and crinoid fragments. Foraminifera and conodonts have also been noted. Collectively, the fauna indicates that the McCraney Formation in southeastern Iowa is Kinderhookian (Early Mississippian) in age. However, the earliest Mississippian record is missing because of an unconformity between the Devonian and Mississippian systems. Over most of southeastern Iowa, the McCraney rests unconformably on the Upper Devonian English River Siltstone.

The McCraney Formation was named for exposures in western Illinois. In

7.8 Mississippian strata exposed along the bluffs of the Mississippi River at Keokuk, Iowa. Drawing by geologist-artist Orestes St. John in 1868, looking north near the upper end of the coffer dam at the base of the Keokuk Rapids. From White 1870.

7.9 Mississippian formations exposed along Flint Creek, near Burlington, Iowa. Drawing by geologist-artist Orestes St. John in 1868, looking north. From White 1870.

southeastern Iowa, the unit ranges in thickness from 0 to 65 feet. McCraney carbonates are absent to the north and west in Iowa. The formation formed in a shallow, restricted to subtidal setting during the first transgression of the Mississippian seas in Iowa (fig. 7.5).

PROSPECT HILL FORMATION

The Prospect Hill Formation derives its name from a site at Burlington, Iowa, now located in Crapo Park. The formation is typically dolomitic quartz siltstone with interbedded shales. The basal contact of the unit is sharp and often displays concentrations of fish bones and teeth. The lag concentrate of fish bones and teeth may indicate an interval of starved sedimentation, prior to widespread transgression of the sea.

Overall, the formation contains a diverse marine fauna, including bivalves, gastropods, cephalopods, scaphopods, brachiopods, conodonts, fragments of bryozoans, and crinoidal debris. Both horizontal and vertical burrows are common. The formation is relatively thin, commonly measuring 3 to 10 feet at most localities in southeastern Iowa. However, thicknesses of 40 to 90 feet are known in Henry, Lee, and Van Buren counties.

The Prospect Hill Formation is the depositional record of the second T-R cycle of the Mississippian seas (fig. 7.5); it represents a marked change in sedimentation from the underlying marine carbonates of the McCraney Formation. The siliciclastic rocks (siltstones and shales) of the Prospect Hill record an influx of land-derived sediments from the east-northeast.

STARRS CAVE FORMATION

The Starrs Cave Formation was designated for exposures along Flint Creek, near Burlington, Iowa. Although the formation is composed typically of fossiliferous oolitic grainstone, other lithologies are known. For example, crinoidal packstones and grainstones are noted, and pelletal and intraclastic grainstones have been observed at some localities. In Washington County, the Starrs Cave is locally absent or difficult to distinguish. The contact of the Starrs Cave with the overlying Wassonville Formation is gradational at many localities in southeastern Iowa. The Starrs Cave Formation is relatively thin and ranges in thickness between 1.5 and 5 feet over much of southeastern Iowa.

Fossils are abundant in the Starrs Cave Formation and include a variety of brachiopods, along with bivalves, gastropods, nautiloid cephalopods, rostroconchs, solitary corals, bryozoans, ostracodes, and trilobites. Crinoid debris is abundant at many localities. The fauna of the Starrs Cave Formation is nearly identical to that of the overlying lower Wassonville Formation. For this and other reasons, deposition of the two formations appears to be linked. Both formations are products of the Wassonville Cycle (fig. 7.5).

WASSONVILLE FORMATION

Named for exposures at Wassonville Mill on the English River in Washington County, the Wassonville Formation is composed primarily of cherty dolomite. However, dolomitic limestones and other limestones also occur within the formation, including grainstones, packstones, wackestones, and lime mudstones. Although the bulk of the formation has been dolomitized, original sedimentary fabrics and primary depositional structures are recognizable at some localities. The presence of flat laminae and hummocky cross-stratification throughout much of the Wassonville implies frequent occurrences of storms during the time of deposition. In addition, the preservation of articulated crinoids at several horizons within the formation is explained by rapid burial of uprooted crinoids during storm events.

An abundant and varied fauna characterizes the Wassonville Formation. The best-preserved fossils occur in skeletal limestones and cherts. The Wassonville fauna includes a variety of brachiopods, such as chonetids, spiriferids, and productids. Articulated crinoids are also known, and trilobites are present in the lower part of the formation. Other fossils include lacy and branching bryozoans, solitary corals, sponge spicules, fish bones and teeth, gastropods, bivalves, scaphopods, rostroconchs, nautiloids, and conodonts.

The Wassonville Formation is probably laterally equivalent to much of the Maynes Creek Formation of north-central Iowa. The Maynes Creek is primarily cherty dolomite, similar in composition and appearance to the Wassonville.

The Wassonville Formation formed during the Wassonville Cycle (fig. 7.5). The initial deposits of the cycle (Starrs Cave Formation) were likely associated with shoaling environments, above fair-weather wave base. Wassonville strata apparently record deepening conditions in the Mississippian seaway, followed by an episode of shallowing. The lower and middle Wassonville beds reflect deposition below fair-weather wave base but above storm wave base. The upper Wassonville probably records shallowing conditions brought about by regression of the sea (fig. 7.5). Discontinuities at or near the top of the Wassonville Formation in southeastern Iowa separate the Wassonville Cycle from the subsequent T-R cycles that produced the Burlington Formation (fig. 7.5).

BURLINGTON FORMATION

The Burlington Formation is one of Iowa's best-known formations. David Dale Owen (1852) mentioned crinoidal strata exposed at Burlington, Iowa, and Hall and Whitney (1858) introduced the term "Burlington limestone" for crinoidal limestones revealed in the bluffs along the Mississippi River at Burlington. Long before the city of Burlington came into existence, Native Americans garnered flint (a variety of chert) for implements from exposures of what is now termed the Burlington Formation. The city that we now call Burlington was known locally as Flint Hills until 1834, when John B. Gray arrived from Vermont, purchased property, and renamed the settlement after his hometown of Burlington, Vermont.

At present, the Burlington Formation is recognized over a wide area of the interior of North America, from Iowa to Arkansas and from Illinois to Kansas. In southeastern Iowa, the Burlington Formation constitutes the lower part of the Osagean Series. The Burlington lies unconformably on the Wassonville Formation and is overlain by the Keokuk Formation. In the Mississippian outcrop belt of southeastern Iowa, the formation ranges from 55 to 80 feet in thickness; the Burlington thickens westward in Iowa's subsurface. Crinoidal limestones with packstone or grainstone textures characterize much of the formation. Cherty, dolomitic, and glauconitic units also occur. These compositional variations have provided a means to subdivide the formation into three members in southeastern Iowa. In ascending order, the subdivisions are the Dolbee Creek, Haight Creek, and Cedar Fork members.

The Dolbee Creek Member was named for a locality in Des Moines County. It is predominantly crinoidal limestone (packstone and grainstone). The *Cactocrinus* zone, proposed by Lowell Laudon (1937), corresponds to this interval. Some beds are dolomitic, and scattered chert nodules may be present. Stylolites (irregular zigzag-shaped surfaces) are common. Attributed to pressure solution, stylolites attest to the solution and removal of substantial thicknesses of calcium carbonate by diagenetic processes.

Low-angle cross-laminae occur in some beds of the Dolbee Creek Member. Multiple hardground surfaces or irregular discontinuities have been recognized in Keokuk and Washington counties. Conodont faunas date the member as Osagean. Spiriferid brachiopods occur and are noteworthy because of their large size. The Dolbee Creek Member ranges between 5 and 17 feet in thickness across most of southeastern Iowa. The Haight Creek Member overlies the Dolbee Creek Member.

The Haight Creek Member, an interval dominated by dolomite and cherty dolomite, comprises the middle portion of the Burlington Formation. The Haight Creek, named for a locality in Des Moines County, was called the *Physetocrinus* zone in earlier literature. Both nodular and bedded cherts occur within the member. The lower portion of the Haight Creek is consistently glauconitic in southeastern Iowa, and abundant glauconite grains are present at or near the base of the unit. The glauconitic zone in the basal Haight Creek is an important horizon marker; it serves as a key reference surface for construction of stratigraphic cross sections and structure contour maps.

Crinoidal limestones (packstones and grainstones), known informally as the middle grainstone interval, occur in the middle portion of the Haight Creek Member throughout much of southeastern Iowa. Equivalent units are recognized in dolomitic facies of the member. Although most of the Haight Creek is sparsely fossiliferous, fossils are abundant in the so-called middle grainstone unit. Fossils include a diverse assortment of crinoid cups, large spiriferid brachiopods, other brachiopods, plus other invertebrates. The Haight Creek Member varies from 28 to 45 feet in thickness in southeastern Iowa and thickens westward in Iowa's subsurface. It is capped by the Cedar Fork Member.

The Cedar Fork Member forms the upper third of the Burlington Formation in southeastern Iowa. Crinoidal limestones (packstones and grainstones) are the most characteristic constituents of the member. The unit is commonly glauconitic. Laudon's (1937) *Dizygocrinus* and *Pentremites* zones occur within the member in southeastern Iowa. The type locality for the Cedar Fork Member is in Des Moines County.

Although the Cedar Fork Member is composed typically of limestone in southeastern Iowa, dolomite and dolomitic limestone occur at some localities. The Cedar Fork dolomites display mud-supported fabrics (wackestones with molds of fossils), whereas the limestone facies of the member (packstones and grainstones) reveal grain-supported fabrics. Dolomite content in the member increases to the north and west in Iowa's subsurface. Nodular cherts occur at most Cedar Fork localities.

The Cedar Fork Member of the Burlington Formation is overlain by the Keokuk Formation. Although there is no evidence of erosional relief at the contact between the two formations, the boundary is distinct. At many locations, the upper 4 to 8 inches of the Burlington Formation preserves significant concentrations of fish bones, spines, and teeth. The widespread distribution of fish parts in the upper Burlington was noted by pioneering paleontologists Charles Wachsmuth and Frank Springer in 1878. Additional bone-bearing horizons occur at other stratigraphic levels within the Cedar Fork Member and within the overlying Keokuk Formation, but they are not as continuous and distinctive as the concentrations at the top of the Cedar Fork. The Cedar Fork Member ranges in thickness between 15 and 30 feet in southeastern Iowa.

Fossils are abundant in the Cedar Fork Member. F. M. Van Tuyl (1925) and Lowell Laudon (1929) recorded diverse benthic faunas from Cedar Fork lithologies in Des Moines and Louisa counties, including 35 species of crinoids, 24 species of brachiopods, 10 species of gastropods, 9 species of corals, 8 species of bryozoans, 4 species of blastoids, 2 species of bivalves, and 2 species of trilobites. More recently, Witzke et al. (1990) reported calcareous foraminifera from the member.

The Burlington Formation originated on a shallow-water carbonate shelf during the Dolbee Creek and Haight Creek T-R cycles (fig. 7.5). As interpreted by Witzke et al. (1990), the Dolbee Creek Member formed in a subtidal, open marine setting. The member preserves several episodes of storm sedimentation.

In addition, hardgrounds are known and probably indicate intervals of non-deposition, most likely in a submarine setting.

The middle member of the formation, the Haight Creek, formed during the Haight Creek T-R Cycle (fig. 7.5). Concentrations of glauconite in the basal Haight Creek beds probably represent a condensed interval brought about by slow rates of sedimentation at the onset of marine transgression. Later deposition reflects a deepening of the sea, and part of the Haight Creek Member likely accumulated below storm wave base. A middle segment of the member, dominated by grainstone lithologies, is interpreted as a product of a shallowing episode. The grainstone interval implies deposition in a wave-stirred setting, above storm wave base. Mudstones and wackestones in the upper Haight Creek suggest origins in quiet water, probably brought about by a phase of relative deepening on the shelf.

The overlying Cedar Fork Member is characterized by crinoidal packstones and grainstones, indicating a return to a depositional position above storm wave base. The presence of glauconite concentrations and bone beds within the Cedar Creek Member implies intervals of relatively slow sedimentation.

Witzke et al. (1990) present a paleogeographic map showing the regional depositional setting of the Burlington Formation (fig. 7.10). Both inner-shelf and middle-shelf settings were present when Iowa's rock record formed.

The carbonate rocks of the Burlington Formation experienced extensive post-depositional changes. Regional dolomitization converted limestones into dolomites, altering most of the original lime muds of the formation. Muds that were originally composed of calcium carbonate were transformed into dolomite, a calcium-and-magnesium carbonate. The mechanism of dolomitization is explainable by the mixing zone model, discussed in chapter 4.

The grainstones of the Burlington Formation are cemented by chemically precipitated calcite ($CaCO_3$). Was the cement formed in a subsea setting, or did it precipitate from groundwater in a terrestrial setting? Geochemical evidence supports the latter hypothesis. The cements appear to be products of a phreatic meteoric system, deposited below the water table in a terrestrial setting.

KEOKUK FORMATION

The Keokuk Formation was named for bluff exposures of cherty carbonate rocks at Keokuk in the southeastern corner of Iowa (fig. 7.8). Prior to the construction of the lock and dam at Keokuk in 1912, the resistant beds of the Keokuk Formation produced a long stretch of treacherous rapids from where the Des Moines River enters the Mississippi River at the lowest point in Iowa below Keokuk to a point upstream near Montrose.

The Keokuk Formation is divided into a lower cherty interval (Montrose Member) and an informal upper Keokuk unit. The contact between the two

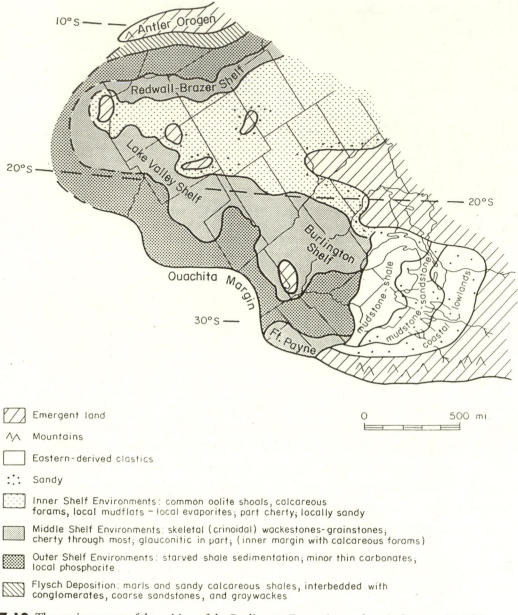

7.10 The environment of deposition of the Burlington Formation and equivalent stratigraphic units. Ancient Iowa was located in inner- and middle-shelf settings during deposition of the Burlington Formation. From Witzke et al. 1990.

members of the Keokuk Formation is transitional. In general, the upper Keokuk is less cherty than the Montrose Member. The Montrose Member is named for exposures near Montrose in Lee County. The member is composed of interbedded fossiliferous limestones, dolomites, and nodular to bedded cherts. Upper Keokuk lithologies are less cherty and include interbedded fossiliferous limestones and shales.

Limestones of the Keokuk Formation are primarily crinoidal packstones that

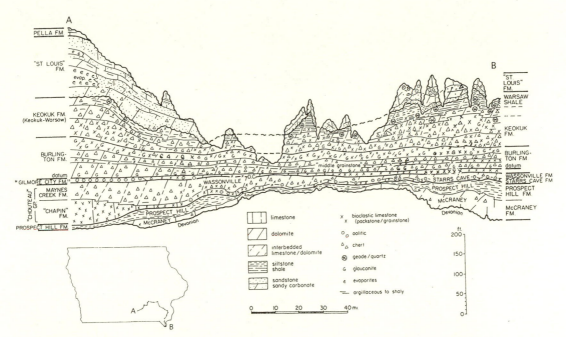

7.11 A stratigraphic cross section across southeastern Iowa (from Monroe County on the left through Mahaska, Keokuk, Washington, Henry, and Des Moines counties, to Lee County on the right). Note the widespread erosional unconformity at the base of the "St. Louis" Formation. From Witzke et al. 1990.

are similar in composition to the limestones of the underlying Burlington Formation. However, the Keokuk limestones are characteristically brown, whereas the Burlington limestones are usually gray. Glauconitic limestones occur in the Keokuk Formation, but they are not as widespread as those of the Burlington Formation. The dolomites of the Keokuk Formation are argillaceous and sparsely fossiliferous. Sponge spicules occur in Keokuk cherts at many localities in southeastern Iowa; this suggests the possibility that siliceous sponge spicules provided the biogenic silica for the development of the abundant cherts of the formation. Quartz crystals are common in some siliceous intervals of the Keokuk Formation, and the upper Keokuk bears geodes.

The Keokuk Formation displays an increase in dolomite and shale content westward in the subsurface of Iowa. These compositional changes make it difficult for geologists to distinguish the Keokuk and overlying Warsaw formations in the subsurface: hence, the use of the term "Keokuk-Warsaw interval" in many subsurface logs. The Keokuk Formation in southeastern Iowa varies in thickness from approximately 45 to 90 feet. Keokuk strata are overlain by the Warsaw Formation at most localities in southeastern Iowa. However, the "St. Louis" Formation rests unconformably on the Keokuk Formation at several sites in Keokuk and Washington counties (fig. 7.11).

The Keokuk and Warsaw formations were deposited during the Keokuk T-R

Cycle (fig. 7.5). This is the sixth of ten T-R cycles that formed Iowa's Mississippian rock record. The bone bed at the top of the Burlington Formation has been interpreted by Witzke et al. (1990) to signify a slowdown in sedimentation, coinciding with the boundary between the Haight Creek and Keokuk T-R cycles. The bulk of the Keokuk Formation apparently formed in a middle-shelf setting. Fossiliferous wackestones and lime mudstones of the formation probably formed in a quiet-water subtidal environment, below fair-weather wave base. Intervals with fossiliferous packstones record times when the seafloor was stirred by storms. The upward increase in clay content (siliciclastics) in the Keokuk stratigraphic sequence implies an influx of sediment from terrigenous (land) sources.

WARSAW FORMATION

The name Warsaw Formation was first used by James Hall in 1857 for exposures at Warsaw, Illinois. Currently, the Geological Survey Bureau recognizes two intervals within the formation: a lower geode-bearing argillaceous dolomite and shale unit and an upper shale-dominated unit. Undolomitized carbonate beds are scarce to absent in the Warsaw Formation of southeastern Iowa, whereas the underlying Keokuk Formation contains some unaltered limestones. In Iowa's subsurface, distinctions between the two formations blur.

The Osagean-Meramecian series boundary has been placed at the top of the Warsaw Formation in Iowa. However, the historic placement of the Osagean-Meramecian boundary in the Mississippi Valley region has varied considerably. The state geological surveys of Illinois and Indiana have found the recognition of separate Osagean and Meramecian series to be impractical and have combined the two in a single Valmeyeran Series.

Warsaw strata are fossiliferous in southeastern Iowa. Common fossils include brachiopods, bryozoans, crinoid parts, and burrows. *Archimedes*, a lacy bryozoan with a distinctive corkscrew-shaped axis, is abundant in the type Warsaw, a few miles downstream from Keokuk, Iowa. The famous geodes of the Keokuk area of southeastern Iowa, western Illinois, and northeastern Missouri occur in the Warsaw Formation. The upper part of the Keokuk Formation is also geode bearing; however, greater varieties of well-developed geodes have been recovered from Warsaw strata.

The Warsaw Formation ranges from 0 to 85 feet in thickness in southeastern Iowa. A prominent unconformity occurs at the top of the Warsaw, and the formation is absent in some areas in southeastern Iowa. For example, the "St. Louis" Formation lies directly on the Keokuk Formation at several sites in Keokuk and Washington counties. Apparently, the Mississippian sea withdrew completely from the midcontinent area at the end of the Keokuk Cycle of deposition (fig. 7.5). A widespread erosional unconformity developed on the Keokuk and Warsaw formations at this time. The underlying cause for the withdrawal of the sea is unclear; it may have been produced by a worldwide drop of sea level or by an uplift of the early North American continent.

The Warsaw Formation was deposited during the upper part of the Keokuk T-R Cycle. It represents a shallowing upward sequence laid down during a time of regression of the Mississippian seaway. Siliciclastic content in the Warsaw records an influx of sediment from terrestrial sources.

SALEM FORMATION

F. M. Van Tuyl (1925) recognized a thin interval in southeastern Iowa that he called the Spergen Formation. These beds pinch out to the northwest in Iowa. The name Spergen has since been replaced by the name Salem in other midwestern states. The Salem Formation of Illinois and Indiana is composed typically of shallow-water grainstones and packstones, oolitic in part. Salem equivalents are poorly known in Iowa, and Salem (Spergen) terminology is applied only in a small area of southeastern Iowa. Salem strata in Iowa may record T-R Cycle 7 (fig. 7.5).

ST. LOUIS FORMATION

The St. Louis Formation is named for exposures near St. Louis, Missouri. As mentioned earlier, the Geological Survey Bureau uses quotation marks with the term "St. Louis" to indicate that the formation terminology used in Iowa only partly correlates with the type St. Louis of Missouri. Furthermore, nomenclatural problems are associated with the use of the term in Iowa, and a comprehensive study of the "St. Louis" interval is needed in both the outcrop and the subsurface of Iowa. The Geological Survey Bureau currently recognizes four members in the "St. Louis" Formation, in ascending order: the Croton, Yenruogis, Verdi, and Waugh. In Iowa, "St. Louis" strata rest unconformably (fig. 7.11) on either the Keokuk, Warsaw, or Salem formations.

The Croton Member is named for the village of Croton in western Lee County. Located along the Des Moines River opposite the Missouri settlement of Athens, Croton is significant in Iowa history. Cannonballs landed in Croton during the Battle of Athens in 1861. This Civil War engagement pitted a group of pro-Confederate Missouri rebels against a combined force from the Second Iowa Regiment stationed at Keokuk and a group of volunteer militia known as the Keokuk Rifles.

The Croton Member corresponds to the "Lower" St. Louis as used by Van Tuyl (1925). The Croton consists primarily of dolomitic limestone and limestone. At many localities, the Croton interval is highly brecciated. The breccias probably formed by solution-collapse processes associated with the dissolution of gypsum and anhydrite. Evaporites (gypsum and anhydrite) occur within the unit in the subsurface of southeastern and south-central Iowa. Mississippian evaporites are currently mined in a subsurface operation in Marion County, and, for a brief time in the 1920s, evaporites were recovered at Centerville in Appanoose County. Although fossils are not abundant in the member, a variety of marine invertebrates occur, including brachiopods, bryozoans, bivalves, trilobites, and various corals. The colonial rugose coral (*Acrocyathus*) is a distinctive fossil in lower "St. Louis" strata. Calcareous microfossils include foraminifera and puzzling varieties that may be of algal origin.

Siliciclastics are common in the lower and middle Croton interval in Keokuk and Washington counties. The upper Croton is composed primarily of dolomite and limestone. Croton strata represent a marine transgression, following an interval of considerable erosion. The unconformity at the base of the member is one of the most pronounced erosional surfaces within the Mississippian System of Iowa.

The Yenruogis Member was proposed by Witzke et al. (1990) for a sandstone unit with minor carbonate beds. It overlies the Croton Member. The name Yenruogis originates from Lake Yenruogis in Yenruogis County Park in Keokuk County. Burrows and sedimentary structures indicate that most, if not all, of the member is of marine origin. The Yenruogis interval represents a progradation of siliciclastic sediments; deposition took place primarily in a nearshore marine setting. Carbonate beds of the Verdi Member cap the Yenruogis.

The Verdi Member derives its name from the former Verdi Railway Station in western Washington County. It consists of a lower marine limestone segment and an upper interbedded limestone and sandstone interval. The lower Verdi bears an abundant and diverse assortment of marine fossils. Microfossils include conodonts, foraminifera, ostracodes, and miscellaneous calcareous constituents. The macrofossil assemblage features a variety of brachiopods, various mollusks, trilobites, crinoid debris, bryozoans, and both rugose and tabulate corals. The fauna of the upper Verdi is less diverse.

The lower Verdi is the product of a marine transgression (fig. 7.5). An ooid grainstone in the upper part of the unit records agitated conditions on the seafloor. The upper Verdi probably formed in very shallow restricted marine settings and nearshore and shoreline environments. The low diversity of faunas of the upper Verdi may reflect stressed environments with high salinities and increased temperatures. The upper Verdi records the final offlap of the "St. Louis" seaway from ancient Iowa. The Waugh Member, a product of various nonmarine environments, overlies the Verdi Member.

The Waugh Member was established in 1987 by McKay et al. (1987). This uppermost member of the "St. Louis" Formation is named for the Waugh Branch of the South Skunk River in southwestern Keokuk County. Five major lithofacies comprise the member: (1) a lower cross-bedded sandstone with thin lime mudstones; (2) sandy shales; (3) massive, brecciated, ostracode- and fish-bearing lime mudstones; (4) thin to thick bedded ostracode- and fish-bearing lime mudstones with interbedded shales; and (5) sandstone with rooted scale trees and paleosols. Two minor facies also occur. Although these facies have a limited areal distribution, they are very significant. Both facies bear fossil amphibians, some of the earliest tetrapods (four-legged vertebrates) to walk on the surface of our planet.

Fossils of the Waugh Member include roots of scale trees, other root molds, carbonaceous plant fragments, ostracods, myriapod arthropods, and small snails. In addition, a wide assortment of vertebrate remains occur: teeth, scales, bones, and questionable spines of sharks; teeth, jaw fragments, scales, spines, plates, and partial body fossils of acanthodian fish; scales, jaw fragments, bones, and partial articulated actinopterygian fish; scales, bones, jaws, partial skulls, and a nearly complete skeleton of lobe-finned (crossopterygian) fish; a toothplate and nearly complete skeleton of lungfish (dipnoans); isolated bones, limb bones, pectoral and pelvic girdles, cranial bones, jaws, teeth, semiarticulated skulls, and partial and nearly complete skeletons of labyrinthodont amphibians; and fragments of other amphibians.

The Waugh Member records a variety of nonmarine environments and preserves an exceptional fauna of amphibians and fish. The discovery of abundant Mississippian amphibians at the Heimstra Quarry near Delta, Iowa, ranks as one of the greatest paleontological discoveries in the state. Globally, the fossil record of early amphibians is quite limited, and the southeastern Iowa discovery represents the oldest well-preserved tetrapod fauna known from North America. The oldest tetrapods in the rock column, known from Upper Devonian strata of East Greenland, document the evolutionary transition of lobe-finned fish to amphibians.

Deposition of the Waugh Member reflects lake, stream, and swamp environments in a coastal lowland setting. Figure 7.12 depicts key facies and environments of deposition for the member.

PELLA FORMATION

The Pella Formation caps the Mississippian sequence of southeastern Iowa (fig. 7.11). The formation consists primarily of fossiliferous calcareous shales and argillaceous limestones (marls). Pella strata reach thicknesses of 30 feet or more in the outcrop belt and approximately 50 feet in the subsurface. The Pella interval marks a marine transgression and unconformably overlies nonmarine upper "St. Louis" beds.

The abundant and diverse fauna of the Pella Formation includes strophomenid, spiriferid, terebratulid, and rhynchonellid brachiopods; at least seven

7.12 Key lithofacies in the Waugh Member of the "St. Louis" Formation in southeastern Iowa and an interpretation of their depositional environments. Facies are shown in stratigraphic position, with lower facies at the bottom and upper facies at the top. After Witzke et al. 1990.

Facies	Depositional Environments
Areally restricted facies of tetrapod-bearing limestone conglomerate, shale, and boulder conglomerate	Preserved in collapse structures; probably produced by dissolution of underlying gypsum beds; preserves tetrapods (amphibians) from lake habitats with fresh to brackish waters
Sandstones with roots of scale trees, also paleosols	Swamp conditions with locally dense growth of scale trees; fluvial influxes of sand eventually buried the swamp
Laminated lime mudstones with ostracodes and fish, also shales	Renewed and continued lacustrine (lake) sedimentation with influxes of fluvial clastics
Massive, fractured, and conglomeratic ostracode-bearing lime mudstone facies	Freshwater to possibly brackish-water lake and lagoonal environments
Planar to cross-stratified sandstone and sandy shale facies	Fluvial to marginal marine, deltaic settings

species of bivalves; several species of gastropods; six varieties of foraminifera; along with bryozoans, corals, trilobites, ostracodes, conodonts, spirorbid worms, and fish teeth. Echinoderms are represented by several species of blastoids, crinoid cups, crinoid debris, and echinoid parts.

The Pella Formation was deposited in T-R Cycle 10 (fig. 7.5), the youngest T-R event in Iowa's Mississippian sequence. The lower Pella consists primarily of oolitic and skeletal limestones and represents a deepening sequence brought about by marine transgression. Clastic sediments of the upper Pella signify progradation of terrigenous muds associated with regression of the sea. A major unconformity separates the Pella Formation from overlying Pennsylvanian rocks in southeastern Iowa.

Mississippian Formations of North-Central Iowa

In 1989, Frederick Woodson and Bill Bunker reported on the stratigraphic framework of Mississippian strata in north-central Iowa. A composite section of the Mississippian sequence at Gilmore City in Pocahontas County in north-central Iowa is illustrated in figure 7.13. In ascending order, the stratigraphic record includes the Prospect Hill, Chapin, Maynes Creek, and Gilmore City formations. As Woodson and Bunker (1989) note, an understanding of Iowa's Mississippian stratigraphic succession has been complicated by inconsistent use of stratigraphic terms, inadequate or poorly representative type sections, the

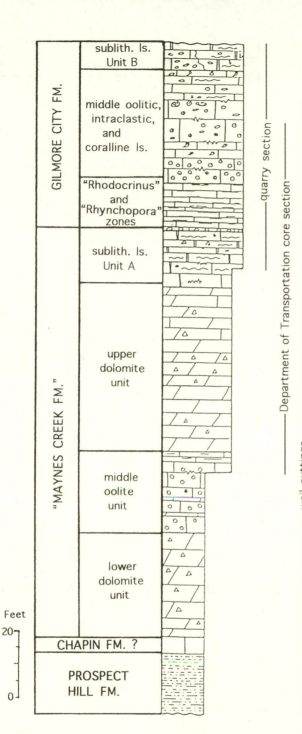

7.13 A composite stratigraphic section of the Mississippian sequence at Gilmore City in north-central Iowa. From Woodson and Bunker 1989.

7.14 The inferred transgression-regression (T-R) cycles of deposition for the Mississippian sequence of north-central Iowa. From Woodson and Bunker 1989.

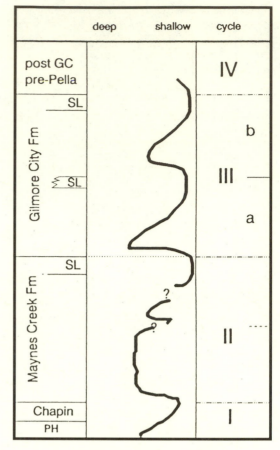

SL - sublithographic limestone
GC - Gilmore City
PH - Prospect Hill

practice of many workers of treating dolomites as laterally persistent units rather than as gradational facies, and the failure of geologists to recognize depositional cycles.

Figure 7.14 shows inferred T-R cycles of deposition for the Mississippian sequence of north-central Iowa. Depositional cycle 1 records the deposition of the Prospect Hill Formation and the Chapin beds. The Maynes Creek interval was deposited in cycle 2, and the Gilmore City beds formed during cycle 3. Both cycles 2 and 3 display fluctuations. Sublithographic limestones at the top of the Maynes Creek, within the Gilmore City, and at the top of the Gilmore City (fig. 7.14) represent very shallow water (tidal flat) deposition. Brief descriptions of the Mississippian formations of north-central Iowa follow. Figure 7.15 summarizes information on the compositions and depositional environments of the formations.

7.15 The Mississippian formations of north-central Iowa and their depositional environments. After Woodson and Bunker 1989 and the Geological Survey Bureau Stratigraphic Column of Iowa (Witzke 1995c).

Series	Formation	Composition	Depositional Environments
Osagean (Middle Mississippian)	"Upper" Gilmore City	Fossiliferous limestone, oolitic limestone, sublithographic limestone, and other limestones	Deposited on a marine shelf during T-R cycle IIIb; oolitic units reflect deposition in wave-agitated settings, above fair-weather wave base; the sublithographic limestone at the top of the formation indicates shallow-water, tidal-flat deposition
Kinderhookian (Lower Mississippian)	"Lower" Gilmore City	Various limestone facies and dolomite	Deposited on a marine shelf during T-R cycle IIIa; consists of a thin transgressive record followed by subtidal and shoal carbonates; a shallow mud-flat facies caps the sequence
Kinderhookian (Lower Mississippian)	Maynes Creek	Dominated by cherty, fossiliferous dolomite; some limestone beds	Deposited on a marine shelf during T-R cycle II; T-R cycle II displays some fluctuations; a shallow-water, mud-flat facies caps the cycle
Kinderhookian (Lower Mississippian)	Chapin	A poorly defined carbonate unit with oolitic, crinoidal, and other varieties of limestone	Deposited on a shallow shelf above wave base during the first Mississippian T-R cycle in north-central Iowa
Kinderhookian (Lower Mississippian)	Prospect Hill	Dolomitic quartz siltstone and silty dolomite	Deposited on a shallow shelf during the initial transgression of T-R cycle I in north-central Iowa; quartz silt indicates influxes of sediment from terrigenous sources

PROSPECT HILL FORMATION

The Prospect Hill Formation was named for exposures near Burlington in southeastern Iowa, but use of the term has been extended into north-central Iowa, where the unit ranges from dolomitic siltstone to silty dolomite. The Prospect Hill is recognized throughout most of north-central Iowa; however, at least two distinct siltstone units occur in Franklin, eastern Hardin, and western Grundy counties, making stratigraphic assignments there more difficult.

CHAPIN FORMATION

The Chapin Formation was named by Van Tuyl (1925) for the village of Chapin in Franklin County. In the type area, the Chapin is represented by 6.5 feet of fossiliferous crinoidal to oolitic limestone and a like amount of dolomite. Unfortunately, neither the lower nor the upper contacts of the Chapin were defined by Van Tuyl. Later, Laudon (1931) transferred the dolomites of the type Chapin section to the overlying Maynes Creek Formation. The term "Chapin" has also been applied to oolitic limestones in the Le Grand area in Marshall County.

MAYNES CREEK FORMATION

Van Tuyl (1925) named the Maynes Creek unit for dolomite exposures in Franklin County, but he did not define lower or upper contacts for the formation. In recent years, use of the term "Maynes Creek" has been expanded by the Geological Survey Bureau to include some limestone strata previously included in the Hampton Formation. For example, the Eagle City Limestone is currently placed in the Maynes Creek Formation.

GILMORE CITY FORMATION

The type section of the Gilmore City Formation was described in the Pennsylvania-Dixie Cement Company quarry in Pocahontas County, near Gilmore City. Laudon (1933) designated the formation, but he defined neither lower nor upper contacts for the Gilmore City. In the type area, the Gilmore City Formation is represented by a variety of limestones, including sublithographic, crinoidal, and oolitic lithologies. The formation has yielded a large number of well-preserved crinoids and some echinoids from quarries near Gilmore City.

The Gilmore City Formation is quarried extensively in north-central Iowa, where it is noted for its high purity. The formation serves as a commercial source of calcium carbonate at locations near Alden and Gilmore City. Proud of its high-quality bedrock, the town of Gilmore City bills itself as Iowa's Limestone Capital (fig. 7.16).

7.16 A sign at Gilmore City, Iowa, proclaiming the town as Iowa's Limestone Capital. Photo by the author.

Geodes

The region around Keokuk, Iowa, is one of the most famous geode localities in the world and is well known to geologists, museum curators, and rock collectors. The geodes are characteristically associated with the lower part of the Warsaw Formation, although a few geodes can also be found in the upper part of the Keokuk Formation. In the Keokuk region, geodes weather free from the Warsaw Formation and occur in abundance in some streambeds. The geodes are highly sought by collectors, for they are objects of great beauty. A variety of minerals have been recognized in the geodes of the Keokuk region (fig. 7.17). Twenty different minerals have been identified in geodes overall, although quartz (silicon dioxide) is the most common composition. Liquid petroleum occurs in geodes near Niota, Illinois, across the Mississippi River from Fort Madison, Iowa.

The geode is a discrete, somewhat spherical body of mineral matter that is commonly hollow and lined inside with clusters and layers of crystals that are aesthetically appealing (fig. 7.18). In general, geodes owe their origin to crystallization of minerals from saturated solutions that filled cavities in the rock. Crystal growth proceeded inward from the outer margin of the cavities toward the center of the cavities, so that the interior crystals are younger than those on the outer margins of the cavities. It is not uncommon for crystals to completely fill a cavity, thereby producing a geode that is solid rather than hollow. Such geodes are not as desirable as the hollow varieties.

7.17 Minerals found in geodes from the Warsaw and Keokuk formations of the Keokuk region of southeastern Iowa, western Illinois, and northeastern Missouri. After Van Tuyl 1916; Tripp 1959; and Sinotte 1969.

Mineral Name	Composition	Mineral Name	Composition
Quartz	SiO_2	Goethite	$HFeO_2$
Chalcedony	SiO_2	Hydrous goethite	$HFeO_2 \cdot nH_2O$
Kaolinite	$2Al_2Si_2O_5(OH)_4$	Chalcopyrite	$CuFeS_2$
Calcite	$CaCO_3$	Malachite	$Cu_2CO_3(OH)_2$
Aragonite	$CaCO_3$	Smithsonite	$ZnCO_3$
Sphalerite	$(Zn,Cd,Fe)S$	Stilpnosiderite	$(Ca,Fe)CO_3$
Iron pyrite	FeS_2	Hematite	Fe_2O_3
Marcasite	FeS_2	Pyrolusite	MnO_2
Barite	$BaSO_4$	Jarosite	$KFe_3(OH)_6(SO4)_2$
Ferroan dolomite	$Ca(Mg,Fe)(CO_3)_2$	Selenite	$CaSO_4 \cdot 2H_2O$

7.18 An 8-inch-diameter geode from the Warsaw Formation makes an attractive set of bookends. Photo by the author.

A geode differs from a concretion, another common body of mineral matter associated with sedimentary rocks, in that a concretion forms by the outward accretion of mineral material, whereas a geode develops by the inward growth of minerals. The term "nodule" is applied to certain concretionary structures of irregular shape that exhibit contrasting compositions from the enclosing rock or sediment. Some hypotheses for the origin of geodes suggest that geodes were derived from calcareous concretions. Other models explain geode formation through a transformation of anhydrite (calcium sulphate) nodules.

Geode formation apparently involved several steps. John B. Hayes (1964) proposed the following sequence. (1) Calcite concretions formed in the fine-grained carbonate sediments of the lower Warsaw and upper Keokuk formations. The concretions probably formed beneath the water-sediment interface on the seafloor while the sediment was still soft. The growth of the concretions may have been localized around bodies of decaying organic matter, such as the soft parts of brachiopods or other invertebrates. (2) Recrystallization of the calcite took place, with the center of the concretions becoming the most coarse grained. (3) A shell of chalcedony (a finely crystalline variety of silicon dioxide) replaced the calcite along the outer margin of the concretions. (4) Chalcedony and quartz replaced calcite inside the concretions. (5) The calcite cores of the concretions dissolved, producing cavities. (6) Later, a variety of crystals precipitated in the cavities from mineral-bearing waters.

Other geologists have recognized relics of evaporite minerals in geodes and have proposed that the precursors of geodes were anhydrite nodules. According to such models, the anhydrite nodules may have been of either primary or of secondary origin. Later, the nodules were silicified and transformed, producing geodes. Siliceous sponge spicules may have provided the source of silica for silicification. See Chowns and Elkins (1974) and Maliva (1987) for additional information.

No matter how they formed, geodes are lovely works of nature (fig. 7.18). This was recognized in 1967 when the Iowa General Assembly made the geode the official state rock of Iowa. The joint Senate-House resolution praised the beauty of the Iowa geode, and the curator of the Department of History and Archives was directed to obtain samples of the Iowa geode and to display them in an appropriate place in the State Historical Library. In addition, the editor of the *Iowa Official Register* was instructed to include a picture of a geode, with appropriate commentary, in the *Iowa Official Register*, along with pictures of the state flower, state bird, and state tree.

Mississippian Life

Iowa has an official state rock but no official state fossil, lagging behind such paleontologically progressive states as Ohio, New York, and Wisconsin. Ohio and Wisconsin proclaim the trilobite as their state fossil, and New York gives recognition to the eurypterid (sea scorpion). If Iowa were to select a state fossil, a Mississippian crinoid would be a logical choice. Iowa legislators considered a resolution in 1996 recognizing the importance of the fossil crinoid, but their action fell short of designating the crinoid as Iowa's official state fossil.

The warm, shallow, carbonate seas of Mississippian time supported an abun-

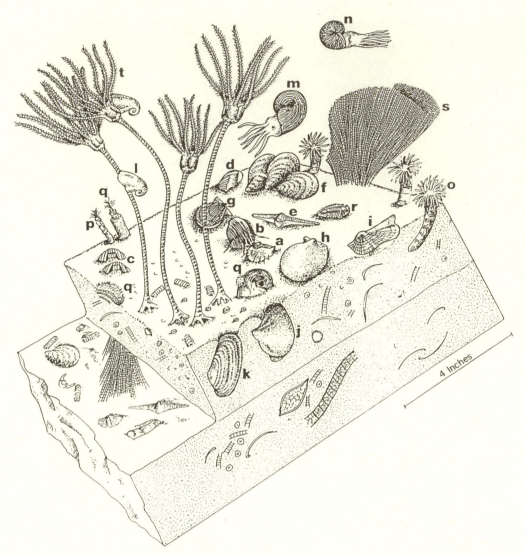

7.19 A reconstruction of life on a Mississippian seafloor. Common Mississippian life forms include: (a–e) brachiopods, (f–k) bivalves, (l) gastropods, (m and n) ammonoid cephalopods, (o) solitary rugose corals, (p) annelid worms, (q) bryozoans, (r) trilobites, (s) lacy bryozoans, and (t) crinoids (sea lilies). From McKerrow 1978.

dant fauna of crinoids and other marine invertebrates (figs. 7.1 and 7.19). A varied assortment of echinoderms populated the seafloors of ancient Iowa, including stalked crinoids and blastoids and mobile starfish and echinoids. Wide-winged spiriferid brachiopods and lacy bryozoans are characteristic fossils in Iowa's Mississippian rock record. Mollusks, corals, arthropods, fish remains, foraminifera, and conodonts are also well known. In addition, a significant terrestrial fauna of amphibians and fish was discovered in 340-million-year-old Mississippian strata near Delta in southeastern Iowa in 1985.

Some representative Mississippian fossils are shown in figure 7.20. Discussions of selected groups follow.

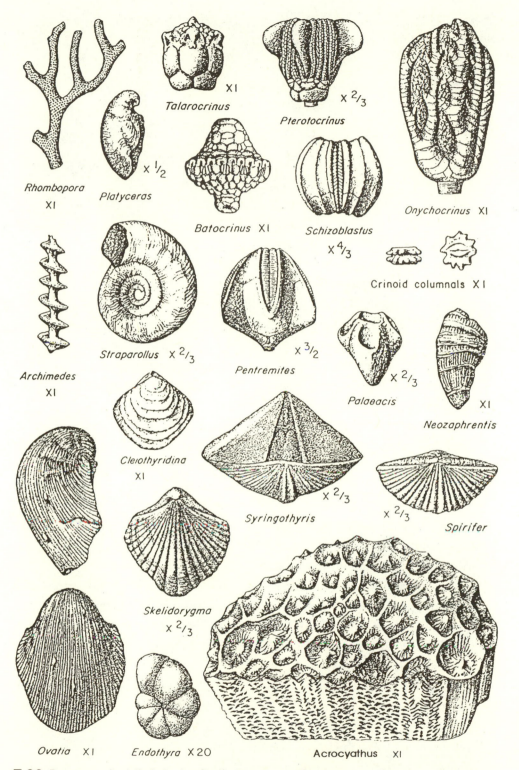

7.20 Representative Mississippian fossils. Bryozoans: *Rhombopora*, *Archimedes*; echinoderms (crinoids): *Talarocrinus*, *Pterotocrinus*, *Batocrinus*, *Onychocrinus*, crinoid columnals; echinoderms (blastoids): *Pentremites*; Brachiopods: *Cleiothyridina*, *Syringothyris*, *Spirifer*, *Skelidorygma*, *Ovatia*; gastropods (snails): *Platyceras*, *Straparollus*; corals: *Neozaphrentis*, *Acrocyathus*; foraminifera: *Endothyra*. From Willman et al. 1975. Courtesy of the Illinois Geological Survey.

7.21 Crinoidal limestone from the Burlington Formation. Photo by the author.

ECHINODERMS

Several different kinds of echinoderms are found in the Mississippian rocks of Iowa. Crinoids occur in the Wassonville, Maynes Creek, Burlington, Keokuk, Gilmore City, and Pella formations in fair abundance, and complete specimens have been recovered from these formations. Crinoids (sea lilies) are delicate, stalked invertebrates that resemble plants more than they do animals (figs 7.1 and 7.19). Pieces of crinoid stems (columnals) are particularly common in the Mississippian limestones of the state (fig. 7.21).

Many complete specimens of crinoids have been recovered from the Maynes Creek and Gilmore City formations. These concentrations of nearly intact crinoid specimens represent storm deposits where crinoids were dislodged or uprooted from the seafloor and quickly buried by sediment. The exquisite Mississippian crinoids from Le Grand, Iowa, merit their worldwide reputation. The remarkable preservation typical of Le Grand crinoids is shown in figures 7.22 and 7.23. A spectacular slab of Le Grand sea lilies is on display at the State Historical Society of Iowa's museum in Des Moines.

Blastoids also occur in the Mississippian rocks of Iowa. Blastoids were stalked echinoderms somewhat similar to crinoids, except that their arms were less well developed than those of crinoids. The arms of blastoids were tiny and are rarely preserved. What is commonly preserved is the tuliplike crown of the blastoid (see *Pentremites*, fig. 7.20).

Echinoids, another division of the phylum Echinodermata, are known from the Mississippian record of Iowa. Echinoids differ from crinoids and blastoids in that they are not stalked. Echinoids are well represented in modern marine environments by such creatures as sand dollars, heart urchins, and spiny sea urchins.

Starfish are among the rarest of the invertebrates in the fossil record. They do not develop discrete plates of calcium carbonate as do their echinoderm cousins,

7.22 A slab from the Le Grand, Iowa, quarry displaying several well-preserved crinoids. A typical crinoid crown (calyx and arms but excluding the stem) measures 1 to 2 inches in length. Courtesy of the Iowa State Historical Society, Des Moines.

the crinoids, blastoids, and echinoids. Therefore, starfish do not have limy hard parts that lend themselves to preservation in the fossil record. It takes unusual conditions of fossilization for starfish to be preserved. These conditions were achieved on the Mississippian seafloor on which the Maynes Creek Formation was laid down. A large slab of Mississippian bedrock, recovered at Le Grand, Iowa, and displayed at the State Historical Society of Iowa's museum in Des Moines, displays dozens of exquisitely preserved starfish (fig. 7.24).

BRACHIOPODS

A variety of brachiopods occur in the Mississippian rocks of Iowa. The spiriferid types are particularly common and are generally characterized by wide hinge lines and winglike margins. Some representative Mississippian brachiopods are illustrated in figures 7.19 and 7.20.

MOLLUSKS

Bivalves and gastropods are present in the Mississippian rocks of Iowa. Although they are not abundant, ammonoid cephalopods have been recovered and serve as important index fossils to establish correlations of Iowa's rock record with Mississippian formations in other parts of the country and the world.

7.23 It is obvious from this photograph why crinoids are known as sea lilies. These beautiful 330-million-year-old specimens were found by Samuel Calvin in 1888 near Le Grand, Iowa. Courtesy of the University of Iowa Museum of Natural History, Iowa City.

CORALS

Both solitary corals and colonial corals are found in the Mississippian rocks of Iowa. *Acrocyathus*, a rather distinctive colonial coral from the "St. Louis" Formation, is shown in figure 7.20. Although corals formed conspicuous biostromes in the Devonian of Iowa, no such features are present in the Mississippian record of the state.

7.24 A magnificent slab of starfish from Mississippian strata near Le Grand, Iowa. Individual starfish measure a few inches in width. Courtesy of the Iowa State Historical Society, Des Moines.

ARTHROPODS

Trilobites and ostracodes occur scattered throughout the Mississippian formations of Iowa. Ostracodes and trilobites are fairly common in the Pella Formation in southeastern Iowa.

FISH TEETH, BONES, AND PARTS

Both bony fish and sharks occupied the Mississippian seas of Iowa. Fish teeth, bones, and other parts are fairly common in Mississippian rocks. In southeastern Iowa, a prominent fish tooth zone occurs at the top of the Burlington Formation. Fish remains are also present in other Mississippian formations, such as the Prospect Hill, Keokuk, and "St. Louis" formations.

In the 1870s, a young Iowan named Orestes St. John started collecting the remains of fossil fish in Iowa. Large collections of fish teeth and other remains were obtained near Burlington and Keokuk and elsewhere in southeastern Iowa. St. John went on to become a leading authority on Paleozoic fish, and he developed great skill at reconstructing these ancient creatures. He contributed to several prestigious monographs on fossil fish, and many forms that he collected in Iowa were described and illustrated in his numerous published works. St. John was also a talented artist whose field sketches illustrated Charles White's two-volume *Report on the Geological Survey of the State of Iowa* in 1870 (figs. 7.8 and 7.9).

FORAMINIFERA

Foraminifera are one-celled organisms, similar to *Amoeba* in terms of complexity. Many foraminifera (forams) secreted a shell of calcium carbonate and are preserved in the fossil record. Forams are generally microscopic in size; they can be obtained from microfossil residues from marine rocks of nearly all ages. *Endothyra*, a common Mississippian foram, is shown in figure 7.20.

CONODONTS

Conodonts can be recovered from nearly all Paleozoic marine rocks, so it is not surprising that they are found in the Mississippian rocks of Iowa. Conodonts are useful index fossils, and they have been used to help establish correlations of the Mississippian formations of Iowa with Mississippian rocks in other parts of North America and Europe.

Ancient Amphibians

Ranking high on the list of fossil finds in North America is the Mississippian amphibian site near Delta, Iowa. A large concentration of fossil amphibian bones was discovered in the spring of 1985 by Robert M. McKay and M. Patrick McAdams. Amphibians are the earliest known four-footed land animals (tetrapods). The discovery of fossil amphibians in Iowa (fig. 7.25) is important for two reasons. First, fossil amphibians are rare worldwide, so the discovery of additional fossils is significant. Second, very few fossil amphibians have been reported from Mississippian strata anywhere. Most fossil amphibians have been recovered from Pennsylvanian or Permian rocks, strata that are younger than the Mississippian. Only a dozen or so localities worldwide reveal fossils of Mississippian amphibians, and most of these localities are in the British Isles. The fossil amphibians from southeastern Iowa are younger than the world's oldest amphibians from the Upper Devonian of East Greenland and Australia, but they are probably older than other Mississippian discoveries in North America. The Iowa locality furnishes significant fossil evidence from a portion of the Mississippian record where amphibians were previously unknown. Overall, the Hiemstra quarry near Delta in Keokuk County serves as an important window to the past and helps document the evolutionary pathway of early land-dwelling vertebrates.

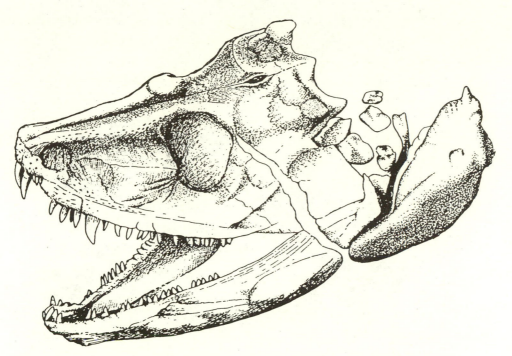

7.25 The skull of an early land-dwelling vertebrate found in Keokuk County, Iowa, in 1985. This 340-million-year-old amphibian skull shows remarkable preservation and is one of Iowa's most significant fossil discoveries. The specimen measures approximately 1 foot, left to right. Drawn by Kay Irelan. From Witzke et al. 1990.

Crinoid Collectors

James Hall, Iowa's first state geologist, was introduced to Midwest geology as a member of David Dale Owen's survey of 1841. Governor James W. Grimes recommended a geological exploration of the state in his first message to the Iowa State Legislature, and he appointed Hall as State Geologist of Iowa in 1855. Hall (fig. 7.26) spent only a few months in Iowa each year, however. He conducted his remote geological survey of Iowa through cleverness and delegation of responsibilities. Josiah D. Whitney, Hall's chief assistant, was designated Chemist and Mineralogist for the Iowa Survey, and he received equal billing with Hall on the titles of state reports and publications. Amos H. Worthen of Warsaw, Illinois, became Hall's official assistant. Worthen's appointment made his magnificent collection of crinoids and other fossils available for study by Hall and for inclusion in the Iowa reports. Hall and Whitney's *Report on the Geological Survey of the State of Iowa* in 1858 has plates illustrating a number of exquisite crinoids and other fossils from the Burlington and Keokuk formations of southeastern Iowa. Included in the report are new species of crinoids, blastoids, brachiopods, and bryozoans named in honor of Amos Worthen. A large spiriferid brachiopod from the Burlington Formation was named for Hall's friend and influential supporter, Governor James Grimes.

7.26 James Hall published exquisite illustrations of Iowa's Mississippian crinoids in 1858 in *Report on the Geological Survey of the State of Iowa*. Hall wore many hats during his long and distinguished career in geology and paleontology. In addition to serving as Iowa's first state geologist, Hall held the title of Foundation Professor of Geology, Zoology, and Natural History at the University of Iowa. There is no evidence that he ever taught a course there, however. From Fischer 1978.

Hall took immense pride in the two-volume publication on Iowa's geology; the volume on Iowa fossils (*Palaeontology of Iowa*) was particularly pleasing to him. Hall and Whitney's report was distributed widely, with copies sent to governments and scientific societies throughout Europe. Amos Worthen went on to a distinguished career as State Geologist of Illinois. Josiah Whitney later served as State Geologist of California, and the highest peak in that state is named in his honor.

Another important crinoid collector was Charles Wachsmuth. Aldo Leopold, the noted conservationist who was born and raised in Burlington, Iowa, wrote in *A Sand County Almanac with Essays on Conservation from Round River* about his fellow townsman: "When I was a boy, there was an old German merchant who lived in a little cottage in our town. On Sundays he used to go out and knock chips off the limestone ledges along the Mississippi, and he had a great tonnage of these chips, all labeled and catalogued. The chips contained little fossil stems of some defunct water creatures called crinoids. The townspeople regarded this gentle old fellow as just a little bit abnormal, but harmless. One day the newspaper reported the arrival of certain titled strangers. It was whispered that these

7.27 Charles Wachsmuth (1829–1896), resident of Burlington, Iowa, became a leading authority on fossil crinoids. His extensive collection of fossil crinoids is housed in the Smithsonian Institution in Washington, D.C. From Keyes 1896.

visitors were great scientists. Some of them were from foreign lands, and some among the world's leading paleontologists. They came to visit the harmless old man and to hear his pronouncements on crinoids, and they accepted these pronouncements as law. When the old German died, the town awoke to the fact that he was a world authority on his subject, a creator of knowledge, a maker of scientific history. He was a great man — a man beside whom the local captains of industry were mere bushwhackers. His collection went to a national museum, and his name is known in all the nations of the earth."

Born in 1829 in Hanover, Germany, Charles Wachsmuth (fig. 7.27) was the only son of a prominent lawyer. He had intended to study law himself, but his failing health forced him to give up his studies, and his physician advised him to pursue a mercantile career.

In 1852, at the age of twenty-three, Wachsmuth traveled from Germany to New York City to become an agent for a Hamburg shipping house. Wachsmuth did not adjust to the climate of New York, and after barely recovering from pneumonia, he moved west to seek a more suitable climate. At the suggestion of friends, he journeyed to Iowa, then a young and promising state. He settled in Burlington, and by 1855 he had married and established his own business. The relocation did not bring improvement in his health, however. Wachsmuth's physician advised him to spend as much time as possible outdoors in the fresh air and suggested that for exercise he could collect fossils. The prescribed treatment agreed with Wachsmuth, and soon his health showed signs of improvement. Wachsmuth became an enthusiastic fossil collector and spent long hours collecting in the ravines and quarries near Burlington while his wife, Bernandina Lorenz, looked after the business. In a few years, he had assembled a fine collection of crinoids from the local limestones near Burlington and was well on his way to becoming a leading authority on fossil crinoids.

Wachsmuth's collection attracted the attention of a number of eastern scientists, including Louis Agassiz of Harvard's Museum of Comparative Zoology. Agassiz encouraged Wachsmuth to continue his collecting and studies and to publish scientific articles on his observations. In 1873, his collection was obtained by Harvard's Museum of Comparative Zoology, and Wachsmuth was hired as an assistant and curator. Wachsmuth's position at the museum, which he held until Agassiz's death, acquainted him with the literature of crinoids on a worldwide basis.

In the late 1870s, Wachsmuth met Frank Springer (fig. 7.28), a native Iowan who was then a young lawyer at Burlington and an enthusiastic student of crinoids. A friendship sprang up between them, and they decided to combine their efforts in the study of crinoids. They worked together, and starting in 1878 their research was published under joint authorship. This cooperative enterprise benefited both men. Wachsmuth was a painstaking worker and, with his wife's able assistance, a tireless collector. Springer was a man of some affluence and was able to supplement their collection by purchasing important collections of European crinoids.

Springer achieved great success as an attorney in New Mexico, where he moved in 1873 and where a town is named in his honor. During the years that Springer resided in New Mexico, he maintained an active interest in crinoids and visited Burlington at frequent intervals to collaborate with Wachsmuth. Wachsmuth and Springer coauthored eighteen principal works on crinoids. Their major work, a monograph on North American crinoids, was in press at the time of Wachsmuth's death in 1896.

Charles Keyes paid special tribute in the *Annals of Iowa* in 1896 to Wachsmuth and Springer for their ambitious monograph on North American fossil crinoids: "Few persons living in the great Mississippi Valley are aware that one of the most important scientific works ever produced in this country has recently been completed in their very midst. Still fewer Iowans there are who will not be greatly surprised when informed that the material which served as the foundations for this

7.28 Frank Springer (1848–1927) collaborated with Charles Wachsmuth, and together they published many important papers on fossil crinoids. From Keyes 1896.

truly great work was obtained largely within the boundaries of their own State. Yet no contribution to the natural history of the State, of the United States, or of the western hemisphere has surpassed it in importance. Few old world undertakings of similar nature rival it. It stands as one of the masterpieces of American science."

No discussion of crinoid collectors or crinoid collecting in Iowa would be complete without mention of B. H. Beane of Le Grand, Iowa (fig. 7.29). Beane, the son of a Quaker minister and farmer, grew up adjacent to a limestone quarry near Le Grand. Fossil stone flowers (crinoids) started to appear in the quarry in reasonable abundance in 1874, some five years before Beane was born. Quarrying in those days was largely a hand operation, and quite naturally the workers noticed the unusual bouquets of stone flowers. The crinoids were described as occurring in a nest because of their concentration in a lens-shaped deposit, perhaps a few tens of feet in diameter.

For the next sixteen years (1874–1890), the nest produced many fine specimens. During these years, a number of local residents accumulated excellent collections of Le Grand crinoids. The Le Grand specimens became well known in

7.29 B. H. Beane (1879–1966) of Le Grand, Iowa, was a well-known collector and preparer of crinoids. Courtesy of the *Des Moines Register*.

paleontological circles, and they were studied and described by the leading crinoid specialists of the day, such as Charles Wachsmuth, Frank Springer, S. A. Miller, and William Gurley.

Burnice Hartley Beane was born in 1879, and the first nest at the Le Grand quarry underwent extensive collecting during his boyhood. He initially became curious and later fascinated by the comings and goings of the learned men who beat a path to the quarry adjacent to his farm home. "These scientists were the idols of my boyhood," Beane is reported as saying in an article by Richard Boyt (1962). "I pestered them with endless questions, and they answered me with inexhaustible patience. I soon became a fossil collector and spent most of my spare time at the quarry."

By 1890, the first nest of crinoids had been completely removed, and crinoid collecting in the Le Grand quarry was meager until 1931, when blasting operations in the quarry uncovered a second nest. This second pocket of sea lilies was located about 100 feet from the earlier discovery and occurred as a lens-shaped deposit about 15 feet in diameter.

Beane spotted the crinoid nest shortly after it was exposed by the blasting operations and began a race with time to save as many specimens as possible from the rock-crushing operations of the quarry. In his examinations of the broken limestone blocks, Beane was amazed to find one block that contained several starfish. A few isolated starfish had been found by previous collectors, but the slab that Beane was examining revealed dozens of well-preserved starfish — rarest of all the fossil animals found at Le Grand. This magnificent slab containing 183 starfish, 12 sea urchins, and 2 trilobites is now on display at the State Historical Society of Iowa's museum in Des Moines (fig. 7.24). In addition to this large slab, numerous smaller slabs with well-preserved crinoids were recovered from the second nest. Several of the slabs contained as many as two hundred remarkably preserved crinoids.

In 1933, another great discovery was made at the Le Grand quarry when blasting revealed a third nest. This nest was about 18 feet in diameter. Several thousand exceptional crinoid specimens were recovered from 1933 to 1937 because of the outstanding cooperation of the quarry operator.

Although the Le Grand quarry remained in operation until 1958, only one additional nest was discovered. Unfortunately, it was consumed by the rock crusher before Beane could save it. No additional discoveries of crinoid nests have been made in recent years, and the general site that contained the earlier finds is now covered by a thick interval of slumped overburden.

Through his many years of collecting, Beane amassed a sizeable collection of Le Grand crinoids. He and his son, Lewis, painstakingly prepared many of the slabs from the Le Grand quarries by removing the stony matrix from around the crinoids with such simple tools as needles and toothbrushes. In 1932, Beane was awarded an honorary degree from William Penn College in Oskaloosa, Iowa, in recognition of his important work.

At first, Beane kept every specimen he collected, unwilling to part with any of his treasures. His house and yard overflowed with exquisite specimens from the local quarry. Professional and amateur paleontologists stopped to visit Beane and to study and admire his fine collection. Eventually, Beane agreed to part with some of his choice specimens, and slabs and specimens were acquired by the University of Nebraska, Simpson College, Augustana College (Rock Island, Illinois), Earlham College, Buffalo Museum of Science, University of Iowa, University of Kansas, University of Oklahoma, University of Minnesota, and University of Alaska. Some specimens were obtained by foreign museums, including museums in London, Paris, Holland, Capetown, and Tokyo.

The State Historical Society of Iowa purchased four of Beane's most spectacular slabs, including the great starfish block. These magnificent materials, along with several specimens that Beane had previously donated, are now part of the collections at the State Historical Society of Iowa's museum in Des Moines. Beloit College in Wisconsin received a sizeable amount of Beane's collection through a donation by a Beloit industrialist and college trustee. The donated materials consisted of two hundred slabs weighing a total of more than 5 tons.

Lowell R. Laudon published several important papers on Iowa's Mississippian paleontology and stratigraphy, including a paper on the crinoid fauna from the Le Grand area coauthored with Beane. Laudon also described crinoid faunas from Burlington and Gilmore City.

After false starts in prelaw at the University of Minnesota and in physical education at Iowa State Teachers College (now the University of Northern Iowa), Laudon found his calling on a geology field trip to the Black Hills conducted by Joe ("Uncle Joe") Runner of the University of Iowa. Laudon went on to complete undergraduate and graduate degrees at the University of Iowa. Upon completion of the Ph.D. degree at the University of Iowa in 1930, Laudon accepted a teaching position at the University of Tulsa. Laudon's stellar teaching career spanned forty-five years, with eleven years at the University of Tulsa, seven years at the University of Kansas, and twenty-seven years at the University of Wisconsin-Madison. He was awarded the Neil Miner Award of the National Association of Geology Teachers in 1986 in recognition of his exceptional contributions to the stimulation of interest in the earth sciences. According to Richard A. Paull, who served as Laudon's citationist for the Neil Miner Award, Laudon probably inspired more people to seek careers in geology than any other educator in North America. Paull states that Laudon possessed the remarkable ability "to make colleagues and students feel it was fun to crawl in the mud during a downpour to collect blastoids, or to wade an icy stream so they could climb an incredibly steep mountain to view a thrust fault."

Harrell LeRoy Strimple (fig. 7.30) was encouraged in paleontology by Lowell Laudon during Laudon's stay at the University of Tulsa. Strimple worked as an accountant for Phillips Petroleum Company and pursued fossil collecting as a hobby after work, on weekends, and during vacations. Strimple attended Laudon's night course in Invertebrate Paleontology and Stratigraphy. That course, along with a high school geology offering at Casper High School in Wyoming, constituted Strimple's formal training in geology and paleontology. Essentially self-trained in paleontology, Strimple launched a full-time paleontological career in 1960. He served as research investigator and curator in the Department of Geology at the University of Iowa from 1962 to 1980.

Strimple authored over three hundred publications in paleontology, primarily on crinoids and related echinoderms. One of his publications includes the description of a new species of crinoid from the Maynes Creek (Hampton) Formation of Iowa in honor of B. H. Beane. Strimple was a major contributor to the *Treatise on Invertebrate Paleontology (Crinoidea)* published in 1978.

Harrell Strimple's collecting prowess was legendary. It was common to hear frustrated and empty-handed colleagues complain on the outcrop: "Harrell

7.30 Harrell Strimple (1912–1983) was one of the most productive self-trained paleontologists of all time. Strimple served as research investigator and curator in the Department of Geology at the University of Iowa from 1962 to 1980. He authored more than three hundred articles on fossil echinoderms and was a major contributor to the *Treatise on Invertebrate Paleontology (Crinoidea)*. Photo by the author.

Strimple must have already been here and cleaned the place out!" Strimple acknowledged his ability as a prolific and skillful collector as follows: "Don't expect to find many echinoderms if you follow me on the outcrop."

Strimple enjoyed great success in working with fossil collectors and amateur paleontologists. His overriding interests were in seeing that an amateur's new material got described and properly curated and that the amateur received recognition for making a significant discovery. The Paleontological Society pays tribute to the memory of Harrell Strimple and his work with its annual Harrell Strimple Award, presented to an amateur paleontologist for outstanding achievement in paleontology. The first award was given in 1984.

Summary

During Mississippian time, shallow seas covered Iowa. Ten T-R cycles are recorded. Carbonate sediments are the principal deposits of these seas. Oolites and sand-size fossil fragments are particularly abundant. Many of the Mississippian carbonates contain an abundance of crinoidal debris. The environment of deposition during part of Mississippian time was similar to that of the present-day Bahama Banks. Sand, silt, and mud washed into the Mississippian seas at various times, and they are now represented by sandstone, siltstone, and shale. Gypsum was deposited in restricted, highly saline areas of the Mississippian seas in southeastern and south-central Iowa. The seas withdrew from Iowa during Late Mississippian time, and the area underwent prolonged weathering and erosion.

Iowa's Mississippian rocks serve as an important source of groundwater in north-central Iowa. Mississippian carbonate rocks are quarried extensively in southeastern and north-central Iowa for road aggregate and other commercial products. The purity of some of Iowa's Mississippian limestones is remarkable.

The Mississippian rocks of Iowa are very fossiliferous and have long attracted the interests of paleontologists. Some exceptional crinoids and starfish have been recovered from these rocks. The discovery of fossil amphibians in Mississippian strata in southeastern Iowa in 1985 ranks as a significant paleontological event.

Acknowledgments

Published works by the following geologists were particularly helpful in preparing this chapter: John R. Bolt, Richard Boyt, Bill J. Bunker, William M. Furnish, Brian F. Glenister, M. Patrick McAdams, Robert M. McKay, Richard A. Paull, Frederick J. Woodson, and Brian J. Witzke. Bill Bunker and Brian Witzke reviewed this chapter and offered constructive suggestions.

The seas being shallow, the conditions were unusually favorable for the formation of extensive coastal swamps which were capable of supporting dense jungles of arborescent plants together with ferns and other vascular cryptograms. The salt waters could easily and frequently invade the swamps or the tropical vegetation could spread out into the quiet bays or sheltered lagoons of the adjacent seas. This may be inferred from the fact that mingled (interbedded) with remains of the plants are the hard parts of numerous marine animals, which evidently swarmed in countless myriads in the more open places.

Charles Keyes (1894) reporting on the coal deposits of Iowa

The great, shallow, carbonate-producing seas retreated from Iowa during Late Mississippian time some 325 million years ago, and the newly formed Mississippian rocks of the craton were subjected to warping, uplift, and erosion. During this time, erosion stripped Mississippian rocks from parts of the craton and removed Lower and Middle Paleozoic strata from large areas of the Transcontinental Arch as well. Thus, strata of the Pennsylvanian System were deposited on rocks of several different ages over the continental interior.

The unconformity between Iowa's Mississippian and Pennsylvanian rocks marks the boundary between the Kaskaskia and Absaroka sequences of L. L. Sloss (1963). The unconformities between the major Sloss sequences are laterally extensive and serve to subdivide the stratigraphic record of the North American continental interior into widespread packages of strata. In Iowa, the Absaroka Sequence is characterized by cyclic deposits of nonmarine and marine Pennsylvanian strata. A variety of nonmarine and shallow marine environments left their mark in the state's Pennsylvanian rock column. The nonmarine settings included stream channels, deltaic coastal plains, and coal swamps (fig. 8.1). Along the eastern and western margins of the ancient North American continent, the Absaroka Sequence includes strata of Permian and Triassic ages. However, Iowa was above the sea and undergoing erosion during Permian and Triassic times, and the state has no record of these geologic systems.

Iowa's Pennsylvanian rocks most commonly lie unconformably on rocks of the Mississippian System, but in eastern Iowa, outliers of Pennsylvanian rocks rest on Devonian and Silurian strata. The basal Pennsylvanian rocks of Iowa were deposited on a rolling plain with a gentle south-to-southwestward slope and a

8.1 An artist's reconstruction of a typical coal swamp like those that covered southern Iowa during Pennsylvanian time. The state's Pennsylvanian record is varied, consisting of alternating layers of marine and nonmarine deposits. Courtesy of the British Geological Survey; copyright NERC.

local relief of up to 200 feet. The plain had been modified by stream erosion and karst activity (sinkhole and solution development) during Late Mississippian and Early Pennsylvanian time.

The Pennsylvanian rocks of Iowa occur in two distinct structural basins. The largest of the two, the Forest City Basin, underlies most of southern Iowa (fig. 8.2) and encompasses approximately 20,000 square miles in the state. The Forest City Basin is centered in northwestern Missouri; Iowa's Pennsylvanian rocks dip southwestward and become progressively younger in that direction. Iowa's Pennsylvanian strata reach a maximum thickness of nearly 1,750 feet in the southwestern part of the state.

Pennsylvanian rocks in Muscatine and Scott counties in eastern Iowa occur in an erosional outlier (fig. 8.2). These strata are physically separated from the Pennsylvanian rocks in the Illinois Basin by the Mississippi River. However, they resemble their Illinois Basin counterparts in composition and fossil content. Therefore, formation terms from Illinois (Caseyville and Spoon formations) are applied to the eastern Iowa exposures.

The Illinois and Forest City basins are separated by a poorly defined structural feature known as the Mississippi River Arch (fig. 8.3). The axis of the arch lies on the Illinois side of the Mississippi River along the southeastern border of Iowa. The Mississippi River Arch is considered to be Middle Pennsylvanian in age, the same age as the maximum subsidence of the Forest City Basin. The Forest City

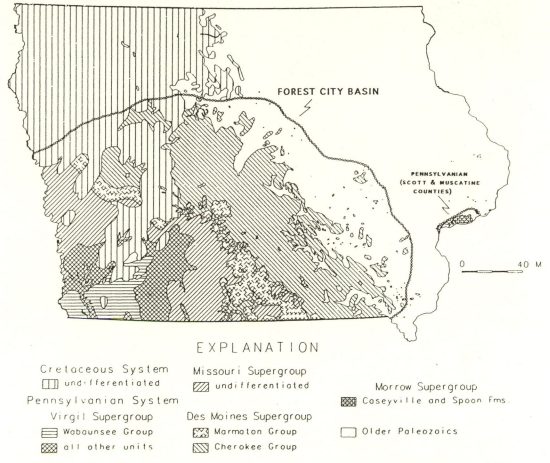

FOREST CITY BASIN

PENNSYLVANIAN
(SCOTT & MUSCATINE
COUNTIES)

0   40 M

EXPLANATION

Cretaceous System   Missouri Supergroup
▢ undifferentiated   ▨ undifferentiated   Morrow Supergroup
                                                      ▦ Caseyville and Spoon Fms.
Pennsylvanian System
   Virgil Supergroup   Des Moines Supergroup
   ▤ Wabaunsee Group   ▨ Marmaton Group   ▢ Older Paleozoics
   ▨ all other units   ▨ Cherokee Group

8.2 The Pennsylvanian and Cretaceous bedrock geology of Iowa and the location of the Forest City
Basin in southern Iowa. Pennsylvanian strata also occur in Muscatine and Scott counties in eastern Iowa.
From Howes 1990.

Basin (fig. 8.3) is bordered on the west by the Nemaha Uplift, a structural feature
that became active during Pennsylvanian time.

Clastic sediments (sand, silt, and clay) were supplied to the Pennsylvanian
seas from the Appalachians on the east and from the Ouachita Mountains and
other uplifts along the southern and western margins of the craton (fig. 8.4). In
addition, the Canadian Shield to the north supplied detritus. The clastic sediments were transported to the seas by streams that drained these bordering
land areas. The streams deposited their sediment loads as complex deltaic
sequences along the shoreline regions of the shallow Pennsylvanian seas that
covered the continental interior. Iowa's Pennsylvanian record includes rocks of
fluvial, deltaic, and marine origins.

During the Pennsylvanian Period, the shoreline shifted back and forth across
ancient Iowa numerous times, causing dozens of shifts from shallow marine
settings to nonmarine environments. Included in the nonmarine settings were

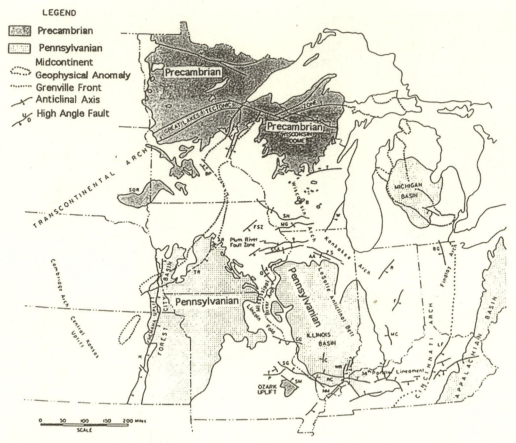

LEGEND

- Precambrian
- Pennsylvanian
- Midcontinent Geophysical Anomaly
- Grenville Front
- Anticlinal Axis
- High Angle Fault

8.3 Major structural features of the north-central midcontinent region of the United States. Note the extensive areas of Pennsylvanian exposures within the Forest City Basin. Exposures of Pennsylvanian strata in Muscatine and Scott counties are structurally isolated from similar Pennsylvanian rocks found in the Illinois Basin. The Mississippi River Arch separates the Forest City and Illinois basins. The Forest City Basin is bordered on the west by the Nemaha Uplift, a structural feature that became active during Pennsylvanian time. Adapted from Bunker et al. 1985.

stream channels and floodplains, deltaic coastal plains, and lush coal swamps. Figure 8.4 shows the approximate position of the shoreline during times of maximum transgression of the seas when the shoreline was located in what is now the Great Lakes region. During times of maximum retreat of the seas, the shoreline extended through what is now the Great Plains region.

As a result of these frequent shore line shifts, Iowa's Pennsylvanian rock record contains an alternating sequence of nonmarine and marine deposits. Such repetitive deposits have been termed cyclothems because of their cyclic nature of repeating in the rock column. Dozens of repetitive cycles of this type characterize the Pennsylvanian rock record of the Midwest. Cyclothems are particularly well developed in the Pennsylvanian rocks of the Illinois Basin to

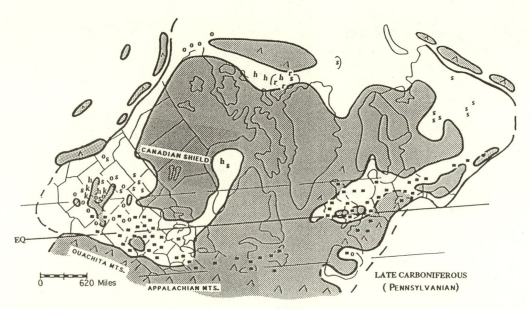

8.4 A paleogeographic map for Middle Pennsylvanian time, approximately 300 million years ago. Ancient Iowa was located in an equatorial region where lush vegetation in coastal swamps grew the year round. Numerous transgressions and regressions of shelf seas produced a varied record of marine and nonmarine rocks in Iowa's Pennsylvanian column. The Appalachian Mountains on the east and the Ouachita Mountains and other uplifts along the southern and western edges of the continent served as source lands for clastic sediments. In addition, the emergent Canadian Shield supplied detritus to the inland seas. Symbols: o = carbonate ooids (oolites) and coated grains, s = sulfate evaporites (primarily gypsum and anhydrite), h = halite (rock salt), k = potash salts, g = glacial deposits, rectangles = coal deposits, EQ = proposed equator, ∧ = mountainous terrains. Continental margins and land areas are shaded. From Witzke 1990b.

the east of Iowa. In the southwestern part of the craton, marine environments were more prominent, whereas swamps and alluvial plains were the predominant settings in the eastern part of the continent in Pennsylvania, Ohio, and West Virginia. Thus, the coal deposits of Pennsylvanian age are thicker and more widespread in states to the east of Iowa.

During Pennsylvanian time, much of the craton was low and relatively flat, perhaps like the Florida Everglades or Louisiana bayous of today. A small rise in sea level or a slight subsidence of the land produced large-scale influxes of marine waters onto the craton. Likewise, a slight drop in sea level resulted in widespread regression. Shoreline shifts of 400 to 500 miles were common during Pennsylvanian time.

Surely many transgressions and regressions of the seas are suggested by Iowa's Pennsylvanian rock record. As we have seen earlier, transgressions and regressions were common during deposition of the Cambrian, Ordovician, Silurian, Devonian, and Mississippian rocks of the state. The transgressions

8.5 The configuration of continental plates during Pennsylvanian (Late Carboniferous) time, with plots of paleoclimatic indicators. Note the presence of glacial deposits in the Southern Hemisphere supercontinent (Gondwana). Symbols: o = carbonate ooids (oolites) and coated grains, rectangles = coal deposits, g = glacial deposits, s = sulfate evaporites (primarily gypsum and anhydrite), h = halite (rock salt), k = potash salts. From Witzke 1990b.

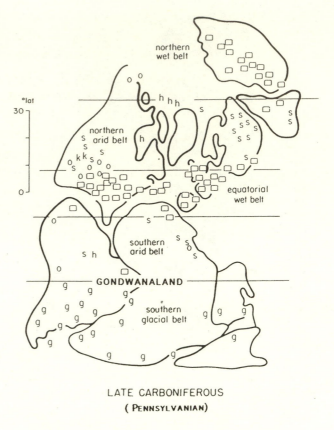

LATE CARBONIFEROUS
(PENNSYLVANIAN)

and regressions of Pennsylvanian time were of much shorter durations, however. Philip Heckel (1986) has estimated the duration of a major Pennsylvanian transgression-regression event for the midcontinent area to be from 235,000 to 400,000 years.

A probable cause for the rather rapid oscillations of the Pennsylvanian seas of North America is found in the rock record of the Southern Hemisphere continents. There, Late Paleozoic glacial deposits are common. The advance and retreat of continental glaciers on the giant Southern Hemisphere supercontinent of Gondwana (fig. 8.5) produced numerous worldwide fluctuations of sea level that had a pronounced influence on the Pennsylvanian rock record globally. Regional and local tectonic movements and shifting of deltas probably exerted some control on sea level changes as well.

The Pennsylvanian Rock Record of Iowa

The older Pennsylvanian strata in Iowa have certain characteristics that suggest cyclic deposition, but these features are not well developed or persistent. Overall, the Pennsylvanian System in Iowa records a shift from nonmarine-dominated deposition in older strata (Lower and Middle Pennsylvanian) to marine-dominated or marine-influenced conditions during deposition of Upper Pennsylvanian rocks. As a result of this transition, a large part of the Pennsylvanian section in Iowa lacks minable coal resources.

The Forest City Basin filled with sediment by the end of Middle Pennsylvanian time, and the region from northern Oklahoma through Kansas, Missouri, Nebraska, and Iowa served as the Northern Midcontinent Shelf during the remainder of Pennsylvanian time (fig. 8.6). The Pennsylvanian record of the Northern Midcontinent Shelf is dominated by cyclic alternations of limestones and shales. Each of these cyclothems is the result of a major rise and fall of sea level.

Figure 8.7 shows a vertical stratigraphic section of a typical midcontinent cyclothem. Note that a standard cyclothem consists of middle (transgressive) limestone, core (offshore) shale, upper (regressive) limestone, and outside (nearshore) shale. The depositional environments and characteristic features of these four key constituents are summarized below.

The middle-limestone interval formed in deepening water brought about by a marine transgression. Dominated by limestones with mud-supported textures (calcilutites in fig. 8.7), these rocks were deposited primarily below effective wave base. Skeletal fossils (shells and invertebrate skeletal materials) are common in this interval. Well-washed carbonate grainstones (calcarenites in fig. 8.7) occur locally at the base of this unit, and they mark shallower water deposition, above wave base.

The core-shale segment of the cyclothem formed during a time of maximum transgression and typically includes gray to black phosphatic shales. These rocks reflect slow deposition in deep water; depths may have exceeded 500 feet. The bottom waters in the deep-shelf setting contained little or no dissolved oxygen, and benthic life was limited to a few varieties of brachiopods and crinoids that were adapted to low-oxygen conditions. In addition, the remains of floating and swimming creatures such as conodonts, fish, and ammonoids are sometimes preserved.

The upper-limestone interval of the cyclothem reflects deposition in shallowing water. Mud-supported textures characterize the limestones (fossiliferous calcilutites in fig. 8.7) and grade upward into skeletal calcarenites. The calcarenites are composed of sand-size fragments of fossils, with some oolitic grains and grains with algal coatings. These rocks formed on the shelf in shallow, agitated waters, above effective wave base. The uppermost part of the upper-limestone interval in Iowa often displays features that indicate deposition in very shallow water, such as lagoonal and peritidal settings. The tops of some of these beds preserve evidence of exposure to the atmosphere and surface weathering processes.

The outside shale is so named because it lies outside of the marine limestone

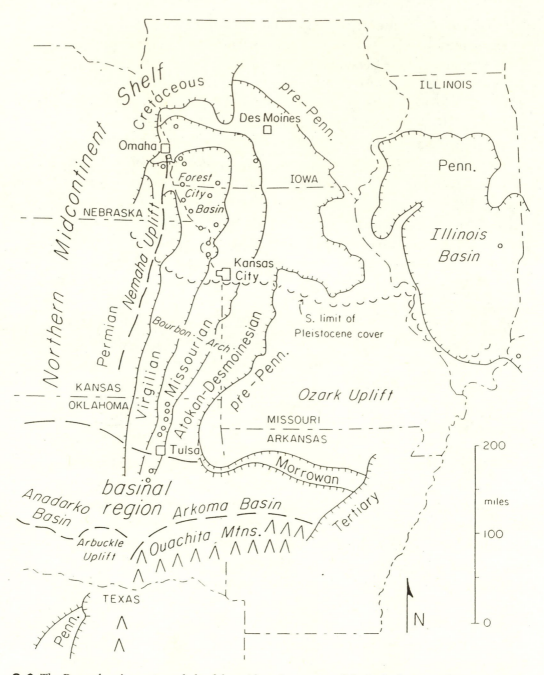

8.6 The Pennsylvanian outcrop belt of the midcontinent area of the United States. The Forest City Basin formed during Early Pennsylvanian time and was largely filled with sediments by the end of Middle Pennsylvanian time. For the remainder of Pennsylvanian time, the region from northern Oklahoma to northern Iowa was occupied by the Northern Midcontinent Shelf and was subjected to numerous transgressions and regressions of the shelf seas. From Heckel 1992.

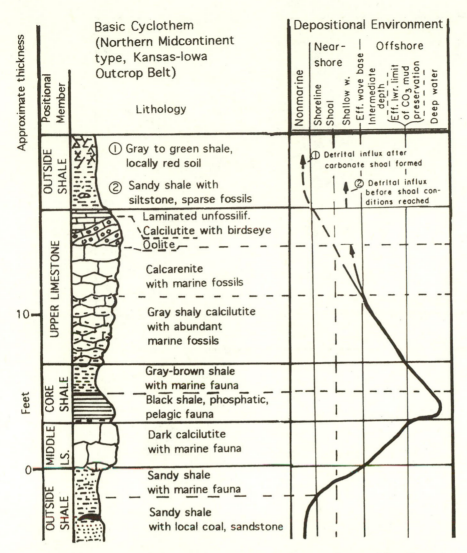

8.7 A standard midcontinent cyclothem displaying the major types of deposits and inferred depositional environments associated with a major transgression-regression (T-R) cycle across the Northern Midcontinent Shelf. Adapted from Heckel 1992.

units of the cyclothem. It formed during lower stands of sea level and includes a wide variety of nearshore and terrestrial deposits. For example, prodeltaic shales are represented, as are delta-front and delta-plain sandstones and coals. The outside-shale segment also contains blocky mudstones that preserve paleosols (ancient soils) and evidence of disturbances by plant roots and animal burrows in terrestrial settings. These blocky mudstones are often overlain by widespread coals. The coals are products of coastal swamps, signifying a rise in sea level and the start of a new cycle of deposition.

Not all cyclothems are as complete as the one just described. A major deepening event was required to produce the black phosphatic shales of the core-

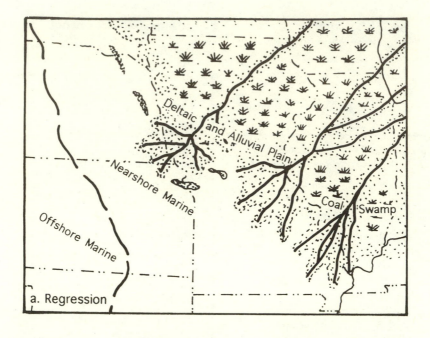

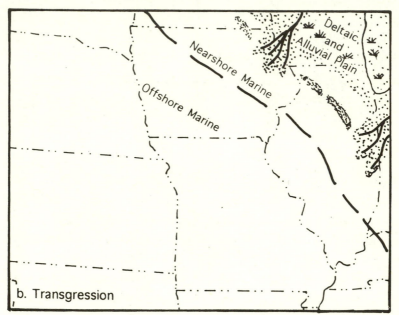

8.8 Generalized paleogeographic settings during regressive (a) and transgressive (b) fluctuations of marine and coastal environments over the central United States during Pennsylvanian time. Adapted from Wanless et al. 1963 and Dott and Batten 1971.

shale segment of a classic cyclothem. T-R cycles of lesser magnitudes produced incomplete, but still recognizable, cyclothems.

Figure 8.8a depicts the paleogeography of ancient Iowa during a time of regression, when low-lying deltaic plains and coastal swamps prevailed. In contrast, figure 8.8b portrays a time of transgression, when Iowa was a marine shelf.

The Pennsylvanian rocks of Iowa are assigned to five series based on age. In ascending order, the divisions are the Morrowan, Atokan, Desmoinesian, Missourian, and Virgilian series. The Morrowan Series represents strata of Early Pennsylvanian age, the Atokan and Desmoinesian series include deposits of Middle Pennsylvanian age, and the Missourian and Virgilian series are of Late Pennsylvanian age.

The Pennsylvanian rock record of the state is complex and varied, consisting of interbedded intervals of shale, siltstone, sandstone, clay, coal, and various limestones. A summary of the major rock types of the Pennsylvanian and their environments of deposition is shown in figure 8.9. Space does not permit a detailed discussion of each of the formations represented in the Pennsylvanian System of Iowa. More than seventy formation names have been introduced to describe the Pennsylvanian rocks of the state. The preponderance of stratigraphic names does serve to illustrate that the Pennsylvanian rock record of Iowa is highly varied. Because formational names are assigned chiefly on the basis of composition, the great variety of formational units indicates that a large number of lithologic changes occur in the Pennsylvanian System in Iowa.

The Pennsylvanian formations of Iowa are assigned to four supergroups or eight groups (fig. 8.10). Groups and supergroups are generally used, instead of formations, when dealing with the Pennsylvanian stratigraphy of Iowa. The rock record can also be classified on the basis of geologic age, and series are appropriate age-related (chronostratigraphic) divisions for the discussion of Iowa's Pennsylvanian history.

MORROWAN SERIES

A large erosional outlier of coal-bearing strata in Muscatine and Scott counties preserves Iowa's oldest Pennsylvanian unit, the Caseyville Formation. The Caseyville Formation is comprised of sandstones, shales, mudstones, a basal conglomerate, and laterally discontinuous coal beds. Fossil spores document the age of the formation as Morrowan. The formation reaches a thickness of nearly 100 feet in the Quad Cities area. The Wyoming Hill exposure in Muscatine County represents the most complete section of the formation in the area (fig. 8.11). Conspicuous facies variations occur in the Caseyville Formation and document lateral changes from stream channel environments through overbank and floodplain settings. The names Wildcat Den Coal and Wyoming Hill Coal are applied to the two principal coal beds of the formation. A variety of well-preserved fossils occur in the Caseyville Formation at the Wyoming Hill site, including stumps of the scale tree *Lepidodendron* preserved in upright position. The overall environment of deposition of the formation was probably fluvial, although a fluvial-deltaic setting cannot be ruled out.

8.9 The major Pennsylvanian rock units and their depositional environments. Adapted from Howes 1990; Ravn et al. 1984; and Van Eck 1965.

Series	Group or Formation	Lithologies (Composition)	Depositional Environments
Virgilian (Upper)	Wabaunsee Group	Primarily shales and limestones with some sandstones, siltstones, and coals	Cyclic changes from deltaic coastal plains to marine shelf settings
	Shawnee Group	Limestone and shale	Cyclic with marine settings predominant
	Douglas Group	Silty shales, siltstones, and limestones; minor coal	Cyclic with marine settings predominant
Missourian (Upper)	Lansing Group	Limestone and shale	Cyclic with marine settings predominant
	Kansas City Group	Limestone and shale	Cyclic with marine settings predominant
	Bronson Group	Limestone and shale; minor coal	Cyclic with marine settings predominant
Desmoinesian (Middle)	Marmaton Group	Alternating layers of shale and limestone with some sandstone and coal	Alternating marine and nonmarine settings; records five T-R cycles
	Upper Cherokee Group	Alternating shale and limestone; includes a number of fairly thick coal beds; well-developed channel sandstones	Alternation of nonmarine settings (deltaic coastal plains and coal swamps) and marine shelves; stream channels
Atokan (Middle)	Lower Cherokee Group (includes channel sandstones of the Spoon Formation in eastern Iowa)	Alternating shale, sandstone, and limestone; coal beds; prominent channel sandstones	Alternating nonmarine and marine environments, with nonmarine settings predominant; includes channel sandstones; possible estuarine settings in low areas of the irregular Mississippian surface
Morrowan (Lower)	Caseyville Formation	Sandstones, shales, mudstones, basal conglomerate, laterally discontinuous coal beds	Fluvial or fluvial-deltaic

ATOKAN SERIES

The Atokan Series has been recognized in Iowa on the basis of palynological (spores and pollen) data, but there is some uncertainty as to the exact placement of the Atokan boundary with the overlying Desmoinesian Series. The Spoon Formation of eastern Iowa and the Cherokee Group of central Iowa may span the Atokan-Desmoinesian boundary (fig. 8.10).

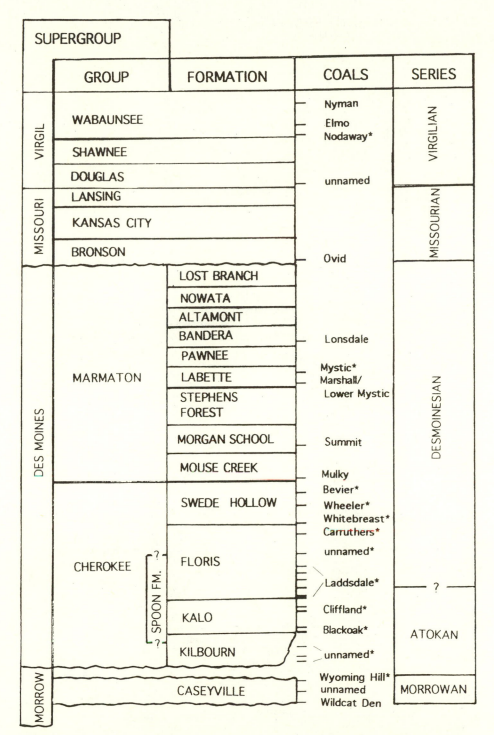

SUPERGROUP				
	GROUP	FORMATION	COALS	SERIES
VIRGIL	WABAUNSEE		Nyman / Elmo / Nodaway*	VIRGILIAN
VIRGIL	SHAWNEE			VIRGILIAN
VIRGIL	DOUGLAS		unnamed	VIRGILIAN
MISSOURI	LANSING			MISSOURIAN
MISSOURI	KANSAS CITY			MISSOURIAN
MISSOURI	BRONSON		Ovid	MISSOURIAN
DES MOINES	MARMATON	LOST BRANCH		DESMOINESIAN
DES MOINES	MARMATON	NOWATA		DESMOINESIAN
DES MOINES	MARMATON	ALTAMONT		DESMOINESIAN
DES MOINES	MARMATON	BANDERA	Lonsdale	DESMOINESIAN
DES MOINES	MARMATON	PAWNEE		DESMOINESIAN
DES MOINES	MARMATON	LABETTE	Mystic* / Marshall/ / Lower Mystic	DESMOINESIAN
DES MOINES	MARMATON	STEPHENS FOREST		DESMOINESIAN
DES MOINES	MARMATON	MORGAN SCHOOL	Summit	DESMOINESIAN
DES MOINES	MARMATON	MOUSE CREEK	Mulky	DESMOINESIAN
DES MOINES	CHEROKEE	SWEDE HOLLOW	Bevier* / Wheeler* / Whitebreast* / Carruthers*	DESMOINESIAN
DES MOINES	CHEROKEE (SPOON FM.)	? FLORIS	unnamed* / Laddsdale*	?
DES MOINES	CHEROKEE (SPOON FM.)	KALO	Cliffland* / Blackoak*	ATOKAN
DES MOINES	CHEROKEE (SPOON FM.)	? KILBOURN	unnamed*	ATOKAN
MORROW		CASEYVILLE	Wyoming Hill* / unnamed / Wildcat Den	MORROWAN

8.10 The generalized stratigraphic nomenclature for the Pennsylvanian System of Iowa. Major units, in ascending order, include the Caseyville Formation of the Morrow Supergroup and the Cherokee, Marmaton, Bronson, Kansas City, Lansing, Douglas, Shawnee, and Wabaunsee groups. Coals of potential economic significance are marked with a *. Adapted from Howes 1990.

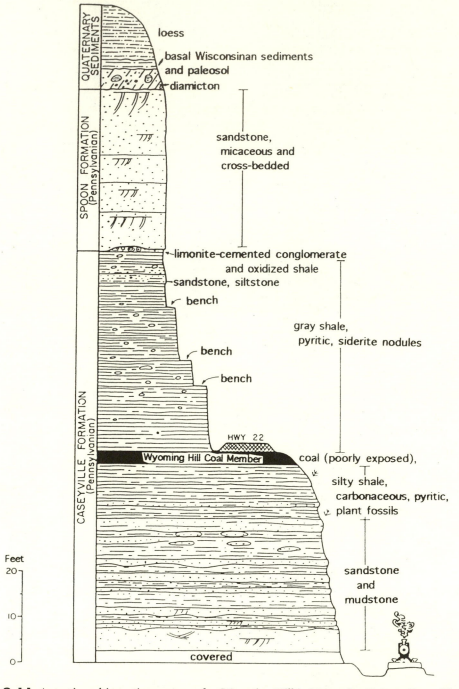

loess

basal Wisconsinan sediments
and paleosol

diamicton

sandstone,
micaceous and
cross-bedded

limonite-cemented conglomerate
and oxidized shale

sandstone, siltstone

bench

gray shale,
pyritic, siderite nodules

bench

bench

HWY 22

Wyoming Hill Coal Member

coal (poorly exposed),

silty shale,
carbonaceous, pyritic,
plant fossils

sandstone
and
mudstone

Feet
20

10

0

covered

QUATERNARY SEDIMENTS

SPOON FORMATION (Pennsylvanian)

CASEYVILLE FORMATION (Pennsylvanian)

8.11 A stratigraphic section measured at Wyoming Hill in Muscatine County, Iowa. The Caseyville Formation is well exposed in the roadcuts along Highway 22 and in the adjacent railroad cuts. The conspicuous cliff-forming sandstones of the Spoon Formation occur higher in the section, well above the grade level of Highway 22. Adapted from Ludvigson and Swett 1987.

The Spoon Formation, a cliff-forming sandstone, is well exposed at the Wyoming Hill locality in Muscatine County (fig. 8.11). It rests unconformably on the Caseyville Formation in Muscatine and Scott counties. Sandstones of the Spoon Formation are interpreted as fluvial deposits; they are highly micaceous, well sorted, and conspicuously cross-bedded. The overall orientation of cross beds suggests that sediments were transported southwestwardly during Pennsylvanian time. The sands of the Spoon Formation contain large amounts of feldspar minerals and grains of metamorphic rocks, in addition to common quartz. This contrasts with the quartz-rich sandstones of the underlying Caseyville Formation.

The sediments that compose the two formations were apparently derived from different source areas. Sands of the Caseyville Formation were likely derived from the recycling of older Paleozoic rocks, whereas the grains in the Spoon Formation probably came from Precambrian terranes to the north and east of Iowa. Spectacular exposures of the Spoon Formation occur at Wildcat Den State Park, where huge blocks of sandstone are creeping downslope on the underlying shales of the Caseyville Formation (fig. 8.12).

The lower part of the Cherokee Group has been assigned to the Atokan Series (fig. 8.10). The Cherokee Group is the thickest of the Pennsylvanian groups in Iowa, averaging 500 feet in thickness in southeastern and south-central Iowa. The group is named for exposures in Cherokee County, Kansas. The major coal resources of Iowa are associated with the Cherokee Group.

The lower Cherokee interval records the infilling of erosional irregularities on the underlying Mississippian surface. The onset of marine transgression during Pennsylvanian time was accompanied by a rise of local base levels. Consequently, alluvial and deltaic sediments were deposited in shoreline areas. Locally, swamps developed in coastal settings, and peat accumulated. Subsequently, the layers of peat were transformed into thin beds of coal.

DESMOINESIAN SERIES

In south-central Iowa, the Desmoinesian Series is comprised of the upper Cherokee Group and the overlying Marmaton Group (fig. 8.10). The Desmoinesian Series derives its name from exposures along the Des Moines River in central Iowa, where coals were mined extensively in the late 1800s and early 1900s. The upper Cherokee Group displays an increase in marine and marginal marine sedimentation in contrast with the largely nonmarine character of the lower Cherokee Group. Coals are present throughout the Cherokee Group, and nine such beds are considered to be potential resources (fig. 8.10). One of these, the Whitebreast Coal, is widespread and occurs over much of the midcontinent region. Most of Iowa's coal resources are in the lower Cherokee interval. Key de-

8.12 Large blocks of Pennsylvanian sandstone that are creeping downslope and away from the cliffs at Wildcat Den State Park in Muscatine County, Iowa. The prominent sandstones are in the Spoon Formation and rest unconformably on shales of the underlying Caseyville Formation. Courtesy of the Calvin Photographic Collection, Department of Geology, University of Iowa.

posits include the Blackoak, Cliffland, and Laddsdale coals, all of which were mined extensively.

The Whitebreast Coal and other Middle Pennsylvanian strata of the Cherokee Group were revealed by the rapid downcutting of floodwaters that surged over the emergency spillway of Saylorville Dam in Polk County during June 1984 and again during the summer of 1993. The rock exposures below the Saylorville Lake emergency spillway represented the most extensive and accessible view of this stratigraphic interval anywhere in Iowa. Two distinct depositional regimes were preserved: a lower fluvial-deltaic assortment of strata and an overlying marine interval displaying transgressive-regressive characteristics.

The variable composition of the Cherokee Group is depicted in figure 8.13. Note the occurrence of prominent channel sandstones within the lower and middle Cherokee. Such channels are well displayed today in Dolliver Memorial State Park in Webster County, in Ledges State Park in Boone County, and at Red Rock Reservoir in Marion County. Primary sedimentary structures in the sandstones indicate a strong southwesterly paleocurrent direction. See figure 8.14 for the locations, trends, and inferred paleodrainage associated with the channel sandstones of the Cherokee Group. Sandstone quarried from erosional outliers of the Cherokee Group in Iowa County was used extensively for the construction of houses in the Amana Colonies.

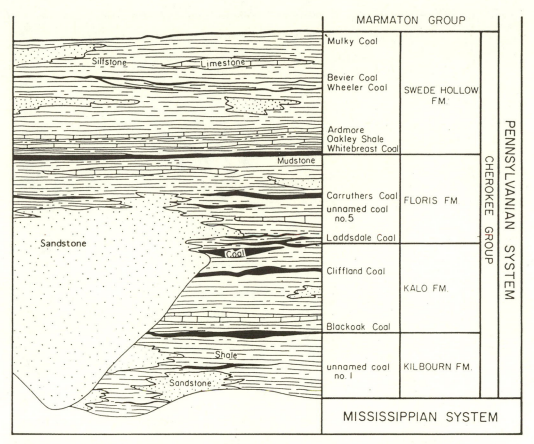

8.13 A schematic cross section of the Cherokee Group of the Desmoinesian Series. Note the variable composition, numerous coal beds, and well-developed channel sandstones. Adapted from Howes et al. 1989.

The Marmaton Group, the uppermost segment of the Desmoinesian Series, records five T-R cycles, probably produced by worldwide fluctuations in sea level. Marine units are well displayed within the Marmaton interval. The Mystic Coal occurs within the Marmaton Group; it has been mined extensively in south-central Iowa.

MISSOURIAN SERIES

Ravn et al. 1984 assigned three groups to the Missourian Series in Iowa. That convention is followed here. In ascending order, the groups are the Bronson, Kansas City, and Lansing (fig. 8.10). Each of the groups preserves alternating shales and limestones. A few sandstones occur, and minor coal seams are known. None of the coals are of economic importance. The depositional environment of the Missourian Series in Iowa was predominantly marine (see figure 8.8 for a generalized environmental setting). Marine fossils such as brachiopods, foraminifera, calcareous algae, conodonts, and fish remains are common in the rocks of the series. The origins of the rocks of the Missourian Series can best be

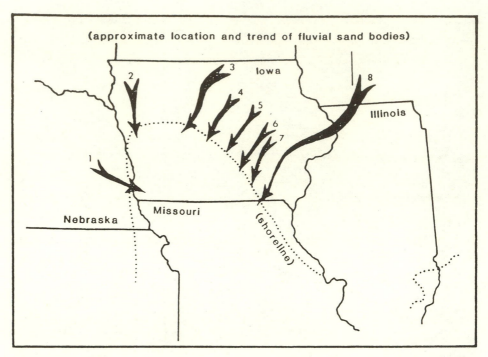

8.14 The location, trends, and inferred paleodrainage of sand bodies within the Cherokee Group. From Lemish, Chamberlain, and Mason 1981.

understood with reference to T-R cycles and the cyclothem model of Heckel (1992) (see figure 8.7). Iowa's record of the series records a complete or partial record of ten cyclothems.

VIRGILIAN SERIES

The Virgilian Series consists, in ascending order, of the Douglas, Shawnee, and Wabaunsee groups. All three groups display cyclic deposition. The Douglas Group is composed primarily of silty shales, siltstones, and limestones. The Shawnee Group contains relatively thick limestones and minor shale beds. The Wabaunsee Group includes shales and limestones with minor amounts of sandstone, siltstone, and coal. Some of the Wabaunsee coals serve as important stratigraphic markers. One coal, the Nodaway Coal, has been mined rather extensively in Adams, Taylor, and Page counties in southwestern Iowa.

The depositional environment for the Virgilian Series was similar to that of the Missourian Series. Rocks of the Virgilian Series reflect cyclic deposition in marine shelf and deltaic coastal plain settings. Coastal swamps produced thin beds of coal at the interface of the land and the sea. Eight cycles of deposition are preserved.

Collectively, the rocks of the Pennsylvanian System provide the state with several important resources. Limestone aggregate is produced from Pennsylvanian quarries in south-central and southwestern Iowa. Pennsylvanian shales are still used in the manufacture of brick and tile, and Pennsylvanian coal had a long history of use in the state. Additional discussion of the state's economic geology is found in chapter 11. A summary of Iowa coal and coal mining follows.

Formation of Coal

Coal is a combustible rock formed by the accumulation and alteration of vegetation. It can be ignited and burned and has many industrial uses. The genesis of coal by the accumulation of plant material has been well established.

Abundant plant remains in European strata gave the Carboniferous System its name. In this country, geologists divide the Carboniferous rock record into the Mississippian and Pennsylvanian systems. Coal deposits of the Upper Carboniferous of Europe and their Pennsylvanian equivalents in the United States provided the energy to power the Industrial Revolution, and they still provide important sources of energy today. Coal swamps of the Late Carboniferous (Pennsylvanian) were some of the most extensive wetlands of all times.

Among the most interesting freshwater wetlands are peat bogs, where environments are highly acidic and depleted of oxygen. In such settings, microorganisms that normally decompose plant materials are absent, and plant remains accumulate as peat. Formation of peat is an early stage in the development of coal. Peat is presently forming in a number of areas, ranging from the tropics to subarctic regions. Two major modern settings for the formation of peat are recognized: sites in the interior of continents and areas in coastal settings.

Peat formation requires that the growth of vegetation exceed the rate of decay and that plant material be allowed to accumulate rather than be removed by erosion. Both of the modern settings for peat formation meet these conditions. Furthermore, both areas are poorly drained and possess high levels of groundwater that protect the plant debris of peat from normal rapid decay. The peat accumulations of coastal swamps are more likely to be preserved than the peat deposits of inland wetlands because they have a better chance of being covered by seas and subsequently buried. Inland peat deposits are more susceptible to destruction by erosional processes. Areas of peat accumulation also need to be protected from rapid influxes of clastic sediment such as silt and clay. Otherwise, the end product is carbonaceous siltstone or shale rather than coal.

The typical ratio of thickness of uncompacted peat to coal is about 10 to 1, so an Iowa coal bed 3 feet in thickness would have required approximately 30 feet of peat. It is clear that the formation of coal beds of commercial value required enormous quantities of ancient vegetation.

For the most part, Iowa's Pennsylvanian coals were associated with river-delta complexes (fig. 8.15). Streams draining the continental areas dumped their sediment loads (clay, silt, and sand) in coastal areas. These deposits usually displayed great lateral variation in composition and thickness.

AREA INFLUENCED BY
MARINE TO BRACKISH WATER

AREA INFLUENCED BY
FRESH WATER

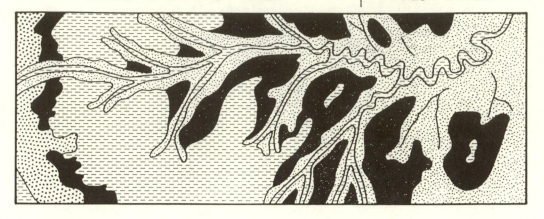

 Fine-grained sand, silt (clay)
[Crevasse splay, overbank deposits]

 Clay (silt, sand)
[Marginal marine, embayments, channels]

 Sand [Beaches, beach ridges]

 Peat [Swamp]

8.15 A depositional model of a peat-forming environment associated with a river delta in a coastal region. The resulting coal beds are often limited in their extent laterally because of the presence of numerous channels. Note the irregular distribution of the peat-forming swamps. Many of Iowa's coal deposits formed in similar environments. From Horne et al. 1978 and Howes et al. 1989.

The tropical climate of Pennsylvanian time in ancient Iowa provided ideal conditions for luxuriant plant growth. Scale trees, ferns, seed ferns, rushes, and other vegetation dominated the coastal swamps and floodplains. The dominant vegetation of Pennsylvanian time was distinctly different from that of today. However, several present-day coastal settings serve as modern analogs for the physical setting of the ancient coal swamps. The Everglades of Florida, the Okefenokee Swamp of Georgia, and the deltaic region of southern Louisiana are examples of such modern coastal swamps. Peat areas adjacent to the northwestern coast of Europe, such as the Dutch Lowlands, also contain features somewhat analogous to the ancient coal-forming region of the interior of the United States during Pennsylvanian time. A reconstruction of a coal swamp as it may have existed during Pennsylvanian time is shown in figure 8.16.

Topographic relief on the ancient coastal landscape was low. Therefore, fluctuations in sea level exerted a strong control on deposition of sediments in coastal settings by raising and lowering the base levels of streams. Sea-level fluctuations also affected the level of the water table in shoreline areas. High water tables during times of marine transgressions provided favorable environments for the conversion of plant material to peat. Later, burial of the peat by influxes of sedi-

8.16 A reconstruction of a Pennsylvanian coal swamp illustrating typical vegetation and a large dragonfly. Constituents include: (a) roots of a lycopod scale-tree (*Stigmaria*); (b) *Lepidodendron*, a common scale tree; (c) tree fern; (d) *Calamites*, an ancient rush; and (e) a large dragonfly. From McKerrow 1978.

ments initiated the changes that ultimately produced coal. The cumulative effects brought about by burial, increased temperature, and time determined the ultimate quality of the coal. Later, circulating groundwater precipitated additional minerals in the coal beds. Among these secondary deposits were pyrite (iron sulfide), calcite (calcium carbonate), quartz (silicon dioxide), and various clay minerals.

The Pennsylvanian coal swamps supported a junglelike growth of vegetation. The principal coal-forming plants were scaly lycopod trees (ancestors of modern club mosses) and tree ferns. Sphenopsids (ancient rushes), seed ferns, and mangrovelike cordaitean trees were present as well. The absence of growth rings and the presence of large, thin-walled cells in the trunks of this vegetation are taken to mean that distinct seasonal changes did not take place in the Pennsylvanian coal swamps. The flora probably grew the year-round under subtropical to tropical conditions. Burial and alteration of the plant material produced peat, and with additional burial, alteration, and compaction, eventually coal was formed.

The kind and relative abundances of swamp plants changed through time, and there is commonly a widespread occurrence of distinctive spore assemblages within a given Pennsylvanian coal bed or seam. Recognition of spores (asexual reproductive bodies of lower plants) and pollen (microscopic reproductive bodies of seed plants) in coal deposits makes it possible for paleobotanists to correlate coal beds. However, during the Pennsylvanian Period, there were relatively few pollen producers, as gymnosperms (cone-bearing plants) were in their early stages of development and angiosperms (flowering plants) had not yet evolved.

Iowa Coal

Coal deposits of economic potential are found in rocks of the Cherokee, Marmaton, and Wabaunsee groups of south-central and southwestern Iowa and in the Caseyville Formation of eastern Iowa (figs. 8.10 and 8.17). Coal seams are not always continuous in Iowa's subsurface. Some coal beds are laterally discontinuous because of the environmental setting under which they formed. For example, the coal seam near region d in figure 8.18 is discontinuous because of the presence of a stream channel in the proximity of the original coal swamp. Other coal seams were removed by subsequent erosion, as fluctuating environments during Pennsylvanian time resulted in streams that cut channels through the deposits of the former coal swamp (see area c in fig. 8.18). Other coal seams are discontinuous owing to modern or preglacial erosion (see areas a and b in fig. 8.18). In addition, the thickness and distribution of coal beds may vary because of depositional factors. For example, many of the coals of the Cherokee Group in the state were probably deposited as lenticular bodies.

Most Iowa coal is classified as high-volatile bituminous, but it just barely makes the bituminous category. In fact, some of the coals from the Fort Dodge area

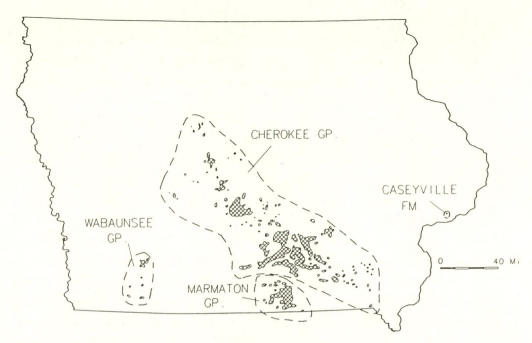

8.17 Generalized areas of coal mining in Iowa. The probable stratigraphic assignment of the coal in each area is indicated. From VanDorpe and Howes 1986 and Howes 1990.

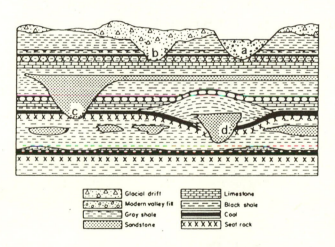

8.18 An idealized stratigraphic cross section showing various factors that result in discontinuity of coal beds: (a) coal removed by modern stream erosion; (b) coal removed by preglacial erosion; (c) coal removed by a stream after coal deposition; and (d) presence of the stream throughout the time of peat accumulation. In addition, coal beds vary in thickness because they were deposited originally as lenticular-shaped bodies; most of the coal beds of the lower Cherokee of Iowa are lenticular in shape. Adapted from Simon and Hopkins 1973.

have been ranked as subbituminous. Subbituminous coals have properties intermediate between lignites (brown coals) and bituminous coals and contain less than 69 weight percent carbon. Bituminous coals are hard, black coals that contain fewer volatiles and less moisture than lignites or subbituminous coals. Bituminous coals contain 69 to 86 weight percent carbon.

Some of the Iowa coals have a relatively high ash and sulfur content. Ash and sulfur are impurities that occur in all coals to some degree, but they occur in sufficient quantities in Iowa coal to restrict its use. The ash originates from sediments (clay, mud, and silt) that were washed into the coal swamps. Sulfur in coal occurs primarily as pyrite (iron sulfide) and small amounts of organic sulfur. Organic sulfur is contained in complex chemical combination with the organic constituents of coal and was probably derived from swamp plants that took in sulfur when they grew.

Pyrite in coal may have originated in part by formation in the swamps at the time of peat production. The presence of microscopic aggregates of pyrite grains dispersed through coal has been offered as evidence that iron sulfide was present at the time the coal formed. A great deal of iron sulfide, however, was likely introduced into the coal seams after the peat was converted to coal. Pyrite coatings along vertical fractures that cut coal seams provide evidence that some pyrite is of secondary origin. Although the origins of high-sulfur coals, such as those in Iowa, are not totally understood, their stratigraphic association with marine strata is suggestive. Sedimentary rocks of marine origin often directly overlie seams of high-sulfur coal. This suggests that marine water from the Pennsylvanian seas may have been the source of much of the sulfur found in the high-sulfur coals of the state.

Iowa coal is similar in quality to the coals of Illinois, Missouri, and Kansas. Analyzed Iowa coals average approximately 12,000 Btu per pound of coal. Most of the Midwest coals are high-sulfur coals. Washing and beneficiation are widely practiced in Illinois to reduce the ash and sulfur content of the coal. Partly because of this, the Illinois coal industry is able to continue to compete in the coal market of the Midwest. In addition, the Illinois coal beds are generally thicker, more laterally continuous, and closer to the surface than those of Iowa.

Funds in Iowa were not readily available to develop modern coal production and washing facilities in the post–World War II era. Consequently, Iowa's coal industry declined significantly, and currently there are no active coal mines in the state.

Iowa's Coal Industry

In 1917, the Iowa coal industry employed approximately 18,000 people, but in 1975 fewer than 100 people were so employed. During the heyday of coal mining in Iowa, some 450 coal mines were active. In 1975, only five small surface (strip) mining operations and two underground mines were operating. By 1981, both of the underground mines had closed. The last year of reported coal production in the state was 1994. When American Coals Corporation of Wintersville, Ohio, declared bankruptcy in September 1994, they closed the two surface mines that they were operating southeast of Knoxville in Marion County. With those closures, an Iowa industry older than statehood came to a quiet end. Although huge quantities of coal still lie beneath the state, no comeback of the coal industry is expected soon.

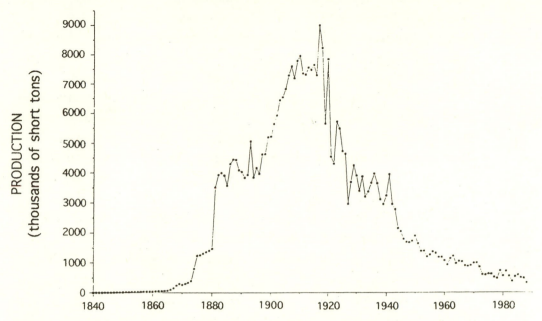

IOWA COAL PRODUCTION 1840-1988

8.19 Annual coal production in Iowa from 1840 to 1988. Peak production in 1917 was just under 9 million tons. The last coal mine in Iowa closed in 1995. From Howes 1990.

Figure 8.19 shows Iowa's annual coal production from 1840 to 1988. Production peaked in 1917 during World War I at just under 9 million tons. The sharp decline in the early 1940s reflects labor shortages and curtailed mining brought on by World War II. The continued decline in coal mining after World War II resulted from competition from cleaner energy sources, such as gasoline, fuel oil, and natural gas.

Estimates of coal reserves in Iowa have been conducted and vary depending on the quality of geologic data available. According to VanDorpe and Howes (1986), the original reserves for coal beds greater than 14 inches thick in thirty-seven Iowa counties were estimated to be nearly 7.4 billion tons. The area of this estimate constitutes only a fraction of the total area of Iowa underlain by Pennsylvanian rocks, however. All of the coal mined in Iowa since 1840 represents only 5 percent of the estimated original reserves for coal beds greater than 14 inches thick. Thus, it seems safe to conclude that the state still has enormous reserves of Pennsylvanian coal.

A Brief History of Coal Mining in Iowa

Iowa was the second state west of the Mississippi River to develop commercial coal mines. Because eastern Iowa was settled before central and western Iowa, coal mining was first developed in that part of the state. The Pennsylvanian rocks and their associated thin coal beds occur as isolated erosional outliers in eastern Iowa in Muscatine and Scott counties and as bedrock exposures in southeastern Iowa in Lee, Van Buren, and Jefferson counties. The early history of Iowa's coal-mining industry is mainly the story of small mines that tapped pockets of coal in these five counties. These small mines generally served only local markets, primarily for home-heating fuel.

A coal mine was operated at Farmington in Van Buren County as early as 1840, and the coal was hauled by team and wagon to Keokuk some 30 miles away. Later, the mine was enlarged to supply coal to steamboats that plied the Des Moines River. During this time, the mine employed from forty to fifty miners. Wages in those days were 5 cents per bushel of coal (80 pounds); coal sold for $2.00 to $2.50 a ton.

The market for Iowa coal was limited in the period from 1840 to 1870. People still preferred to use wood for heating because of its cleanliness and availability; they were reluctant to experiment with an unproven substitute in the form of dirty, dusty, smoky coal.

By 1870, all the major railroads had reached the Missouri River. Development of the railroads increased the demand for coal as fuel for the locomotives. The growth of the railroads in Iowa provided a ready market for Iowa's coal resources and stimulated coal mining in the state. In that decade, it was of great significance that Iowa's coal mines represented the last place where railroads could secure adequate supplies of coal for their locomotives before they started their long haul westward across the Great Plains. Iowa was the leading coal producer west of the Mississippi River until the coalfields of Colorado were developed.

One of the first large mining operations in Iowa was located in Des Moines south of the Raccoon River, near what is now the Seventh Street Bridge. At this site, Wesley Redhead developed the Black Diamond Mine (later called the Pioneer Mine). This mine (fig. 8.20) recovered coal from a 4.5-foot seam of coal, 125 feet below the surface. In 1876, the Black Diamond Mine employed 150 persons and produced 200 tons of coal per day.

Other mining ventures followed in the Des Moines area. Although there are no mines operating in the city today, there are extensive underground systems of tunnels under the city to attest to coal-mining activities of the past. In particular, the area just east of the state capitol was once the site of an extensive subsurface-mining operation. During the summer of 1978, a large hole appeared suddenly in the front yard of a residence on East Capitol Street. Certainly, the subsidence and collapse of old mine tunnels would have to be a prime suspect to explain this surprising event.

Another early coal-mining venture of significance was the Whitebreast Coal and Mining Company's operation in Lucas County. The company was named for Whitebreast Creek near Lucas, where coal exposures occur in western Lucas

8.20 The Pioneer Mine, Des Moines, Iowa. This was one of the first large surface coal-mining operations in Iowa. From Olin 1965.

County. The original Whitebreast Mine produced coal from a coal bed 5 feet in thickness at a depth of about 250 feet. In 1880, the mine employed 360 people and produced 650 to 750 tons of coal per day. The Whitebreast Company later developed other mines in Lucas County and in surrounding counties as well.

Thomas Lewis was employed by the Whitebreast Company at its mine near Lucas. He had immigrated to Iowa from Wales in the late 1870s. Having experienced the back-breaking work, unsafe environments, and stark poverty associated with coal mining in Wales, Lewis hoped to find improved working conditions and higher wages in the coalfields of Iowa. Instead, he found many of the same conditions. Because of this, Lewis proceeded to organize his fellow miners, and he led them in a strike in 1882. Although the bitter strike was eventually settled, Lewis was denied re-employment and was put on a blacklist that prevented him from being hired as a miner elsewhere.

As a result of Lewis's banishment from the coalfields, the Lewis family experienced a long period of wandering and instability, during which Lewis worked a variety of jobs, including night watchman and city jailer in Des Moines. In 1897, the blacklists were nullified, and the Lewis family returned to Lucas, where Thomas and his two sons, John and Thomas, found employment with the Big Hill Coal Company.

John L. Lewis (1880–1969), a high-school dropout, got his start in mining as a mule driver at the Big Hill Mine at Lucas. He later ascended to the presidency of the United Mine Workers of America. Displays, photographs, and videos at the John L. Lewis Museum of Mining and Labor in Lucas tell the story of the fiery, Iowa-born labor leader.

According to Olin (1965), during the 1880s and 1890s, the Lucas operation employed 2,000 miners, and Lucas's Main Street featured twenty-seven saloons. The last coal mine in the Lucas area, the Nebraska-Iowa mine, shut down in 1923. Today, the population of Lucas is less than 250.

Lucas is not the only Iowa town that fell on hard times when coal mining started to decline after World War I. A number of coal boomtowns that sprang up in Iowa in the late 1800s and early 1900s no longer exist. One such boom-to-bust town was Muchakinock in Mahaska County. Muchakinock was located about 3 miles south of Oskaloosa. A 6-foot-thick coal bed at Muchakinock was worked in seven openings — one vertical shaft and six horizontal drifts. The Chicago and North Western Railway was the chief consumer of the Muchakinock coal.

Coke ovens were constructed at the Muchakinock site in 1877 to convert the fine screenings of coal into a useful product — coke. Until that time, the fine screenings had been treated as a waste product. These coke ovens were probably the first coke ovens to be operated west of the Mississippi River. The high sulfur content of the coal resulted in a rather unsatisfactory product, however. Consequently, the coking operation was soon discontinued.

Black laborers were recruited from Virginia and other southern states to work in the coal mines of Muchakinock. In the early 1880s, of the 500 persons employed in the coal mines of Muchakinock, approximately 350 were black. By 1900, most of the mining had ceased at the Muchakinock coal camp. The equipment and personnel were moved south to Monroe County, where the Chicago and North Western Railway had purchased several thousand acres of ground to develop a new coal mine. The town of Buxton was established at the site of this new mining venture; it was named in honor of the mine superintendent, J. E. Buxton.

Coal was plentiful near Buxton, and several mines opened in the area in the early 1900s. One mine, the Consolidated Coal Company Mine Number 18, was probably the largest coal mine ever to operate in Iowa. The mine had a huge engine room nearly 200 feet long that was filled with dynamos, steam turbines, and hoisting equipment used in the operation of the mine.

Buxton grew rapidly and developed into a prosperous community of 9,000 people (fig. 8.21). The town had a substantial population of African Americans because large numbers of black workers had been recruited from the South to work in the coal mines of the Muchakinock and Buxton mining districts. Buxton could also claim a significant population of black professionals — doctors, lawyers, teachers, ministers, and others. In addition, the town was unique in having the nation's first industrial YMCA, which became the center of activity for black miners of the community.

By the late 1920s, the coal of the Buxton mines was running out, and the town's

Looking S. E. from Water Tower, Buxton, Ia.

8.21 Buxton, Iowa, in the early 1900s, when the area around Buxton was one of Iowa's principal coal-mining centers. Note the uniform appearance of the company-owned frame houses. Courtesy of the Iowa State University Archaeological Laboratory, from the collection of Wilma Stewart.

population started to decline. Buildings were torn down and sold for lumber. Today, Buxton is a ghost town, and crumbling foundations and a couple of dilapidated buildings are all that remain of what was once one of Iowa's principal coal-mining communities.

Another landmark of Iowa's coal-mining past that has long since vanished is the Ottumwa Coal Palace (fig. 8.22). The Coal Palace was built in 1890 as an exposition building to publicize the coal resources of the state. A corn palace already existed in Sioux City to pay tribute to one of Iowa's chief agricultural products, so it seemed appropriate to residents of the coal-mining region of southern Iowa to honor coal in a similar manner. President Benjamin Harrison visited the Coal Palace during its first season and commented on its significance: "If I should attempt to interpret the lesson of this structure, I should say that it was an illustration of how much that is artistic and graceful is to be found in the common things of life."

The Coal Palace was 120 feet high and 130 feet wide. It was veneered with blocks of coal and featured large display rooms for exhibits and an auditorium that seated 6,000 persons. The key attraction of the facility was undoubtedly its coal mine, which was located below the auditorium. Visitors entered the mine by way of the gallery, where they were lowered by car down a dusty, coal-lined shaft. Below, a mule waited, hitched to a train of pit cars. Visitors were hauled in these cars into the model mine, where they observed miners digging into veins of coal.

8.22 The Ottumwa Coal Palace in 1890. From Kneedler 1890.

The Coal Palace enjoyed only two successful seasons (1890–1891) as a tourist attraction and exhibition hall. After the 1891 season, the Coal Palace lost its drawing power, and attendance and use plummeted. Its glory gone, the empty structure was allowed to stand for several years before it was eventually torn down. Today, a city park occupies the site that once housed Ottumwa's fabulous Coal Palace.

Iowa's coal mines drew large numbers of job-seeking immigrants to the state in the early 1900s. The coal camps of south-central Iowa were largely filled by European immigrants (French, Germans, Belgians, and Croatians) who had heard of the opportunities in the Iowa coalfields from friends and relatives. There was little work available in Europe then, and the immigrants were drawn to Iowa in search of a better life.

Coal production peaked in Iowa in 1917 when the total tonnage reached just under 9 million tons. Since then, coal production in the state has declined significantly. In the early 1940s, the railroads switched to diesel engines and no longer needed coal to power their locomotives. About this time also, homes and industries began using oil and gas for heat. The last operating coal mine in Iowa closed in 1994.

Types of Underground Mines

Several different types of mining operations were used in Iowa. If the coal beds were horizontal and outcropping along a hillside or slope, a horizontal entry was driven directly into the coal seam. An operation of this type is called a drift mine (fig. 8.23). Where the coal beds were below the surface but generally less than 100 feet underground, a sloping tunnel was driven down to intersect the coal. This type of operation is called a slope mine (fig. 8.23). If the coal was well below the surface, approximately 100 feet or more, vertical shafts were sunk to intersect the coal layers, and horizontal entries were developed off the shafts to mine the coal. Such an operation is known as a shaft mine (fig. 8.23). After the coal was encountered in an underground mine (drift, slope, or shaft), either of two methods was used to remove the coal — the longwall method (fig. 8.24) or the room-and-pillar method (fig. 8.25). The method used depended on the thickness of the coal, the depth of the coal, the type of rock that formed the roof of the mine, and other factors.

The New Gladstone Mine in Appanoose County serves as an example of a mine that used the longwall method. This mine (a slope mine) was located about 8 miles west of Centerville; it closed in 1971. The mining method used in the New Gladstone Mine was the advancing longwall method. The miners utilized special machines similar in appearance to huge chain saws to cut the soft rock from underneath the coal bed. Once the layer of rock beneath the coal had been undercut, the weight of the beds overlying the coal bed caused the coal to break off. The coal was then handloaded into carts and pulled to the surface.

At the New Gladstone operation, ponies were used to pull the loaded coal cars through the mine tunnels to the mine entrance. During the late 1800s and early 1900s, it was common to use ponies or mules to haul coal from the mines. Pony mines, as they were called, were abundant in Iowa. Many of these mines had low roofs, about 5 feet high, so only small draft animals like ponies or mules could maneuver easily through the mine tunnels (fig. 8.26).

The New Gladstone Mine was featured in a documentary film produced by the Iowa State University Film Production Unit. The film, *The Last Pony Mine*, depicts a day in the life of the miners of the New Gladstone Mine. The New Gladstone Mine was apparently the last mine in the United States to use ponies for underground haulage of coal, hence the title of the film.

The miners in the New Gladstone Mine began to mine along the coal bed at the bottom of the main entrance slope of the mine. As coal was removed, the mine face (wall) receded, and the mine started to take on a fanlike shape, with the entrance located at the handle of the fan. The length of the mine face determined the number of workers that were needed in the mine. Generally, each miner worked a 40-foot section of the face.

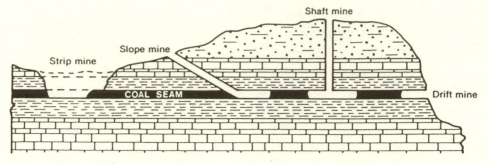

8.23 A comparison of drift, slope, shaft, and strip mines. From Schwieder and Kraemer 1973.

8.24 A plan view of a longwall mine with a vertical shaft entry. Skips and roadways were branch entries through the coal seam to the road-head area where coal was mined. From Howes et al. 1986.

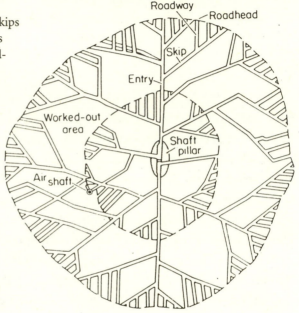

As the mine enlarged, it was necessary to erect supports to reinforce the mine roof to keep it from sagging or collapsing. The miners used timbers to support the mine roof near the working face, and they used rock debris from the mine to build pack walls about 4 feet back from the working face. The pack walls provided additional support for the mine roof. Added supports were placed at points of stress, such as where two tunnels came together. These supports, called cribs, were square-shaped wooden structures packed with rock debris. The New Gladstone Mine was one of the last advancing longwall mines in the United States.

The room-and-pillar method of mining (fig. 8.25) was also used in Iowa. In the room-and-pillar system, at least two parallel main entries are driven into the coal seam at the base of the mine shaft or slope. Cross entries are then developed off the main entries, and the coal is worked in 30-foot-wide areas, called rooms. An 8-to-10-foot pillar of coal is left between adjacent rooms to provide support for the mine roof. Although the room-and-pillar system became the standard

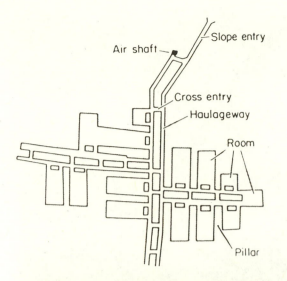

8.25 A plan view of a room-and-pillar coal mine. Coal was removed from the rooms, but pillars were left to support the roof of the mine. Consequently, only 40 to 60 percent of the coal was recovered from a typical room-and-pillar operation. From Howes et al. 1986.

8.26 A so-called pony mine, where ponies or mules were used to haul coal from the mine. Such operations were common in southern Iowa and northern Missouri. From Olin 1965.

method of underground mining, it was not as efficient as the advancing longwall method. In the longwall method, 85 to 95 percent of the coal was removed. In the room-and-pillar method, the coal in the pillar areas was not recovered; it was left to support the roof of the mine. Typically, only 40 to 60 percent of the coal was recovered initially in room-and-pillar operations. However, it was common practice to remove additional coal from the pillars, just before abandoning a mine.

Strip Mining

Surface mining, or strip mining, was used in Iowa where the coal was close to the surface. The rock above the coal layer (overburden) was dug away to expose the coal. A large dragline with a bucket (fig. 8.27) was generally utilized to re-

8.27 A strip-mining operation near Oskaloosa, Iowa, in the early 1970s. From Schwieder and Kraemer 1973.

move the overburden. In Iowa, a typical strip mine also used front-end loaders, bulldozers, and trucks to remove the coal. Conventional road construction equipment was utilized on an experimental basis during the 1970s in the Iowa State University Demonstration Mine near Oskaloosa.

Strip mining can have a damaging effect on the land unless the land is properly restored after the mining operations have been completed. It is essential that careful replacement of top soil and leveling of spoil banks take place after the mining operations have been completed.

Coal Mining and the Environment

From time to time, Iowans living over Pennsylvanian bedrock experience that sinking feeling. It might result from a collapsed front yard in Des Moines, a subsiding house in the Centerville area, or a mired combine in a cornfield near What Cheer. Most of Iowa's coal mines were underground operations, perhaps as many as 6,000 mines in an 80,000-acre area. Potential long-term detrimental effects of underground mining include subsidence of land and the drainage of acid water from mines. Subsidence involves the sinking of the land surface brought about by the collapse of an underground mine roof or the failure of support pillars. Subsidence over abandoned coal mines (fig. 8.28) has caused damage to buildings and property in both rural and urban areas in the state. A total

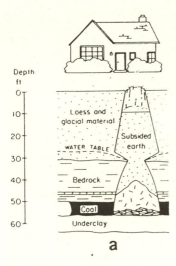

8.28 The effects of collapse over an underground coal mine. After DuMontelle et al. 1981 and Howes et al. 1986.

a

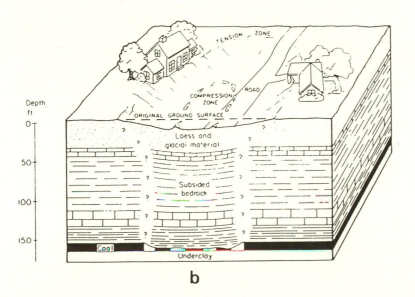

b

of 222 coal mines operated in the city of Des Moines alone, and that city is underlain by an extensive system of underground mine tunnels. The Geological Survey Bureau has documented the extent of underground coal mines in several areas of the state, including Des Moines in Polk County, Centerville in Appanoose County, and What Cheer in Keokuk County.

In the 1940s, a shift occurred in coal mining in Iowa. Surface (strip) mines began to replace underground mines. In surface mining, overburden (top soil, glacial drift, shale, and other rock material) is removed in order to extract the underlying coal. The overburden was left in the mines as spoil piles. Unfortunately, the shales in the spoil piles contained undesirable amounts of pyrite (iron sulfide, or fool's gold). Iron sulfide oxidizes when exposed to surface water and oxygen, and the resulting weathering products include sulfuric acid. Sulfuric acid contributes to acid runoff and is toxic to most wildlife and vegetation. The lack of vegetation on the spoil piles of abandoned coal mines contributes to erosion, exposing additional shales with iron sulfide to oxidation. Thus, more acid runoff is produced and the environment is further degraded. According to Van-Dorpe and Howes (1986), approximately 14,000 acres of abandoned strip-mine lands have been identified in Iowa, and some 9,500 acres of these lands were judged to have health, safety, or general environmental problems.

Since the middle and late 1970s, Iowa and federal laws have required that coal-mined lands be restored so that environmental problems are minimized. In addition, new mining regulations cover the protection of local groundwater supplies and the control of surface runoff and sedimentation. Limited funding has been provided through a portion of coal taxes to reclaim abandoned mine lands that were identified as hazardous. Reclamation efforts in Iowa have been moderately successful but so far have been limited to approximately 1,000 acres.

Pennsylvanian Nonmarine Life

Iowa's Pennsylvanian rock record reveals a varied assortment of both non-marine and marine fossils. The nonmarine record will be discussed first. The Pennsylvanian swamplands of the midcontinent supported an abundant growth of scale trees and other plant species that have long since become extinct. Today's common deciduous trees, flowering plants, and modern conifers had not yet evolved. The tangled junglelike growths in the Pennsylvanian coal swamps were dominated by the giant predecessors of the present-day club mosses and horsetail rushes. Ferns and *Cordaites* (an ancestor of the modern conifers) were also widespread. The common plant fossils of the Pennsylvanian of Iowa can be assigned to a few major taxonomic groups: scale trees (lycopods), rushes (sphenopsids), ferns and seed ferns, and cordaitean trees.

SCALE TREES (LYCOPODS)

Scale trees (lycopods) were abundant during the Pennsylvanian Period and were important contributors to Iowa's coal beds. *Lepidodendron* and *Sigillaria* were two of the best-known lycopod genera. Both have distinctive leaf scar patterns on their trunks and branches (fig. 8.29).

The ancient lycopods are distant relatives of the present-day club mosses. Although today's club mosses are diminutive in size, their ancient counterparts often grew to heights of more than 100 feet. *Lepidodendron* had a long, tapering trunk which in some cases measured more than 3 feet in diameter. The trunk ended in a spreading crown that formed by repeated dichotomous branching (fig. 8.29). The leaves of *Lepidodendron* were linear or awl-shaped and ranged in length from 1 inch to in excess of 30 inches. The leaves of *Sigillaria* were also long

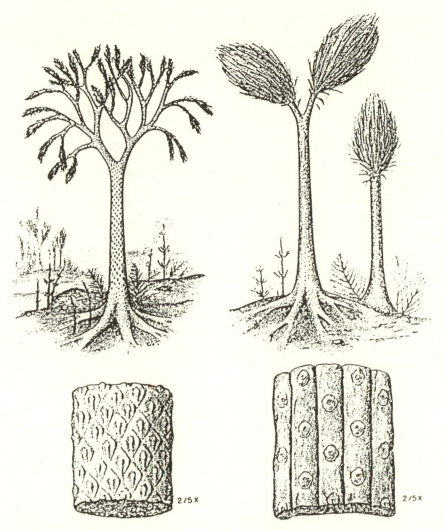

8.29 Common fossil scale trees from the Pennsylvanian of Iowa: reconstruction of the scale tree *Lepidodendron* and fossil of a trunk of *Lepidodendron* showing the distinctive pattern of leaf scars (left); reconstruction of the scale tree *Sigillaria* and fossil of a trunk of *Sigillaria* showing the characteristic seal-like leaf scars (right). From Collinson and Skartvedt 1960.

and bladelike. Therefore, it is generally difficult to tell the leaves of *Sigillaria* and *Lepidodendron* apart. The scale trees reproduced by spores, which were borne in cones at the tips of their branches. Lycopod cones are sometimes preserved as fossils, but the most common remains of scale trees are molds and casts of trunks and branches.

RUSHES (SPHENOPSIDS)

The ancient rushes (sphenopsids) of Pennsylvanian time are related to the modern horsetail rushes. The ancient rushes grew to tree size and were widely

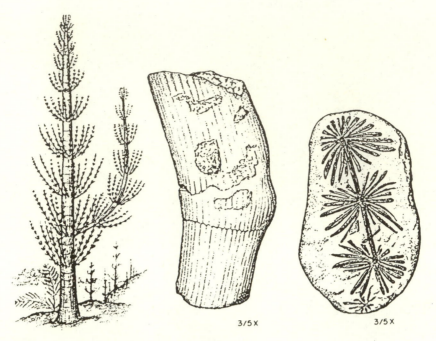

8.30 Fossil rushes: *Calamites*, showing distinctive segmentation (left and center); whorled leaves of the ancient rush *Annularia* (right). From Collinson and Skartvedt 1960.

distributed in the Pennsylvanian lowlands and coal swamps. Some of the ancient rushes reached heights of over 40 feet, but on the average they measured about 20 feet in maximum height. The trunks of the rushes were jointed and bore whorls of branches at the joints (nodes). Small leaves grew in whorls along the nodes of the smaller branches. *Calamites* (fig. 8.30) was one of the most common of the ancient rushes and is a common Pennsylvanian fossil.

FERNS AND SEED FERNS

Ferns were common in the Pennsylvanian deltaic forests and swamplands. They contributed primarily to the undergrowth of the forests, but some species were treelike and reached heights of 30 to 40 feet (fig. 8.16). Seed ferns resembled true ferns, but they reproduced by seeds rather than by spores. The leaves (fronds) of seed ferns and true ferns are similar, and it may not be possible to distinguish the two without the presence of spores or seeds. Most seed ferns contributed to the undergrowth of the Pennsylvanian coal forests, but some achieved treelike dimensions. Impressions of fern leaves (fig. 8.31) are present in Iowa's fossil record at a number of localities.

CORDAITEAN TREES

Cordaites was an early gymnosperm (cone-bearing plant). It is considered to be the forerunner of the modern conifers. Cordaitean trees were among

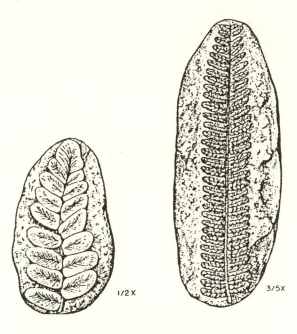

8.31 Impressions of ferns in concretions from Pennsylvanian rocks. From Collinson and Skartvedt 1960 and Collinson 1964.

1/2 X

3/5 X

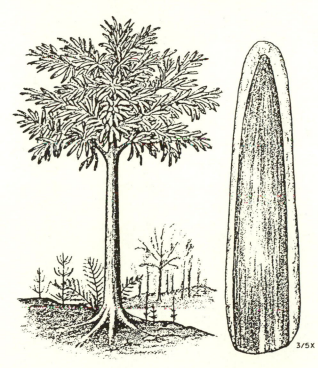

8.32 A cordaitean tree (left) and leaf (right). Cordaitean trees were forerunners of the modern conifers. From Collinson and Skartvedt 1960.

3/5 X

the tallest plants of Pennsylvanian time, and some reached heights of more than 100 feet. The typical cordaitean tree trunk was unbranched for three-fourths of the height of the tree and was topped by numerous branches with large, straplike leaves (fig. 8.32). Logs and trunks of *Cordaites* are known from southern Iowa.

Pennsylvanian Marine Life

Marine fossils are present in Iowa's Pennsylvanian rock record and are particularly abundant in Upper Pennsylvanian strata of the Missourian and Virgilian series (fig. 8.9). All of the major invertebrate groups are known, as are foraminifera, fish, algae, and conodonts. As mentioned previously, conodonts are excellent index fossils for the entire Paleozoic. Their use in dating and correlating Pennsylvanian strata is significant. Because the major groups of marine fossils have been discussed in previous chapters, discussion here will be quite brief.

FUSULINID FORAMINIFERA

Tiny fossils the size of kernels of wheat are common is some of the marine Pennsylvanian strata of south-central and southwestern Iowa. These fossils, fusulinid foraminifera, represent the limy skeletons of a one-celled organism of the complexity of the amoeba. The fusulinid foraminifera were bottom-dwellers and lived on the floor of the Pennsylvanian sea. They constructed an intricate skeleton of calcium carbonate (fig. 8.33). Fusulinids serve as good index fossils for Pennsylvanian marine strata.

BRACHIOPODS

Brachiopods are fairly common in Iowa's marine Pennsylvanian rocks. Productid brachiopods are articulate brachiopods with a strongly curved pedicle valve and a brachial valve that is either flat or curved in the same direction as the pedicle valve (fig. 8.34). Productids were bottom-dwellers. Apparently, the strongly curved pedicle valve helped anchor them on the seafloor, and their pedicle valve probably sank part way into the substratum. Some productids had spines on their pedicle valves to assist with anchorage. Although productid brachiopods are somewhat characteristic of Iowa's Pennsylvanian strata, they are also found in the state's Devonian and Mississippian rocks.

CALCAREOUS ALGAE

Algae were abundant in the seas during Pennsylvanian time. They conducted photosynthesis and deposited calcium carbonate on and within their tissues as part of their life processes. Calcareous algae are abundant in many of Iowa's Pennsylvanian limestones, and some of these algal limestones have a distinctive texture that resembles oatmeal (fig. 8.35). Algae are associated with both transgressive and regressive limestones of Pennsylvanian cyclothems (fig. 8.7) and are indicators of shallow, sun-lit environments. Calcareous algae are absent in the offshore shale facies of cyclothems; such beds formed in deeper water, where photosynthesis was absent.

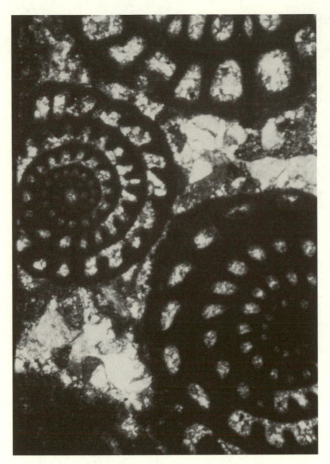

8.33 A thinly sliced section of Pennsylvanian limestone showing cross sections of fusulinid foraminifera. These wheat-shaped microfossils are excellent index fossils for Iowa's Pennsylvanian rocks. The specimens shown are from .04 to .12 inch in diameter. Photo by the author.

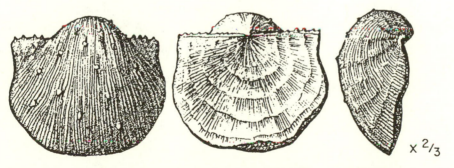

x ²/₃

8.34 *Linoproductus*, a common Pennsylvanian productid brachiopod. Three views show pedicle valve (left), brachial valve (center), and side view (right). Note the presence of broken spines on the pedicle valve. From Willman et al. 1975.

8.35 The texture of an algal limestone from Pennsylvanian strata of south-central Iowa resembles oatmeal. Photo by the author.

Summary

Alternating marine and nonmarine environments dominate in Iowa's Pennsylvanian record. These cyclic deposits were produced by numerous fluctuations in sea level, probably controlled by the waxing and waning of continental glaciers in the Southern Hemisphere supercontinent of Gondwana. During times of nonmarine conditions, Iowa was a low-lying coastal swamp and deltaic plain. Streams draining the ancient landscape left thick deposits of channel sands. These beds are now represented by prominent exposures of cliff-forming sandstone at Dolliver Memorial, Ledges, and Wildcat Den state parks, at Red Rock Reservoir, and elsewhere.

The coastal swamps and deltaic plains of Pennsylvanian time supported a lush growth of ferns, rushes, and scale trees. Cordaitean trees and seed ferns are known also. The vegetation was buried to form peat. After burial under the weight of additional peat and sediments, the peat was later changed to coal.

Iowa's coals were mined extensively during the late 1800s and early 1900s. Iowa still has huge reserves of coal. However, the relatively high sulfur content of the coal has restricted its use. Currently, there are no active coal mines in the state.

Marine conditions prevailed in Iowa during part of Pennsylvanian time, and marine limestones and shales are well represented in the state. These rocks contain a varied fossil record of marine invertebrates, calcareous algae, fish remains, and various microfossils.

Pennsylvanian limestones of the state have been used for aggregate and in the manufacture of cement. Brick and drainage tile have been produced from Pennsylvanian shales in southern Iowa, and locally near the Amana Colonies in eastern Iowa, Pennsylvanian sandstones were utilized for building purposes.

Acknowledgments
Philip H. Heckel's work provided key information on depositional environments. The publications of Mary R. Howes and Paul E. VanDorpe furnished data on Iowa's coal resources and Pennsylvanian stratigraphy. In addition, published works by the following were particularly useful: Charles W. Collinson, Romayne Skartvedt, Lon Drake, Ron Neimann, John F. Baesemann, Matthew H. Culp, Helen Greenburg, E. R. Landis, Orville J. Van Eck, John D. Lemish, Daniel R. Burggraf Jr., Howard J. White, Greg A. Ludvigson, Keene Sweet, Hubert L. Olin, Robert L. Ravn, John W. Swade, J. L. Gregory, Raymond R. Anderson, Dorothy Schweider, Richard Kraemer, Marsha J. Miller, and Susan J. Lenker. Mary Howes reviewed this chapter and provided helpful suggestions. I appreciate her constructive review.

Throughout the whole region, the Fort Dodge gypsum has the exact
appearance of a sedimentary deposit. It is arranged in layers like the regular
layers of limestone, and the whole mass from top to bottom is traced with fine
horizontal laminae of alternating white and gray gypsum, parallel with the
bedding surfaces of the layers, but the whole so intimately blended as to form
a solid mass. The darker lines contain almost all the impurity there is in the
gypsum, and that impurity is evidently sedimentary in character. From these
facts, and also from the further one that no trace of fossil remains has been
detected in the gypsum, it seems not unreasonable to entertain the opinion
that the gypsum of Fort Dodge originated as a chemical precipitation in
comparatively still waters which were saturated with sulphate of lime
and destitute of life.

    C. A. White (1870) describing Fort Dodge gypsum beds

The Mesozoic Era (245 to 65 million years ago) consists of the Triassic, Jurassic,
and Cretaceous periods. Mesozoic means middle life, and at the time it was
named it was thought to be the middle part of earth history. We now know the
great age of the Earth (4.6 billion years) and that the Mesozoic is but a part of the
late history of the Earth.

The rock record of the Mesozoic is not well developed in Iowa. Triassic rocks
are absent, and Jurassic strata (fig. 9.1) are found only in a small area near Fort
Dodge in Webster County. Cretaceous strata constitute the bedrock in northwestern and parts of west-central and southwestern Iowa and occur as erosional
outliers in central and northeastern Iowa. The Cretaceous System, however, is
poorly exposed and tends to be covered by glacial till and loess throughout most
of its outcrop belt in Iowa.

Iowa was apparently above sea level during the latter part of Paleozoic time
and during the early part of Mesozoic time. Deposits of Late Paleozoic (Permian)
age and Early Mesozoic (Triassic) age are not represented in the state. During
Permian and Triassic time, Iowa was a low-lying plain with an arid to semiarid
climate. Paleoclimatic and paleomagnetic data suggest that the equator passed
through North America during Permian and Triassic time. The climate in Iowa
then was probably similar to that of parts of northern Africa today.

Iowa apparently remained emergent through most of Jurassic time. In Late
Jurassic time, about 155 million years ago, a sea covered a large portion of the

9.1 Prominent laminations in rock gypsum from the Fort Dodge Formation suggest a sedimentary origin. The gypsum deposits of the Ft. Dodge, Iowa, area are some of the purest in the United States and have been mined for commercial purposes for well over a century. Photo by the author.

present-day Great Plains and Rocky Mountain regions. Embayments and lagoons of this sea probably extended into what is now central Iowa. Subsequent erosion has removed these deposits, except for a small area in Webster County near Fort Dodge. Iowa's Jurassic record was probably deposited contemporaneously with the well-known dinosaur-bearing strata of the Morrison Formation and the marine units of the Sundance Formation of the western United States and Canada (fig. 9.2). The gypsum deposits and red beds at Fort Dodge are erosional remnants of ancient Jurassic sedimentary environments that were once much more extensive. Other examples include the Hallock Red Beds of Minnesota and North Dakota and the Gypsumville Formation of Manitoba (fig. 9.2).

Iowa's Cretaceous record is also incomplete. The gap in Iowa's record between the Jurassic Fort Dodge Formation and the Cretaceous Dakota Formation represents approximately 50 million years. Cretaceous deposits are now restricted primarily to western Iowa. Erosional outliers of Cretaceous rocks, however, occur in central and northeastern Iowa and as far to the east as western Illinois and western Wisconsin. The occurrence of outliers such as these indicates that Cretaceous deposits probably covered all of Iowa at one time. Post-Cretaceous erosion has since removed nearly all of the Cretaceous rocks from central and

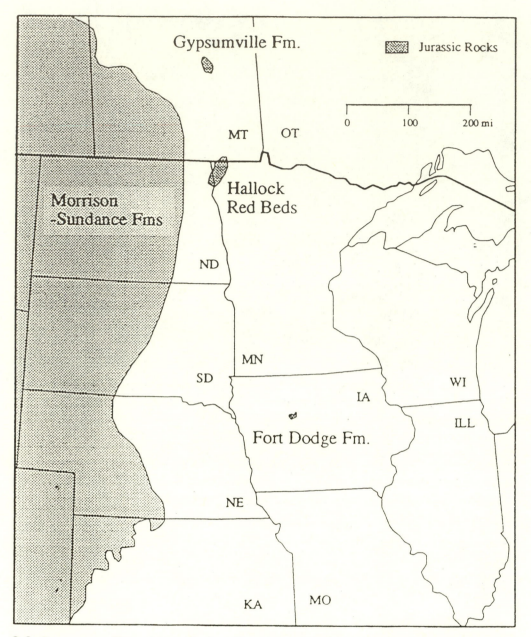

9.2 Occurrences of Jurassic strata in the midcontinent area of North America. Note the presence of erosional outliers of Jurassic rocks in north-central Iowa, northwestern Minnesota, and east-central Manitoba. In addition, gypsiferous shales of probable Jurassic age occur in the Michigan Basin. From Cody et al. 1996.

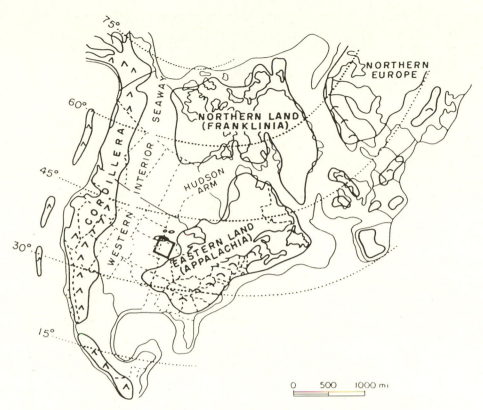

9.3 The generalized paleogeography of North America and northern Europe during Cretaceous time about 90 million years ago, when ancient Iowa was located near the edge of the vast Western Interior Seaway. A rugged mountainous terrain (Cordillera) occupied western North America. Less rugged land areas (Appalachia and Franklinia) were located in the eastern and northern parts of the continent. The box encloses the area of the type Dakota Formation in eastern Nebraska and western Iowa and the area of Dakota outcrops in central and western Iowa. From Witzke and Ludvigson 1994.

eastern Iowa. Figure 9.3 depicts the paleogeography of Late Cretaceous time approximately 90 million years ago, when Iowa was located near the eastern edge of the vast Western Interior Seaway.

Mesozoic Rock Record of Iowa

The Jurassic rock record of Iowa consists, in ascending order, of unnamed basal units, the Fort Dodge Gypsum, and clastic strata referred to informally as the Soldier Creek beds. The Cretaceous column of the state includes, in ascending order, the Dakota Formation, Graneros Shale, Greenhorn Formation, Carlile Shale, Niobrara Formation, and Pierre Shale. However, the Niobrara and Pierre are very limited in their distributions. They are known primarily from the subsurface in the Manson Impact Structure area of north-central Iowa. Lateral equivalents of the Dakota, known as the Windrow Formation, occur in northeastern Iowa, southeastern Minnesota, and southwestern Wisconsin.

9.4 Iowa's Mesozoic formations and their depositional environments. From Anderson and Witzke 1994; Cody et al. 1996; and Witzke and Ludvigson 1987, 1994, and 1996a, b.

Geologic Age	Formation or Unit	Composition	Probable Environment of Deposition
Late Cretaceous	Pierre Shale	Gray shales, noncalcareous	Muddy marine shelf with low levels of oxygen in bottom waters
Late Cretaceous	Niobrara Formation (unconformity)	Silty shales, slightly calcareous; some mudstones and siltstones; fish debris and fragments of inoceramid bivalves	Muddy and limy marine shelf
Late Cretaceous	Carlile Shale	Gray shales with ammonoids, fish debris, and inoceramid bivalves	Muddy marine shelf with low levels of oxygen in bottom waters
Late Cretaceous	Greenhorn Formation	Gray calcareous shale, argillaceous (muddy) limestone, and chalk; fish debris and inoceramid bivalves (clams), calcareous remains of floating algae (nannofossils)	Muddy and limy marine shelf with low levels of oxygen in bottom waters
Late Cretaceous	Graneros Shale	Light gray calcareous shale and mudstone; siltstone at base; fossil mollusks and fish debris; pelagic algae and foraminifera	Coastal setting, followed by a muddy marine shelf with low levels of oxygen in bottom waters
Early–Late Cretaceous	Dakota Formation (Windrow Formation as erosional outliers in northeastern Iowa, southeastern Minnesota, and southwestern Wisconsin) (unconformity)	Predominantly sandstones and mudstones but also includes shales, siltstones, and conglomerates; siderite (iron carbonate) pellets and concretions are common; plant debris and low-grade coals (lignites) mark intervals of nonmarine deposition; limonitic iron ores in northeastern Iowa and southeastern Minnesota	The upper Dakota preserves a variety of depositional environments indicating estuarine, coastal, and nearshore marine settings; the lower Dakota reflects fluvial deposition with westward and southwestward transport directions; the iron-rich deposits in the Windrow Formation probably represent bog iron ores formed in wetland soils on flood plains

Geologic Age	Formation or Unit	Composition	Probable Environment of Deposition
Late Jurassic	Fort Dodge Formation (Soldier Creek beds)	Red, buff, and green shales and red sandstones and siltstones; some shales bear nodules of celestite (strontium sulfate); one locality reveals reworked fossils of Pennsylvanian age	Probably fluvial; possibly deltaic
Late Jurassic	Fort Dodge Formation (Fort Dodge Gypsum)	Crystalline gypsum with a maximum thickness of about 33 feet for the formation; very pure with only minor impurities; contains 89% to 96% pure $CaSO_4 \cdot H_2O$; slight impurities impart a distinctive banding to the formation	Precipitation of gypsum from highly saline waters of an evaporite basin in an arid climate where evaporation exceeded precipitation; probably represents the erosional remnant of a restricted marine evaporite basin or a large evaporite lake, marginal to the Jurassic seaway
Late Jurassic	Fort Dodge Formation (unnamed basal units) (unconformity)	Conglomerate with clasts of limestone and dolomite, some with Pennsylvanian-age fossils; thin claystone layers at top and bottom; often overlain by dark gray shale, claystone, or clay	Probably fluvial; known only from a few localities near Fort Dodge; some of the upper clay may be residual, derived from dissolution of the overlying gypsum
Triassic	No rock record		

Figure 9.4 summarizes the environments of deposition of the Mesozoic formations of Iowa. A discussion of each of the major stratigraphic units follows.

TRIASSIC

There are no Triassic rocks in Iowa's rock record. The Permian System of the Paleozoic is also absent in Iowa. In states to the west of Iowa, the Permian and Triassic systems are represented by red beds and gypsum deposits. The gypsum beds and associated rocks of the Fort Dodge Formation were once considered to be of Permian age because of their similarity in composition and appearance to Permian beds in Kansas and Nebraska. However, the Fort Dodge Formation is now dated as Jurassic in age, based on the remains of fossil plant materials.

The Jurassic System is represented in Iowa by exposures of the Fort Dodge Formation in Webster County. The distribution of the gypsum-bearing Fort Dodge Formation is shown in figure 9.5. Originally, the formation was probably continuous over a much larger area in the state, but the gypsum beds are now known to occur only in the northern part of Webster County. The Fort Dodge Formation (fig. 9.4) is composed of unnamed basal units consisting of conglomerate, clay, claystone, and shale; a middle unit of crystalline rock gypsum; and an upper sequence of red, buff, and green shales and red sandstones and siltstones. Only the gypsum unit will be discussed in detail.

Gypsum exposures in the Fort Dodge area have been known for a long time (fig. 9.6). The occurrence of gypsum in the vicinity of Fort Dodge was noted in 1852 by geologist David Dale Owen in his geological survey of Wisconsin, Iowa, Minnesota, and part of Nebraska. Iowa's first gypsum mill was erected at Fort Dodge in 1872. Since then, the gypsum industry in the Fort Dodge area has grown to its present production of more than 1.65 million tons a year. Moreover, the Fort Dodge area consistently ranks as one of the nation's leading producers of commercial gypsum, and the state of Iowa, thanks to the production at Fort Dodge, generally ranks from second to fourth among the states in the annual production of gypsum. The occurrence of gypsum in the Fort Dodge area is limited, however (fig. 9.5). Recent estimates indicate that the gypsum resources will be depleted in the next thirty years.

United States Gypsum Corporation, National Gypsum Company, Georgia-Pacific Corporation, and Celotex Corporation all conduct operations in the Fort Dodge area. The gypsum is used as a retarder in the manufacture of portland cement and in the manufacture of plaster, plaster board, and related building products. Gypsum also finds use as soil conditioner and as an ingredient in classroom chalk.

Gypsum from the Fort Dodge Formation is noted for its remarkable purity, making the rock a valuable commodity for industrial use. Rock gypsum in the Fort Dodge area contains between 89 and 96 percent $CaSO_4 \cdot H_2O$ and very small amounts of impurities. Mineral impurities impart the distinctive banding found in the Fort Dodge gypsum. The main impurities are quartz, iron oxide, clay, calcite, and dolomite.

As noted by geologist Charles A. White in the epigraph at the beginning of this chapter, the alternating bands of white and dark laminations are striking features in the Fort Dodge gypsum. The laminations typically are irregular, and many are contorted. The dark banding is predominantly blue-gray in color, but yellow, red, or brown colors occur also. The dark horizons are apparently associated with mineral impurities, but only discontinuous stringers of accessory minerals are visible in thin sections of the gypsum. Larger gypsum crystals occur in the

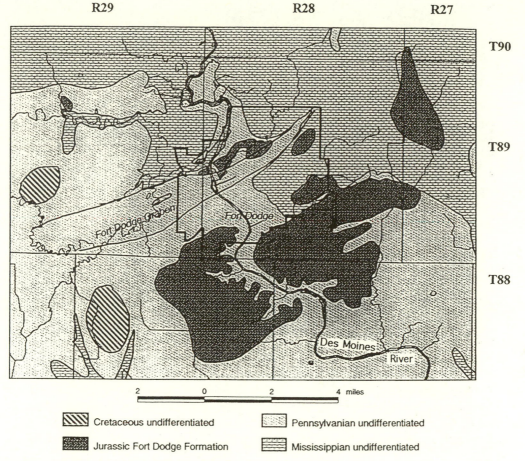

9.5 The distribution of the gypsum-bearing Fort Dodge Formation in the vicinity of Fort Dodge, Iowa. Boundaries of the city are outlined. From Cody et al. 1996.

dark laminations, compared to smaller gypsum crystals that comprise the adjacent nodular white bands. Since larger crystals scatter less light and allow greater light penetration than do smaller crystals, the bands composed of larger crystals appear darker. The white bands consist of three types: nearly horizontal laminations composed of vertically oriented, elongate gypsum crystals; contorted lamina that may have resulted from the coalescence of small nodules; and intercalations of marl (a mixture of clay and calcium carbonate) within the gypsum.

In addition to containing prominent laminations, the Fort Dodge gypsum reveals nodules, ripples, small folds, joint-controlled channels, and veins of fibrous gypsum. Gypsum breccias occur at some locations. Four varieties of gypsum are recognized based on crystal size and crystal shape. Alabastrine gypsum is the most common variety and consists of small crystals less than 0.002 inch in diameter. Alabastrine gypsum occurs in nodules and in irregular bands of coalescing nodules. Relict crystal textures are common in the alabastrine gypsum and suggest that it had a previous history as anhydrite ($CaSO_4$). Gypsum con-

9.6 A drawing by geologist-artist Orestes St. John showing the Cummins Quarry as it appeared in 1868. The area is now part of Snell Park in Fort Dodge, Iowa. The Cummins Quarry was the first recorded gypsum quarry in the Fort Dodge area and was utilized to extract dimension stone. Note the irregular surface at the top of the gypsum. Gypsum is very soluble, and the upper surface of the gypsum is often irregular and marked by solution features such as shown here. From White 1870.

verts to anhydrite under burial conditions as temperature increases. Conversely, under the right conditions, anhydrite will combine with water and hydrate to form gypsum.

Recrystallization and other types of postdepositional changes have altered the original gypsum deposits significantly. This makes it difficult to interpret the depositional environment of the Fort Dodge gypsum in detail. However, the general pattern is agreed upon. The gypsum deposits apparently formed by precipitation from highly saline waters in a closed basin or similar restricted setting. The climate was arid, and the setting was in an area where evaporation exceeded precipitation. The high purity of the Fort Dodge gypsum and the absence of lateral changes in composition within the unit suggest that the gypsum was once more widespread than its current distribution. Was the gypsum deposited in a marine setting or in a large lake? We cannot be entirely certain, but the overall purity of the gypsum suggests a marine origin. The distinctive banding in the gypsum may result from the periodic deposition of wind-blown silt.

The Fort Dodge gypsum contains no visible fossils. The only reported fossils are those noted by Aureal Cross (1966), who listed the presence of palynomorphs (microscopic plant remains such as spores and pollen). Based on these fossil remains and other considerations, the Fort Dodge gypsum is placed in the Upper Jurassic. It may have been deposited approximately 155 million years ago during a time of marine transgression that produced the Sundance Formation of Wyoming and South Dakota.

At some localities, the gypsum beds of the Fort Dodge Formation are overlain by red, buff, and green shales and red sandstones and siltstones. These clastic rocks, known informally as the Soldier Creek beds, reach a maximum thickness of approximately 50 feet. The origin of red coloration in sedimentary rocks, such as the red clastics of the Fort Dodge Formation, is somewhat controversial. Red coloration in sedimentary rocks has often been considered to indicate deposition under unique paleoclimatic conditions, but exactly what those climatic conditions were is still being debated.

Red coloration in sedimentary rocks is an indication of thorough oxidation of iron in the sediments, and only a small amount of iron (1 to 6 percent) is needed to produce a bright red coloration. The chemical environment needed for the formation and preservation of red beds must be one that is strongly oxidizing. Nonmarine settings such as broad alluvial plains and low-lying coastal plains are often sufficiently oxidizing so that red beds could form there. In contrast, most marine environments are oxygen poor (reducing) below the sediment-water interface, so the red pigment of red beds would generally be altered to a drab gray coloration in such environments.

The red coloration in red beds is formed chiefly by the combination of iron (ferric iron) with oxygen to produce the iron-oxide mineral hematite (Fe_2O_3). Hematite forms as a red pigment in the soil horizons of humid tropical regions. If this pigment is later transported and deposited in an oxidizing environment, red beds can be produced.

Red beds also form by the alteration of iron-bearing sediments after deposition. Oxygen-bearing groundwaters that percolate through the sediments are capable of oxidizing the iron in the sediments to produce a hematite coating around the grains. Such a coating is sufficient to give a red coloration to sediments and rocks. Most red beds probably form by this method. The red coloration in some of Iowa's Jurassic rocks may have been produced by ancient soil-forming processes.

The red shales, siltstones, and sandstones of the Fort Dodge Formation are not bright red in color as are some of the Mesozoic red beds in the western states. They do contain oxidized iron, however, which suggests that they probably formed by deposition or alteration in a nonmarine setting. The environment of deposition of the red clastics of the upper Fort Dodge Formation is interpreted as fluvial or deltaic (fig. 9.4).

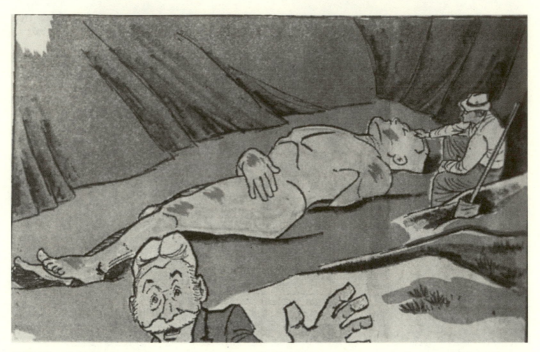

9.7 The Cardiff Giant was carved from a large block of Fort Dodge gypsum in 1868 and buried near the village of Cardiff in upstate New York. Proclaimed by some to be a petrified man and by others to represent a statue from an ancient civilization, the giant attracted thousands of viewers. Today, the gypsum giant rests peacefully in a museum in Cooperstown, New York — a reminder of one of the greatest hoaxes of all time. From a cartoon by Jack Bender, courtesy of the *Waterloo Courier*.

Fort Dodge Gypsum and the Cardiff Giant

A novel use was found for a block of gypsum from the Fort Dodge area. The likeness of a giant man — 10 feet and 4.5 inches long, 3 feet wide, and weighing 2,990 pounds — was sculpted from the gypsum block and later buried near Cardiff, New York, as a hoax. The giant, termed the Cardiff Giant, was discovered and identified as either a petrified prehistoric man or an ancient stone statue. The original Cardiff Giant (fig. 9.7) is now owned and displayed by the New York State Historical Association in Cooperstown, New York. A replica is on display in the reconstructed historical fort in Fort Dodge, Iowa.

The idea for the hoax originated in 1866 when George Hull was visiting his sister in Ackley, Iowa. While in Ackley, Hull heard the Reverend Turk preach and disagreed with the minister over the meaning of the biblical passage: "There were giants in the earth in those days." Hull was irked by Reverend Turk's literal interpretation of the scripture, so he decided to construct a giant as a practical joke to demonstrate how gullible people are. To that end, Hull had a large block of gypsum quarried in the Fort Dodge area. The block was shipped to Chicago, where Hull had a stonecutter carve it into the likeness of a sleeping giant. Special hammers and needles were used to impart porelike indentations on the statue to simulate pores in skin. The statue was then rubbed with sand, water, ink, and sulfuric acid to make the giant look ancient. Subsequently, the giant was shipped to New York and secretly buried one dark night in 1868, near Cardiff, on the farm of George Hull's brother-in-law, William C. "Stub" Newell. One of the greatest hoaxes of all time was on.

On a Saturday morning in October 1869, two men were digging a well on Stub Newell's farm near Cardiff. Newell had told them exactly where he wanted the well dug. At a depth of 3 feet, the workers' shovels struck a rocky object. Soon the form of a huge human foot was revealed, and in short order a gigantic human figure came into view. The right arm and hand lay across the body, while the left arm was pressed against the back, directly opposite. The lower limbs were slightly contracted as if by pain; the left foot rested partially on the right.

Stub Newell soon erected a tent over the trench that contained the mysterious stone giant and charged an admission fee of 50 cents to the many visitors who came to see it. Fifty thousand sight-seers are reported to have visited the Newell farm to see the Cardiff Giant.

According to Ruth Gallaher (1921), the following description was given by a visitor to the Newell Farm: "The roads were crowded with buggies, carriages, and even omnibuses from the city, and with lumber-wagons from the farms — all laden with passengers — a gathering which at first sight seemed like a county fair. In the midst was a tent, and a crowd was pressing for admission. Entering, we saw a large pit or grave, and, at the bottom of it, perhaps five feet below the surface, an enormous figure — it was a stone giant, with massive features, the whole body nude, the limbs contracted in agony. It had a color as if it had lain long in the earth, and over its surface were minute punctures like pores. An especial appearance of great age was given it by deep grooves and channels in its underside, apparently worn by the water which flowed in streams through the earth and along the rock on which the figure rested. Lying in its grave, with the subdued light from the roof of the tent falling upon it, and with limbs contorted as if in a death struggle, it produced a most weird effect. An air of great solemnity pervaded the place. Visitors hardly spoke above a whisper."

Among those who examined the giant was James Hall, State Geologist of New York, who some years earlier had directed one of the first geological surveys of Iowa. Hall was quite certain that the figure was a statue carved from crystalline gypsum rock, but he believed that the statue was of great antiquity and that it had been buried in the earth for many generations. Hall proclaimed the Cardiff Giant to be "the most remarkable object yet brought to light in this country."

Other explanations were offered. According to one interpretation, the giant was the petrified body of an ancient Native American prophet. Supposedly, age-old tales of the Onondaga Indians told of such giants. Dr. John F. Boynton, a local lecturer, believed the statue was "of Caucasian origin and designed by the artist to perpetuate the memory of a great mind and noble deeds." A prominent clergyman pronounced: "This is not a thing contrived of man, but is the face of one who lived on the earth, the very image and child of God." Alexander McWhorter, a graduate student at Yale, studied the Cardiff Giant in great detail and was convinced that he recognized an inscription in an ancient language. McWhorter concluded that the gypsum statue depicted the Phoenician's god, Baal.

After initially being displayed on William Newell's farm, the Cardiff Giant was exhibited in Albany, Syracuse, New York City, and Boston. While the giant was in Boston, Oliver Wendell Holmes examined it and bored a hole in back of the left ear, seeking to substantiate the statue theory of origin. Ralph Waldo Emerson visited the Cardiff Giant in Boston, too. He pronounced it beyond his depth, "very wonderful, and undoubtedly ancient."

Eventually, the Cardiff Giant was exposed as a hoax, but not before George Hull's initial investment of a few thousand dollars netted him a handsome profit. Hull also proved his point, rather convincingly, that people are gullible.

Prior to finding a permanent home in Cooperstown, New York, the Cardiff Giant resided for a time during the 1930s in the Des Moines home of publisher Gardner "Mike" Cowles. Cowles bought the giant from a bankrupt circus in Texas for $4,500, plus $1,500 for shipping. He displayed the gypsum carving in the family den, complete with theatrical lighting and posters. To Cowles's dismay, the giant was involved in an embarrassing incident at the hands of Cowles's son and two of the son's playmates. The mischievous lads decided that it would be great fun to smash the giant's penis with a hammer, and they succeeded in breaking off the tip. Eventually, Cowles found an artisan who performed the delicate procedure of cementing the tip back in its original position.

Cretaceous

A variety of depositional environments are recorded in Iowa's Cretaceous rock record (fig. 9.4). Sandstone, mudstone, shale, siltstone, conglomerate, and lignite occur in the Dakota Formation of western Iowa. These rocks represent sediments that were laid down in and adjacent to the advancing Cretaceous sea in nonmarine and nearshore marine settings (fig. 9.3). Fluvial conditions dominated in western Iowa during deposition of the lower and middle Dakota Formation (fig. 9.8). Later, coastal and nearshore marine conditions prevailed when the upper Dakota strata were laid down (fig. 9.9). At one time, many Dakota sediments deposited at the marine-nonmarine interface were interpreted as fluvial-deltaic facies (fig. 9.9). It now appears that these so-called fluvial-deltaic facies represent estuarine sediments, deposited in drowned river valleys. Witzke

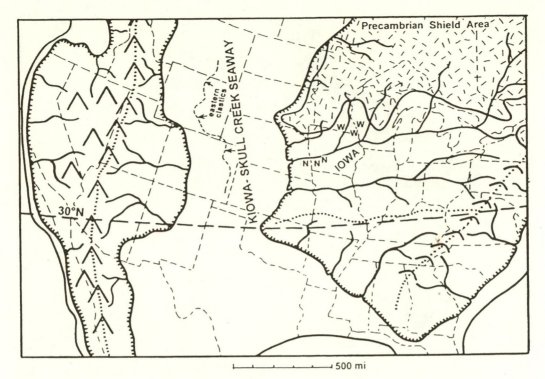

9.8 A paleogeographic reconstruction for Cretaceous time when the Dakota Formation was deposited. Hypothetical river drainages are shown. The maximum westward distribution of eastern-derived sandstones is shown by a dashed line. The Kiowa–Skull Creek Seaway accumulated marine deposits, whereas the lower Dakota Formation of Iowa represents fluvial deposits. Estuaries extended to the Iowa-Nebraska line. The general outcrop area of the lower Dakota (Nishnabotna Member) is shown by N. W locates erosional outliers of the Dakota Formation that are termed the Windrow Formation. From Witzke and Ludvigson 1996b.

and Ludvigson (1996b) have recognized tidally influenced sedimentation within estuarine sandstones of the Dakota Formation at Homer, Nebraska, within a few miles of the Iowa border.

Erosional outliers of Dakota facies in northeastern Iowa, southeastern Minnesota, and southwestern Wisconsin have been termed the Windrow Formation (fig. 9.4). The rest of the Cretaceous record in Iowa (Graneros, Greenhorn, Carlile, Niobrara, and Pierre) is of marine origin. A description of each of the Cretaceous formations of Iowa follows.

DAKOTA FORMATION

The Dakota Formation was named by Fielding B. Meek and Ferdinand V. Hayden in 1862 for exposures along the Missouri River valley in Dakota County in northeastern Nebraska. These exposures, along with Cretaceous exposures across the Missouri River valley in the vicinity of Sioux City, Iowa, were among

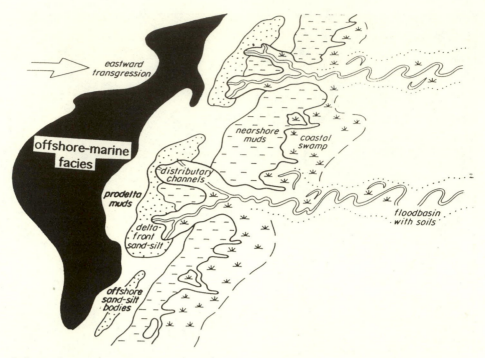

9.9 A schematic representation of upper Dakota deposition, which took place in coastal and nearshore marine settings. The model for upper Dakota deposition is presently being reformulated, and the deltaic facies shown here may actually represent estuarine deposition. From Witzke and Ludvigson 1994.

the first Cretaceous rocks to be studied in North America. Captain William Clark of the historic Lewis and Clark Expedition described rocks of the Cretaceous Dakota Formation when he recorded "yellow, soft sandstone" in his 1804 journal, referring to the bluffs of the Missouri River near the Omaha Indians' Blackbird burial grounds.

The term "Dakota" has been used as a stratigraphic name over a wide geographic area of the north-central and western United States. In the 1870s, Hayden expanded usage of the Dakota terminology southward into Kansas and westward outside the Missouri Valley to include sandstones exposed along the flanks of a number of mountain ranges from Wyoming to New Mexico. Subsequently, the stratigraphic names Dakota Sandstone, Dakota Formation, or Dakota Group were used by other geologists at various times to describe lower to mid-Cretaceous sandstones and mudstones in the following states and provinces: Nebraska, Iowa, Minnesota, North Dakota, South Dakota, Kansas, Oklahoma, Texas, New Mexico, Arizona, Colorado, Wyoming, Montana, Alberta, Saskatchewan, and Manitoba. Although alternate terminology has been used as of late in Canada, Texas, the northern Great Plains, and parts of the western United States, usage of Dakota still persists in Colorado, New Mexico, Utah, and

Arizona. For example, a prominent ridge of Cretaceous sandstone in the foothills of the Front Range of Colorado is termed the Dakota Hogback because it is underlain by a formation that Colorado geologists refer to as the Dakota Sandstone.

According to Witzke and Ludvigson (1994), the current usage of Dakota Group or Dakota Formation as a rock-stratigraphic unit in Colorado and elsewhere in the western United States is confusing and ill advised. There is no evidence that the Cretaceous sandstones and mudstones termed "Dakota" in the western states are continuous to the east and merge with rocks of the same composition in the type area of the Dakota in eastern Nebraska and western Iowa. Furthermore, the clastic materials (pebbles, granules, sand, silt, and clay) that comprise the Dakota of the western states were derived from western source lands and transported eastward to their sites of deposition. In contrast, the clasts in the rocks of the type Dakota area were carried by westerly-to-southwesterly flowing streams that drained eastern source lands.

Strata of the lower half of the type Dakota sequence are assigned to the Nishnabotna Member. The Nishnabotna Member unconformably overlies Paleozoic or Precambrian rocks and includes coarse-grained sandstones, pebbly sandstones, and conglomerates. Minor mudstones occur, some of which preserve evidence of ancient soil formation. The Nishnabotna is restricted to the subsurface in the type Dakota area, but exposures occur in western Iowa and eastern Nebraska. The member was deposited by fluvial systems with a westerly-to-southwesterly transport direction. The sandstones of the Nishnabotna Member are porous and permeable, making the lower Dakota a significant aquifer in western Iowa and eastern Nebraska.

The upper half of the type Dakota sequence is primarily shale and mudstone; sandstone, siltstone, and lignite occur also. These strata are placed in the Woodbury Member and are well exposed in the type area, where they are overlain by marine beds of the Graneros Shale. In the type area, the Woodbury Member includes a basal mudstone interval with well-developed soil features, abundant small nodules of iron carbonate, and local sandstone channels. A middle interval is represented by sandstone, shale, and lignite beds — the deposits of an ancient deltaic distributary channel. An upper unit of shale and siltstone, with minor sandstone and lignite, displays ripple laminations and hummocky stratification. Locally, a molluscan fauna occurs in the upper unit and signifies a marine to brackish-water environment of deposition.

The siderite (iron carbonate) nodules in the Woodbury Member record a geochemical record from which ancient climates can be interpreted. Data from the Dakota Formation of Iowa, Nebraska, and Kansas document warm, uniform Cretaceous climates.

Overall, the sandstones of the type Dakota are high in quartz, with minor feldspar, mica, and metamorphic rock fragments. Conglomerates in the Dakota Formation contain rounded pebbles and granules of quartz and chert. Most of the Dakota sediment probably originated from deeply weathered eastern source lands that were composed of Precambrian igneous and metamorphic rocks. Clasts of chert in the Dakota conglomerates suggest a secondary source of sediment from Paleozoic carbonate rocks that contained chert beds or chert nodules.

The type Dakota sequence is capped by an interval of calcareous marine shale that has been termed the Graneros Shale. However, the type Graneros Formation of Colorado is noncalcareous silty shale and siltstone. Therefore, it appears that the so-called Graneros Shale of western Iowa and eastern Nebraska is inappropriately labeled and that alternative nomenclature is needed. The term "Graneros" will be used in this text because alternative terminology has not yet been proposed.

WINDROW FORMATION

In northeastern Iowa, iron-rich residual deposits and alluvial sands and gravels accumulated during Cretaceous time. These deposits were originally assigned to the Windrow Formation, although they are now known to correlate with the Dakota Formation of western Iowa. The Windrow Formation was named for the locality of Windrow Bluff in Monroe County in southwestern Wisconsin. The formation is discontinuous and is known only from isolated exposures in western Wisconsin, southeastern Minnesota, and northeastern Iowa (fig. 9.10). The Windrow Formation was divided into an Iron Hill Member (lower) and an East Bluff Member (upper). In Minnesota, the East Bluff interval is termed the Ostrander Member. A generalized composite section of the Windrow Formation is shown in figure 9.11.

The Iron Hill Member is named for exposures at Iron Hill, about 1 mile north of Waukon in Allamakee County, Iowa. The member is characterized by masses of limonite, an amorphous mixture of various hydrates of iron. Limonite at this locality occurs in various shades of brown, yellow, and reddish brown. The limonite deposits at Iron Hill are generally porous and have a cellular texture. Chert fragments are common in the limonite beds, and rocks and fossils from the underlying carbonate beds of the Ordovician Galena Group occur in the basal Windrow beds at the Iron Hill locality near Waukon.

The Waukon Iron Company was formed in 1900 to develop the iron deposits near Waukon. A concentration plant, consisting of a crusher and washer, was built. The company eventually folded because of the high cost of hauling ore from the Iron Hill locality to the railroad at Waukon. In addition, the physical nature of the ore made the washing treatment for concentration ineffective. In

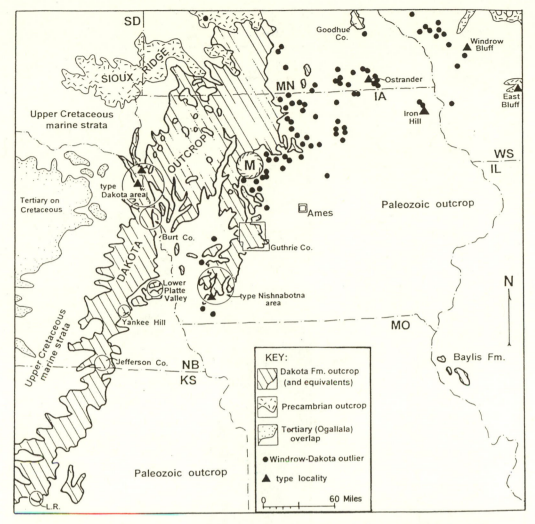

9.10 The regional outcrop area of the Dakota Formation and equivalent strata (Windrow Formation). Cross-ruled pattern denotes the Dakota Formation in western Iowa, eastern Nebraska, southern Minnesota, southeastern South Dakota, and eastern Kansas. Outliers of equivalent strata (Dakota and Windrow) shown by solid dots. Note the location of the type Dakota and the Nishnabotna Member of the Dakota. Guthrie County is the location of Iowa's sole dinosaur discovery, and that is tentative, based on one piece of bone. The type localities of the Windrow Formation and its members are also shown (Windrow Bluff, East Bluff, Ostrander, and Iron Hill). Also shown is the Late Cretaceous Manson Impact Crater. From Witzke and Ludvigson 1996b.

1909, the financial interests of the Waukon Iron Company were acquired by the Missouri Iron Company of St. Louis, which constructed a larger and more efficient treatment plant. Over 27,000 tons of ore were mined, processed, and shipped from the Waukon area from approximately 1913 to 1918. The Waukon deposit was reinvestigated in 1943 during World War II, but no further mining and processing took place. However, the iron-bearing beds of the Windrow For-

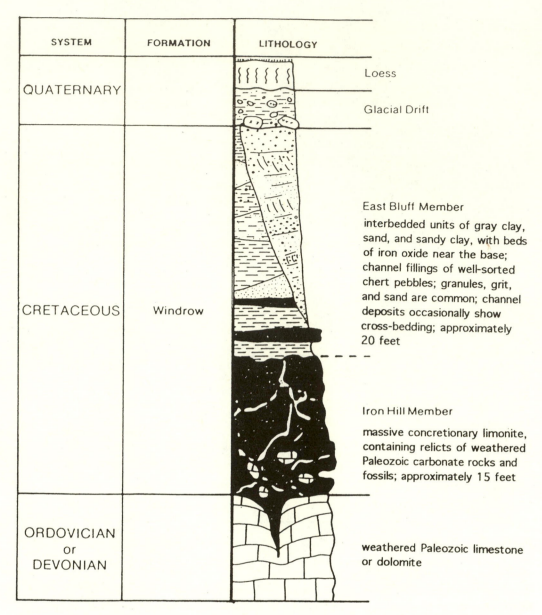

SYSTEM	FORMATION	LITHOLOGY	
QUATERNARY			Loess
			Glacial Drift
CRETACEOUS	Windrow		East Bluff Member
interbedded units of gray clay, sand, and sandy clay, with beds of iron oxide near the base; channel fillings of well-sorted chert pebbles; granules, grit, and sand are common; channel deposits occasionally show cross-bedding; approximately 20 feet
			Iron Hill Member
massive concretionary limonite, containing relicts of weathered Paleozoic carbonate rocks and fossils; approximately 15 feet
ORDOVICIAN or DEVONIAN			weathered Paleozoic limestone or dolomite

9.11 A generalized stratigraphic section of the Windrow Formation. Adapted from Andrews 1958 and Anderson 1983.

mation in southeastern Minnesota were mined commercially from 1942 to 1969, with a cumulative production of approximately 8 million tons of iron ore.

Although the origin of the iron-bearing interval of the Iron Hill Member has been debated, it appears to have formed in part by the replacement of underlying Paleozoic carbonates. In northeastern Iowa, the limonite deposits lie on weathered Ordovician bedrock. In southeastern Minnesota, the Iron Hill Member rests on either Devonian or Ordovician carbonate bedrock. Some geologists suggest that the iron deposits formed by a process whereby bog iron ores formed on the landscape during Cretaceous time and later soaked into the underlying carbonate bedrock, filling in porous areas and solutional cavities and/or replacing the upper weathered zone of the carbonate bedrock. Others propose that the iron-bearing deposits formed by a process involving iron-rich soils that formed on the Paleozoic bedrock during Cretaceous time and accumulated as a residuum. Later, iron was leached from the residual soils and deposited in the underlying bedrock, which was deeply weathered, highly fractured, and marked by solutional features.

The upper member of the Windrow Formation (the East Bluff or Ostrander) has been recognized in extreme northeastern Iowa, southeastern Minnesota, and southwestern Wisconsin. The member varies in composition, although it is usually characterized by conglomerate with peanut-size clasts and coarse sandstone. The pebbles in the conglomerates are almost entirely quartz or chert. The distribution of the member and its textural and compositional features are all consistent with a fluvial origin.

GRANEROS, GREENHORN, AND CARLILE FORMATIONS

The Dakota Formation in Iowa is overlain by offshore marine deposits of the Graneros Shale, followed by offshore marine units of the Greenhorn and Carlile formations. The contact between the Graneros Shale and the Dakota Formation, however, has been placed at different positions by various geologists over the years. For this reason, Witzke and Ludvigson (1987) recommend that the Graneros-Dakota contact be placed at the base of the calcareous shale–dominated unit in the lower Graneros.

Noncalcareous or weakly calcareous shales, mudstones, and siltstones comprise the upper Dakota in the Sioux City, Iowa, area. The basal Graneros reveals siltstone laminae and ripples. Tiny white specks (less than 0.04 inch in diameter) are common in the shales of the formation and represent calcareous foraminifera and fecal pellets. Fish scales, coccoliths (microscopic calcareous plates of pelagic algae), and planktonic (floating) foraminifera occur throughout the formation. Inoceramid bivalves (clams) and ammonoid cephalopods occur but are rare. Bones of plesiosaurs (marine reptiles) have been recovered from Graneros strata in the Sioux City area. Deposition of the Graneros Shale of western Iowa was initiated in a prodelta setting. (The prodelta is the part of a delta that lies below the effective depth of wave erosion; it is located seaward of the delta front.) Later, offshore shelf conditions dominated as the Cretaceous seaway advanced eastward over Iowa.

The Greenhorn Formation caps the Graneros in western Iowa and forms

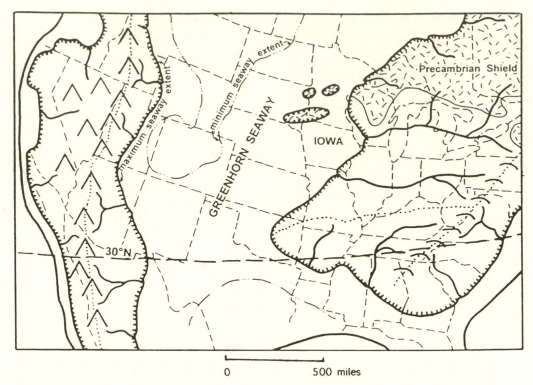

0   500 miles

9.12 The inferred paleogeographic setting during deposition of the Cretaceous Greenhorn Formation. At the time of its maximum extent, the Greenhorn Sea probably reached western Wisconsin and western Illinois. The record of Greenhorn strata in northwestern Iowa reflects deposition in an offshore marine setting largely removed from influxes of clastic sediments from eastern source lands. Chalk and shaley chalk are common in the formation and represent the calcareous remains of pelagic (floating) algae and foraminifera. A generalized shoreline position marking the minimum extent of the Greenhorn Seaway is also shown (dashed and dotted lines). From Witzke and Ludvigson 1996b.

resistant cliffs and ledges above Graneros strata in surface exposures, such as those near Stone State Park in Woodbury County. The Greenhorn is composed predominantly of impure limestone, shaley chalk, and calcareous shale. Microscopic remains of calcareous foraminifera and coccoliths are abundant in the chalks. Macrofossils in the formation include inoceramid bivalves, fish parts, and shark remains. During deposition of the Greenhorn, the Cretaceous shoreline was near the eastern edge of Iowa (fig. 9.12). Consequently, the influx of clastic sediment from eastern source areas was greatly reduced in western Iowa.

The Greenhorn Formation is overlain by the Carlile Shale at some localities in northwestern Iowa. The Carlile is primarily silty shale. Calcareous content decreases upward in the formation. Fossils in the Carlile include planktonic microfossils such as foraminifera and coccoliths. Fish debris, bivalves, and ammonoids are also known. During deposition of the Carlile, siliclastic sediment (silt and clay) accumulated at the edge of the Cretaceous seaway, building the shoreline westward. Conditions on the seafloor were generally inhospitable for most organisms, probably because the bottom waters contained very little dissolved oxygen. Similar oxygen-deficient conditions also characterized the Cretaceous seafloor during deposition of the Greenhorn and Graneros formations.

NIOBRARA AND PIERRE FORMATIONS

The Niobrara and Pierre formations are both well known in Nebraska and South Dakota. These Upper Cretaceous formations were apparently once present over much of western and central Iowa, but they have mostly been eroded back beyond the state's western boundary. Niobrara strata are confirmed in the subsurface of northwestern Lyon County, based on the occurrence of coccolith-bearing sandy marls in a water well. In addition, the Niobrara and Pierre formations are tentatively identified in Iowa from wells within the ring graben (an area of down-dropped blocks) of the Manson Impact Structure in north-central Iowa. According to Anderson and Witzke (1994), nearly 1,100 feet of Upper Cretaceous rocks were probably deposited in the Manson area originally; part of that record is preserved within the ring graben area of the Manson crater.

The thickest Cretaceous sequence in the Manson area is from the Erling Malmin well (fig. 9.13). The 630 feet of Cretaceous strata in the well represent the thickest penetration of Cretaceous bedrock anywhere in the state and include from top to bottom about 70 feet of so-called Pierre Shale, 110 feet of probable Niobrara Formation, 120 feet of Carlile Shale, 30 feet of Greenhorn Formation, 85 feet of Graneros Shale, and 210 feet of Dakota Formation. The rocks assigned to the Niobrara Formation are primarily silty shales, with an upper interval characterized by interlayered silty shale and calcareous mudstones and siltstones. Fossils in the Niobrara interval include fish debris, fragments of inoceramid bivalves, and pieces of other mollusks.

The so-called Niobrara beds from the Manson area compare favorably in composition to the Niobrara Formation of western Minnesota. The Niobrara Formation at its type area in northeastern Nebraska contains abundant chalk. Overall, the Niobrara Formation of Nebraska and Kansas is chalky and contains less silt and clay than equivalent strata in Iowa and Minnesota. This difference in composition is explained by the offshore setting of Nebraska and Kansas during Late Cretaceous time, whereas Iowa and Minnesota were in closer proximity to sources of eastward-derived clastic sediments (silt and clay).

About 70 feet of shale in the Malmin well is assigned to the Pierre Shale. Samples from the Pierre interval are highly weathered and of poor quality. The dominant rock is gray, silty, noncalcareous shale with abundant secondary

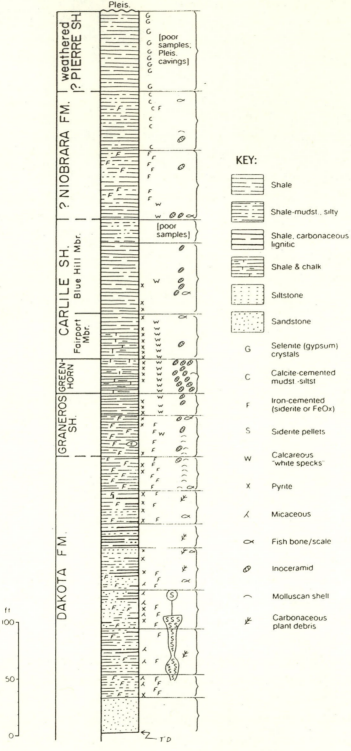

9.13 A lithostratigraphic log of Cretaceous strata in the Erling Malmin well near Manson, Iowa. Adapted from Anderson and Witzke 1994.

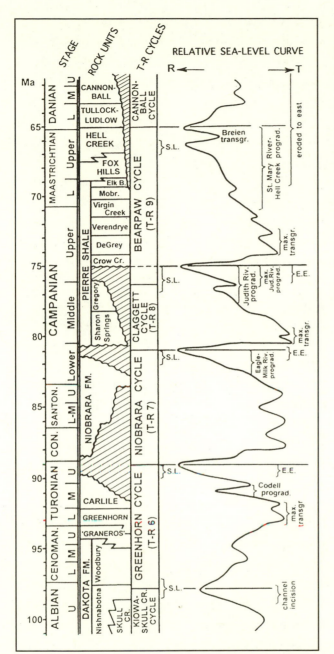

9.14 A relative sea-level curve for the eastern-margin area of the Western Interior Cretaceous Seaway in South Dakota and western Iowa. Ages in millions of years before present are shown on the left. Iowa's Cretaceous rock record includes the Dakota, Graneros, Greenhorn, Carlile, Niobrara, and Pierre formations, although the Pierre may only be preserved in a structurally depressed area of the Manson Impact Structure. The uppermost portion of the Cretaceous record is absent in Iowa, and a huge gap exists between the Mesozoic and Cenozoic systems. Adapted from Witzke, Hammond, and Anderson 1996.

gypsum (selenite crystals). The composition of the Pierre strata in the Manson area is similar to that of better-known exposures of the formation in South Dakota.

Additional Cretaceous formations are known in states to the west of Iowa (fig. 9.14), but these units are not found in Iowa. Iowa's Cenozoic rock record is composed primarily of deposits of the Pleistocene Ice Age. Consequently, there is a huge gap in the state's rock record between the Mesozoic and Cenozoic.

The Manson Impact Structure

Early attempts to date the Manson Impact Structure in the late 1980s yielded an age of approximately 65.7 million years, essentially indistinguishable from the accepted age of the Cretaceous-Tertiary (K-T) boundary. However, more accurate age determinations in the early 1990s based on analyses of argon gas in sanidine feldspar (a product of recrystallization from impact melt) demonstrated that the Manson Impact Structure formed 73.8 million years ago. This date, now well accepted by the geological community, indicates that the impact at Manson occurred 8 to 9 million years prior to the end of the Cretaceous Period. Even though the Manson Impact Structure no longer coincides with the K-T boundary as was once thought, it is still an important geological feature and ranks as the largest intact, onshore meteorite crater in the continental United States. Koeberl and Anderson (1996b) describe many of the significant features of the Manson Impact Structure.

The surfaces of the rocky planets and moons of our solar system are pocked with craters, indicating that impact cratering is a significant process in sculpting the surfaces of planetary and lunar bodies. Even though 150 impact structures have been identified on Earth to date, the importance of impact cratering as a geological process may not be fully appreciated by the general public. The huge amounts of energy released during impact events shatter rocks, blast out craters, generate heat, and disrupt life over wide areas. Impact-generated shock waves impart irreversible changes in the crystal structures of common rock-forming minerals such as quartz and feldspar.

Shock metamorphism in minerals is produced primarily by the enormous pressures associated with impacts and to a lesser degree by the elevated temperatures. Some of the best indicators of shock metamorphism are microscopic features such as planar deformation features (PDFs) in quartz, feldspar, and other rock-forming minerals (fig. 9.15). Other key products of impact are shock-formed glass, high-pressure minerals, and mineral and rock melts. On a larger, or macroscopic, scale, the occurrence of brecciated rocks, as crater fill or ejecta, provides additional evidence for impact. Ejecta often display an inverted stratigraphic position, with older rocks overlying younger units.

The presence of meteorites or meteoritic material at impact crater sites is very rare. Why is this so? Apparently, the shock waves produced by the impact process completely vaporize the original impacting bodies. This appears to be the case at the Manson Impact Structure. An impacting body (possibly a stony meteorite) with a diameter of about 1.2 miles is inferred from the size of the Manson crater, but no meteorites or meteorite fragments have been recovered in the area.

The Manson Impact Structure (fig. 9.16) is nearly circular and has a maxi-

9.15 Planar deformation features (PDFs) in a quartz grain from the Crow Creek Member of the Pierre Formation, near Yankton, South Dakota. Shock-produced features such as these are considered prime evidence of impact origin. Such microscopic features are common in rocks of the Manson Impact Structure and in the Crow Creek Member of the Pierre Formation. Shocked quartz and feldpar grains within the Crow Creek Member in northeastern Nebraska and southeastern South Dakota are interpreted as ejecta from the Manson Impact Structure. The field of view is approximately .025 inch. Photographed through crossed polarizers. Courtesy of Raymond R. Anderson, Geological Survey Bureau, Iowa City.

mum diameter of about 23 miles. Centered just north of the town of Manson near the border of Calhoun and Pocahontas counties, the Manson structure is a well-preserved, nearly circular impact structure with a large central peak within a central depression (fig. 9.16). The crater displays no surface expression because it is covered by approximately 65 to 295 feet of Pleistocene glacial drift. The structure developed in a region covered by a thick sequence of shale-dominated Cretaceous marine rocks overlying about 2,600 feet of Paleozoic strata and a wedge-shaped body of Middle Proterozoic clastic rocks that ranges in thickness from 0 to more than 9,800 feet. The sedimentary rocks at Manson rest on Early and Middle Proterozoic basement rocks composed primarily of gneiss and granite.

The Manson area has been known as an area of anomalous subsurface geology since 1912, when Norton et al. (1912) reported on the underground water resources of the region. The water from the Manson town well proved to be unusually soft when compared to water in other Iowa wells. The Manson town well produced water from broken and brecciated Precambrian igneous and

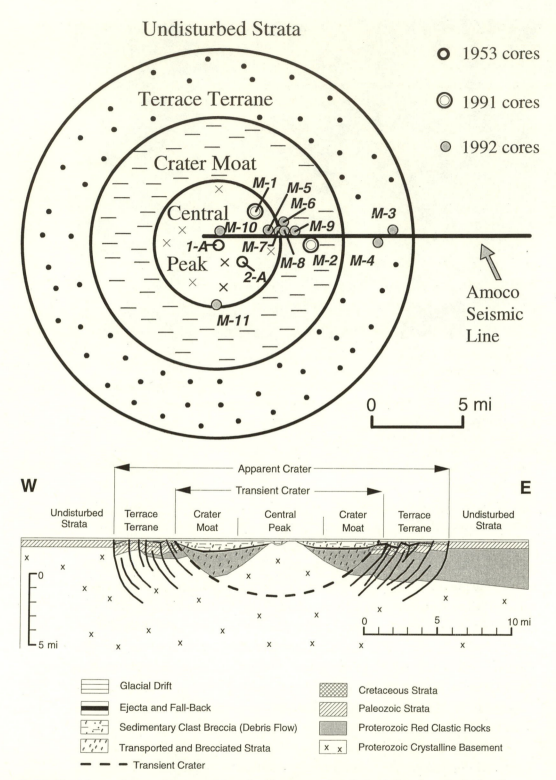

9.16 The Manson Impact Structure (above) with cross section (below). Based on Koeberl and Anderson 1996a; updated version courtesy of Raymond R. Anderson, Geological Survey Bureau, Iowa City.

metamorphic rocks, whereas most of the state's bedrock aquifers are porous and permeable sedimentary rocks such as sandstone, limestone, or dolomite.

Hoppin and Dryden (1958) suggested that the Manson structure was a cryptovolcanic feature. The term "cryptovolcanic structure" is used for highly deformed, strongly brecciated, circular features that are believed to have been produced by blasts of volcanic gases. Typically, these structures lack conclusive evidence of volcanic action, such as the presence of volcanic rock, contact metamorphism, or hydrothermal action. Such features are thought to be produced by completely concealed volcanic activity.

The Manson structure was first recognized as an impact structure in 1966 when Short (1966) identified PDFs in granitic rocks from the central peak of the crater (fig. 9.15). In the 1980s, a team of scientists led by Nobel Laureate physicist Luis Alvarez and his son, Walter, a geologist, proposed a connection between large impacts and the extinctions at the K-T boundary (Alvarez et al. 1980). This sparked an increase in research on impact craters in general and on the Manson structure in particular. Because its age was presumed to be 65.7 million years, the Manson structure was linked with the numerous extinctions (including dinosaurs) at the K-T boundary. As mentioned, it is now known that the impact at Manson took place 8 to 9 million years prior to the K-T boundary, and instead the Chicxulub crater in the subsurface of Mexico's Yucatan Peninsula is believed to be the site of a massive meteorite impact that contributed to the demise of the dinosaurs and other Mesozoic creatures. The size of the Chicxulub crater is estimated to be 110 miles in diameter, nearly five times larger than the Manson crater.

Although the extraterrestrial impact at Manson is no longer seen as the smoking gun that contributed to extinctions at the K-T boundary, it was significant nevertheless. The impact must have caused a huge splash in the Cretaceous sea over north-central Iowa. Impact ejecta, including grains with PDFs, have been recovered from the Crow Creek Member of the Pierre Shale in South Dakota and Nebraska, some 125 to 185 miles northwest of Manson (fig. 9.17). Shocked grains of quartz and feldspar occur throughout the Crow Creek Member and comprise up to 3 percent of the rock volume of the basal unit of the member. Witzke, Hammond, and Anderson (1996) interpret the ejecta grains within the Crow Creek as reworked material derived from a regional Late Cretaceous unconformity surface. According to them, the Manson impact event preceded Crow Creek deposition.

Others, however, interpret the Crow Creek ejecta material as tsunami sedimentation, triggered by the Manson impact event. A tsunami is a gravitational sea wave produced by any large-scale, short-duration disturbance of the ocean floor. Tsunamis are usually associated with shallow submarine earthquakes, but rapid submarine earth movements, subsidence, or volcanic eruptions can also

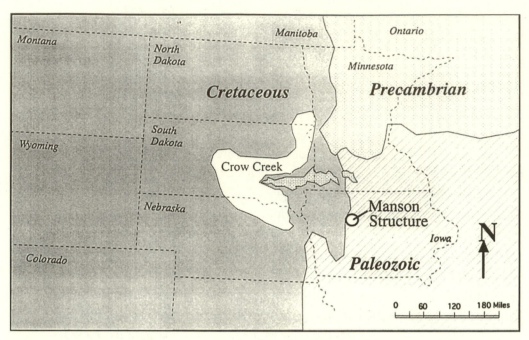

9.17 The distribution of the Upper Cretaceous Crow Creek Member of the Pierre Formation. Grains of quartz and feldspar with planar deformation features (PDFs) occur within the member and are interpreted as ejecta from the Manson Impact Structure, located 125 to 185 miles to the southeast. Courtesy of Raymond R. Anderson, Geological Survey Bureau, Iowa City.

generate such waves. Steiner and Shoemaker (1996) hypothesize that the Crow Creek Member is a tsunami deposit. Under such a scenario, the Crow Creek sediments were laid down in a few days or less.

Life of the Mesozoic

The Mesozoic Era is often called the Age of the Dinosaurs. The era saw the diversification of dinosaurs and other reptiles and the first appearance of birds, mammals, and angiosperms (modern flowering plants). Birds and mammals are unknown from Iowa's Mesozoic rock record, although bird tracks are reported from the Cretaceous of Kansas. Evidence for dinosaurs in the state is scant, consisting of a single bone fragment from Guthrie County. Witzke and Ludvigson (1996b) tentatively identify the bone as dinosaurian based on its size and microstructure. In Nebraska, duck-billed dinosaurs have been documented, and their presence in Iowa is plausible in that suitable terrestrial environments were present. The Dakota Formation of north-central Kansas has yielded a partial skeleton of an ankylosaur (plated dinosaur). Fossils of plesiosaurs (marine reptiles) have been reported from the Sioux City area. Mosasaurs (another type of marine reptile) occur in the Niobrara and Pierre formations in states to the west of Iowa. The skeleton of a 20-foot-long mosasaur from the Cretaceous of northeastern

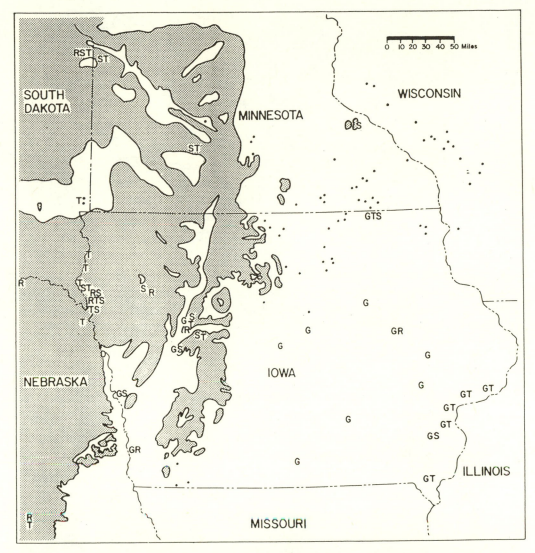

9.18 The distribution of Cretaceous strata and vertebrate fossils in Iowa and adjacent states. Erosional outliers of Cretaceous strata (Dakota equivalents) in northeastern Iowa, southwestern Minnesota, and western Wisconsin have been termed the Windrow Formation. Patterned area depicts the present-day distribution of Cretaceous rocks. Dots show Cretaceous outliers (Dakota or Windrow formations). Cretaceous vertebrate fossil occurrences are shown. Symbols: T = teleost (bony) fish, S = sharks, R = marine reptiles (plesiosaurs, mosasaurs, or turtles), D = dinosaurs, G = Cretaceous fossils recovered from Pleistocene glacial deposits and gravels. From Witzke 1981a.

Nebraska is on display at Augustana College, Rock Island, Illinois. The remains of giant sea turtles are known from Cretaceous strata in southeastern Nebraska. Fossil remains of teleost (bony) fish and sharks occur in a number of Cretaceous formations in Iowa and adjacent states. The occurrences of vertebrate fossils from Cretaceous strata in Iowa, Minnesota, eastern Nebraska, and eastern South Dakota are shown in figure 9.18. Larger vertebrate animals found in Cretaceous strata of Iowa and adjacent states are depicted in figure 9.19.

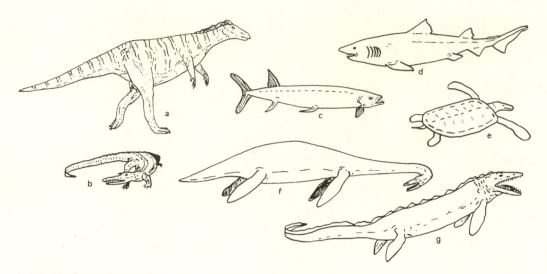

9.19 The larger Cretaceous vertebrate animals known from Iowa or adjacent states: (a) ornithopod (duck-billed) dinosaur, Dakota Formation; (b) crocodilian reptile, Dakota and Windrow formations; (c) large teleost (bony) fish, Graneros and Greenhorn formations; (d) large shark, Greenhorn and Windrow formations; (e) giant sea turtle, Graneros Formation; (f) plesiosaur, Graneros Formation; and (g) mosasaur, Niobrara and Pierre formations. From Witzke 1981a.

Iowa's Mesozoic record also includes an assortment of plant remains. Fossil spores and pollen from the Fort Dodge Formation help date the formation as Upper Jurassic. Conifer pollen is abundant in the formation, but primitive gymnosperm pollen and spores of ferns are also known.

The Dakota Formation preserves altered plant materials in low-grade brown coals (lignites). Dakota strata have also yielded petrified wood, casts of tree trunks, and compressions of angiosperm leaves such as those of willows, poplars, and magnolias. Leaf fragments of sycamore-like angiosperms also occur. In addition, pine needles and sequoia-like foliage and branches are known. According to Witzke and Ludvigson (1996b), the compression flora and the palynoflora (spores, pollen, and related materials) of the Nishnabotna Member of the Dakota Formation provide a glimpse of one of the plant communities that inhabited the broad flood-basin setting of the eastern continental lowlands during the Late Cretaceous. The diverse assemblage includes bryophytes, lycopods, ferns, gymnosperms, and angiosperms. The vegetated lowlands included forests of pines, cypresslike trees (possibly in swamps), and other trees. Smaller trees, bushes, and vines of early angiosperms were also present. Cycads were locally abundant. The understory of the floodplain forests and the forest breaks were likely carpeted by a diverse array of ferns.

Ongoing research by Robert L. Ravn, Brian J. Witzke, and others suggest that the palynoflora of the Dakota Formation is exceptionally rich and diverse. This flora will likely become the standard to which other Cretaceous palynofloras are compared.

The marine formations of Iowa's Cretaceous preserve a variety of marine mollusks, including bivalves, gastropods, and ammonoid cephalopods. Microscopic marine fossils include foraminifera and calcareous nannofossils (floating algae).

Summary

Ancient Iowa was a land area undergoing erosion during most of Mesozoic time. Iowa has no Permian (Late Paleozoic) or Triassic (Early Mesozoic) rock record. Iowa's Jurassic record is limited, consisting of the Fort Dodge Formation, a sequence composed primarily of rock gypsum with some associated red, green, and gray clastic rocks. These strata were probably deposited in both restricted marine and nonmarine environments along the eastern edge of an inland sea. Deposition of the gypsum beds apparently took place in restricted marine embayments or lagoons. Gypsum provides the raw material for an important industry at Fort Dodge, and Iowa annually ranks high nationally in the commercial production of gypsum. The clastic units of the Fort Dodge Formation reflect deposition in nonmarine settings.

Cretaceous strata of the state are best represented in western and northwestern Iowa, although erosional outliers occur in central and northeastern Iowa. Iowa's Cretaceous record includes the Dakota Formation and its lateral equivalent, the Windrow Formation. The Dakota Formation was deposited under fluvial, coastal, and marginal marine settings. The Dakota is composed primarily of sandstone and mudstone, with some associated shale, siltstone, conglomerate, and lignite. The Windrow Formation consists of limonitic beds of possible residual or bog origin and sandstones and conglomerates of fluvial origin. The Graneros Formation overlies the Dakota and marks a transition to marine deposition. The rest of Iowa's Cretaceous record (Greenhorn, Carlile, Niobrara, and Pierre) formed on a muddy marine shelf, often with low levels of oxygen in the bottom waters. In Iowa, the Pierre Formation is preserved only in the graben area of the Manson Impact Structure.

Approximately 73.8 million years ago, during the Late Cretaceous, a meteorite smashed into the Cretaceous seaway over north-central Iowa. The resulting impact produced the Manson Impact Structure, one of the largest and best-preserved impact craters in the world. Ejecta from the impact occur in the Crow Creek Member of the Pierre Formation in eastern South Dakota and northeastern Nebraska, 125 to 185 miles northwest of Manson, Iowa.

Iowa does not have a rock record of the uppermost formations of the Cretaceous System, and the state's Cenozoic record consists primarily of Pleistocene deposits. Consequently, there is a huge gap in Iowa's rock record between the Mesozoic and Cenozoic.

Acknowledgments

Published works by the following were particularly helpful in preparing this chapter: Raymond R. Anderson, Brian J. Witzke, Robert L. Brenner, Richard F. Bretz, Bill J. Bunker, Derrick L. Iles, Greg A. Ludvigson, Robert M. McKay, Donald L. Whitley, Robert D. Cody, Ruth A. Galhaher, James T. Dunn, Gardner Cowles, Jack B. Hartung, Christian Koeberl, Robert L. Ravn, George W. Shurr, Richard H. Hammond, and J. R. Poppe. Greg Ludvigson read a draft of this chapter and provided helpful comments.

To Iowa the scientific world has learned to look for the standard American section of the deposits of the great ice age, for here the record of successive glacial invasions of the Pleistocene period are most complete and have been most satisfactorily deciphered.

   Samuel Calvin (1906) remarking on Iowa's glacial record

The trail of the ice monster has been traced, his magnitude measured, his form and even his features figured forth, and all from the slime of his body alone, where even his characteristic tracks fail.

   W. J. McGee (1891) commenting on the Pleistocene history
   of northeastern Iowa

In general the rocks transported by the Iowan ice were brought in large masses, ten, fifteen, or twenty feet in diameter. . . . The largest of the many large boulders seen in this county is that known as Saint Peter. . . . Saint Peter is fully twenty feet in height and more than eighty feet in circumference.

   Samuel Calvin (1902) discussing the glacial deposits
   of Chickasaw County

The Cenozoic Era is divided into two periods, the Tertiary and the Quaternary. The Tertiary Period represents about 65 million years, and the Quaternary Period includes the last 1.65 million years. The Quaternary consists of the Pleistocene Epoch and the Holocene (Recent) Epoch. The Holocene represents the last 10,000 to 10,500 years, essentially the time since the continental glaciations in North America and Europe.

   Iowa was above sea level during the Tertiary and subjected to extensive weathering and erosion. States to the west of Iowa, however, have a well-preserved record of Tertiary rocks in the form of stream-laid sediments. These Tertiary sediments represent the deposits of stream systems that flowed eastward from sources in the Rocky Mountain area. The ancient stream deposits contain a diverse fauna of mammals and other land vertebrates. The Tertiary strata of the Badlands of South Dakota and Nebraska are particularly well known for their abundant and well-preserved vertebrate faunas. A few Tertiary vertebrate remains have been recovered in southwestern Iowa from alluvial sediments. Among these are parts of rhinoceroses and three-toed horses in Pleistocene sand-and-gravel deposits. The excellent preservation and relatively unabraded condition of these fossils suggest they were not transported far. Perhaps they were derived from nearby Tertiary strata that remain undetected in the subsurface of western Iowa.

So-called salt-and-pepper sands are common in western Iowa. They are similar in appearance and characteristics to deposits of the present-day Platte and Missouri rivers to the west of Iowa. The "salt" in the sands is common quartz, but the "pepper" is more distinctive. Many of the dark or pepper-colored grains are fragments of volcanic glass, indicative of volcanic source lands such as those in the Rockies. These telltale grains in western Iowa strongly suggest the presence of an eastward-flowing stream system with headwaters in a volcanic source land. Such a stream system apparently existed in western Iowa prior to the establishment of the present course of the Missouri River.

Most of the salt-and-pepper deposits are older than the start of glacial deposition in western Iowa some 2.5 million years ago. The tooth of an ancient Pliocene elephant, recovered from gravels in Plymouth County, helps date some of these western-derived alluvial sediments. Successive glaciations in western Iowa probably diverted the southeastward-flowing ancestral drainage system to the south, forming the present Missouri River system.

During the long episode of erosion in Tertiary time, Iowa's Mesozoic rock record was largely removed by erosion, and streams carved their valleys into the Paleozoic bedrock of the state. This topography had some control on the path of the Late Pliocene and Pleistocene continental ice sheets that invaded Iowa. The ice sheets tended to follow low topographic trends. Glacial deposits and glacially related deposits constitute a significant part of Iowa's Cenozoic record, as this chapter's opening quotation from Samuel Calvin implies. However, the understanding of these deposits was not as complete as Calvin envisioned back in 1906. Since then, geologists have learned a great deal more about this intriguing record. The past three decades in particular have seen major revisions in Iowa's Ice Age history.

The climate in Iowa during the Late Pliocene and Pleistocene was considerably different than it had been previously. Whereas Paleozoic and Mesozoic climates were warm and tropical or subtropical, conditions during the late Pliocene and Pleistocene were temperate and punctuated by glacial ages.

The Pleistocene Epoch started about 1.65 million years ago. The placement of the end of the Pleistocene is somewhat arbitrary, but most North American geologists place the last 10,000 to 10,500 years in the Holocene (Recent) Epoch. This corresponds approximately to the time that continental glaciers retreated from the midlatitude regions of North America and Europe. Obviously, the polar and alpine regions of the Earth today are still in an ice age. Iowa last emerged from glaciation about 12,000 years ago, but portions of the Great Lakes region of the United States and Canada were still under the influence of glacial ice as late as 8,000 years ago.

Nature of the Pleistocene Record

When early naturalists in Europe and North America first recognized the presence of striated and polished bedrock and the overlying poorly sorted deposits of clay, silt, sand, pebbles, cobbles, and boulders, they interpreted these features very differently than how they are interpreted today. Most observers recognized that some exceptional carrying agent must have been responsible for transporting the boulders in particular, for some of the rocks are immense (fig. 10.1). Still, it did not occur to them that the northern part of Europe and North America had been covered by continental ice sheets. Some explained the poorly sorted deposits as being caused by the great biblical flood. These deposits were considered to have formed from melting icebergs that drifted southward during the time of the biblical deluge. It was logical, with this interpretation, to call the poorly sorted deposits drift. It was not until the mid-1800s that these deposits were correctly attributed to continental glaciers.

Glacial Drift

Geologists still use the old term "drift," but its meaning is different. Today, geologists use the term "drift" for materials that were carried and deposited by glaciers and for sediments that have been transported by running water emanating from glaciers. Drift includes poorly sorted material, such as till, and stratified-and-sorted deposits that form from meltwater within and adjacent to glaciers.

Till

Till (fig. 10.2) is material transported and deposited directly from glacial ice. It is usually poorly sorted and unstratified. Because of its high viscosity, glacial ice carries the largest and smallest fragments together, just as a dump truck may carry a load of soil and concrete slabs mixed in the same load. When a glacier dumps its load, the mixture of particles of different sizes is deposited without evidence of sorting or stratification.

Geologists also use the term "diamicton" for poorly sorted, unstratified deposits. Diamicton is a nongenetic term. It does not imply any specific mode of deposition, whereas the term "till" signifies material that was deposited directly by glacial ice. Diamicton is an appropriate term to use if the origin of a poorly sorted deposit is unclear. It is also useful in describing tills that have been disturbed or modified by slumping or other gravitational processes.

Associated with glacial till are rocks that are foreign to the local area. These rocks, called erratics, are usually of compositions totally different from the rocks found in the local drainage basins where they occur. Most of the large erratics in Iowa are of igneous or metamorphic origins and were derived from the Precambrian bedrock of areas to the north of Iowa, such as Minnesota and Canada. Granite is a common erratic, but a wide variety of igneous and metamorphic

10.1 St. Peter rock in Chickasaw County is one of Iowa's largest glacial erratics, standing 20 feet in height and measuring approximately 80 feet in circumference. Photo by the author.

rocks are known. Copper-bearing erratics, probably from the Lake Superior area, have been recovered from glacial till in Iowa, and minor amounts of gold are associated with some glacial tills. Some erratics were used as stone for building construction during the early days in Iowa; others found use in fieldstone fences, building foundations, and stone fireplaces.

Some of the rocks that occur in glacial till have distinctive markings that attest to their journey within glacial ice; these rocks are scratched and flattened where they rubbed together. The bedrock surface under glacial till is often scratched and abraded, too. These markings were produced by the scraping action of rocks frozen within the glacial ice. Scratches on bedrock produced by the action of such frozen debris are called glacial striations. Deeper markings are called glacial grooves. Generally, striations and grooves on bedrock are parallel to the direction of ice movement.

Geologists often get a view of striations or grooves in quarries — on the upper bedrock surface, just below the glacial till. The public can observe glacial markings on bedrock at Stainbrook Geological Preserve in Johnson County.

10.2 Glacial till, illustrating the absence of stratification and the poor sorting that is characteristic of such deposits. Photo by the author.

Glaciofluvial Deposits

Sand-and-gravel deposits (fig. 10.3) are commonly associated with melting glaciers. Called glaciofluvial deposits, such sediments form from meltwaters that descend through openings in the ice or flow away from the margin of the ice. Glaciofluvial deposits are often referred to as outwash. Outwash can form under the ice, at its margin, or beyond the border of the ice margin. Glaciofluvial deposits are important sources of aggregate in Iowa, and they also serve as groundwater aquifers.

Loess Deposits

Strictly defined, loess (fig. 10.4) is a fine-grained deposit of wind-carried sediment. Loess typically contains particles of silt and clay size, although some loess contains particles of very fine sand. The source of loess in the Midwest was fine rock powder (rock flour) that was produced by the grinding processes of continental glaciers. This fine-grained material was deposited on the floodplains of streams that developed near the margins of glacial ice. After the rock powder dried, it was picked up by the wind and deposited as an all-covering blanket over the land surface. The floodplains of the Missouri and Mississippi rivers and their tributaries were the principal sources of the midwestern loess deposits. Local

10.3 Outwash sediments in Dickinson County, Iowa. Note the cross-bedding in these sand-and-gravel deposits. Photo by the author.

10.4 Loess exposures, such as these deposits in western Iowa, often display steep slopes in roadcuts. Photo made from a slide taken in 1941 in Harrison County by Charles Gwynne.

source areas were present along the floodplains of the Des Moines, Skunk, Iowa, and Cedar rivers, too. Loess deposits are thickest in the proximity of their source areas because of the limited carrying power of wind. For this reason, the loess deposits of Iowa are thickest in the western part of the state, near their principal source area — the Missouri River valley.

The word "loess" is of German origin (Löss) and is difficult for most Americans to pronounce correctly. The name was applied to massive silt deposits exposed along the Rhine River valley in Germany. Löss was derived from "loose," so named by peasants and bricklayers because of the unconsolidated character of the material. In addition to its German pronunciation (*luehss*), loess has been pronounced lurse, less, and luss (rhymes with bus). The latter is the most common pronunciation in this country and is considered acceptable by the geological community. The loess deposits of western Iowa were referred to as siliceous marl, or the Bluff Deposit, in the early literature.

Irregular-shaped nodules of calcium carbonate occur in some of Iowa's loess exposures. These configurations of lime somewhat resemble tiny dolls, and for that reason they have been called loess *kindchen* (loess babies). They are formed by the irregular deposition of calcium carbonate by infiltrating water. Occasionally, tube-shaped bodies of iron-bearing minerals are concentrated in loess, where plant roots were once present. Called pipestems, these cylindrical features were deposited by groundwater.

Many early geologists and naturalists thought that loess was a water-laid sediment that formed in a quiet lake setting. Bohumil Shimek, the renowned nineteenth-century Iowa naturalist for whom Shimek State Forest is named, was the first to recognize loess as wind-deposited material. Shimek's interpretation was based on his documentation of fossil, terrestrial, air-breathing snails within the loess deposits of western Iowa.

Interglacial Deposits

Weathering, erosion, and landscape development were dominant processes during interglacial times. The records of such times are known chiefly from the soil horizons and weathered zones that developed on exposed glacial sediments during the interglacial intervals. Some of these horizons were later buried under younger glacial deposits and preserved in the Pleistocene record. In addition to containing ancient soil horizons and weathered zones, the interglacial record sometimes preserves alluvial, lacustrine, or bog deposits.

Ancient soil horizons are particularly useful in determining the history of interglacial times. These old soils are called paleosols. The prefix "paleo" means ancient or old, and the term "sol" is from the Latin solum, or soil. Paleosols, designated as formal soil-stratigraphic units, are also referred to as geosols.

Iowa's Quaternary Record

Iowa's Quaternary record includes deposits of several glacial and interglacial stages that formed during the Pleistocene Epoch. These deposits include glacial drift, tills, glacial fluvial sediments, loess, and paleosols. In addition, the Quaternary record consists of Holocene alluvium and eolian sand.

The Pleistocene deposits of Iowa afford one of the most extensive and complete sections of continental glacial deposits to be found in North America. Studies conducted in Iowa during the late 1800s and early 1900s provided the basis for several of the major subdivisions of the Pleistocene. For example, reference sections of the so-called Nebraskan and Kansan Glacial stages were established in southwestern Iowa, and the Aftonian Interglacial Stage was named for exposures near Afton Junction in southwestern Iowa. In addition, the type area of the Yarmouthian Interglacial Stage was based originally on material from two hand-dug wells at Yarmouth in southeastern Iowa. The names Afton and Yarmouth were used for two of the major subdivisions of the Pleistocene chronology for North America, and these names were subsequently used in the geologic literature nationally and internationally. Figure 10.5 shows the divisions that were used to describe Iowa's Pleistocene record prior to the late 1970s.

10.5 The classification of Pleistocene and Holocene events in Iowa prior to the late 1970s. Adapted from Ruhe 1969.

Glacial Stage	Interglacial Stage	Evidence
	Holocene	Erosion, landscape development, postglacial wetlands, eolian sands, alluvial and other sediments
Wisconsinan		Cary drift and moraines, Lake Calvin terraces, Iowan Erosion Surface, Wisconsinan loess, Tazewell drift
	Sangamonian	Soils, weathered zones, peat, and erosion
Illinoian		Loveland Loess (western Iowa), weathering and landscape development (southern Iowa), Illinoian drift (southeastern Iowa)
	Yarmouthian	Soils, peat, weathered zones, and erosion
Kansan		Glacial drift
	Aftonian	Soils, peat, weathered zones, and erosion
Nebraskan		Glacial drift

THE PRE-ILLINOIAN GLACIAL AND INTERGLACIAL STAGES

Additional glacial deposits and interglacial features (not shown in fig. 10.5) are now known from the subsurface of southwestern Iowa and adjacent states. According to our present understanding, more than seven separate till sheets comprise the Pre-Illinoian sequence of western Iowa. For this reason, terms such as "Nebraskan Glacial," "Aftonian Interglacial," and "Kansan Glacial" that were once historically significant have been abandoned. The term "Pre-Illinoian" is now used collectively for all of the glacial and interglacial deposits older than the Illinoian Glacial Stage. The youngest of these deposits is at least 500,000 years old.

Layers of volcanic ash, referred to collectively as the Pearlette Family of volcanic ash deposits, occur interlayered with Pre-Illinoian glacial tills in western Iowa and provide evidence for the age of Pre-Illinoian deposits. The chemistry and mineralogy of the volcanic ash are characteristic of ancient volcanoes that erupted in the Yellowstone area of northwestern Wyoming. Through the technique of fission-track dating, the ash beds from the Yellowstone source area have been given ages of 0.6, 0.7, 1.2, and 2.2 million years old. These dates establish the age of the Pre-Illinoian glacial deposits of western Iowa. The oldest of the ash-fall deposits overlies Pre-Illinoian till, indicating that glaciation in western Iowa began prior to 2.2 million years ago. This date falls within the Late Pliocene (fig. 10.6).

Separate Pre-Illinoian deposits can be recognized on the basis of their mineral composition and physical properties. Two such units, the Wolf Creek Formation and the Alburnett Formation (fig. 10.6), comprise the glacial sequence of much

10.6 A stratigraphic classification of Iowa's Late Pliocene, Pleistocene, and Holocene record. Adapted from Prior 1991 and Bettis et al. 1996.

Geologic Age	Stratigraphic Evidence	Approximate Age in Years Before Present
Holocene	DeForest Formation (alluvium, slope deposits, and pond and lake sediments), local deposits of wind-blown sand	Present to 10,500
Wisconsinan Glacial	Dows Formation (glacial drift), Noah Creek Formation (glaci-ofluvial sediments), Wisconsinan loesses and eolian sand deposits of the Peoria Formation, Sheldon Creek Formation (glacial drift), Pisgah Loess	Dows drift (12,000–15,000), Noah Creek Fm. (11,000–14,000), Peoria Loess (12,500–25,000), Sheldon Creek drift (26,000–40,000), Pisgah Loess (26,000–55,000)
Sangamonian Interglacial	Sangamon Soil (paleosol) and alluvium	55,000–140,000
Illinoian Glacial	Kellerville Till Member of the Glasford Formation (glacial drift), Loveland Loess	140,000–300,000 for the Illinoian Glacial, overall; Iowa's record is incomplete, 140,000–160,000
Yarmouthian Interglacial	Yarmouth Soil (paleosol) and alluvium	300,000–500,000
Pre-Illinoian Glacial and Interglacial Stages (Pleistocene)	Wolf Creek Formation (glacial drift), volcanic ash, Alburnett Formation (glacial drift), volcanic ash	500,000–720,000 720,000–1,650,000
Additional Pre-Illinoian Glacial and Interglacial Stages (Late Pliocene)	Unnamed glacial and inter-glacial deposits, volcanic ash	1,650,000–2,500,000

of eastern Iowa. The current stratigraphic terminology for the Pleistocene of Iowa is shown in figure 10.6.

Loess-mantled Pre-Illinoian drift forms the upland landscape throughout most of southern Iowa. The Pre-Illinoian drift is overlain by Illinoian drift in southeastern Iowa and by Wisconsinan glacial deposits in north-central Iowa. The Iowan Surface, a Late Wisconsinan–age erosion surface, developed on Pre-Illinoian drift over a wide area in northwestern, north-central, and northeastern Iowa. The limits of the major advances of the Pre-Illinoian, Illinoian, and Wisconsinan glacial advances are shown in figure 10.7.

YARMOUTHIAN INTERGLACIAL STAGE

The Yarmouthian Interglacial Stage represents the interval between the last Pre-Illinoian glaciation and the initiation of the Illinoian Glacial Stage, approx-

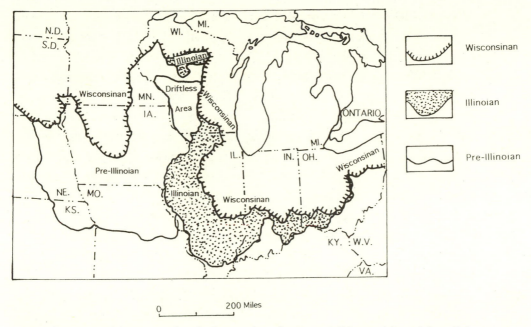

10.7 The limits of the major glacial advances into the Upper Midwest. Adapted from Prior 1991.

imately 300,000 to 500,000 years ago. The major streams of southern Iowa were established during this time.

The classic type location for the Yarmouthian Interglacial Stage was based on Frank Leverett's (1898) interpretation of the sediments recovered from two dug wells near Yarmouth in Des Moines County, Iowa. George Hallberg and Richard Baker obtained and analyzed cores from the same area and discovered that the sediments on which the Yarmouthian Interglacial was based are actually closely associated with glacial deposits of the Illinoian Glacial Stage. Pollen and wood from the cores indicated that the climate was full glacial rather than interglacial. For this reason, Hallberg and Baker (1980) redefined the type section for the Yarmouthian Interglacial. The new type section was designated as the well-developed paleosol underlying the Illinoian Glacial sediments in the Yarmouth cores. A stream cut near Mediapolis, Iowa, that exposes a Yarmouth paleosol between the Illinoian and Pre-Illinoian tills was designated a principal reference section.

ILLINOIAN GLACIAL STAGE

The Illinoian drift (Kellerville Till Member of the Glasford Formation) in Iowa is limited to a narrow belt paralleling the Mississippi River in the southeastern corner of the state (fig. 10.7). The Illinoian ice entered Iowa from the east and northeast and is believed to have originated in the Labrador Ice Field east of Hudson Bay. The Illinoian drift in Iowa contains numerous rocks and fossils of Pennsylvanian age, constituents that were apparently derived from the bedrock of eastern and central Illinois. In contrast, the Pre-Illinoian ice advances into the state came from the north-northwest and contain abundant rocks of Tertiary, Cretaceous, and Precambrian ages.

As the Illinoian ice advanced from the east, it displaced the ancestral Mississippi River westward. The Illinoian ice probably temporarily dammed the Iowa and Cedar rivers in eastern Iowa. Terraces, sedimentary features, and topographic features that suggested the former presence of a glacial lake in the lower reaches of the Iowa and Cedar River valleys were recognized by some geologists. Early geologists attributed these features in Johnson, Washington, Cedar, Louisa, and Muscatine counties to a glacial lake. Called Lake Calvin, the presumed lake was named for the prominent Iowa geologist Samuel Calvin.

The origin of Lake Calvin is somewhat of a puzzle. Terraces of the Lake Calvin area have been dated by carbon-14 methods and yield ages that are within the Wisconsinan Glacial Stage. This demonstrates that the Lake Calvin features are younger than Illinoian Glacial time and that they cannot be explained by Illinoian glacial events. The Lake Calvin features are probably not of lacustrine origin at all. Rather, they appear to be complex alluvial features of the Cedar and Iowa river systems. They formed at approximately the same time that the complex Iowan Surface was developing in northern Iowa.

Throughout most of southeastern Iowa, the Illinoian drift rests on a paleosol that formed on Pre-Illinoian drift. At a few localities, the Illinoian drift lies on Paleozoic bedrock. The Illinoian drift is overlain by a thin sequence of Wisconsinan loess. The average thickness of the Illinoian drift in southeastern Iowa is about 30 feet. However, deposits of the Illinoian ice advances are thicker and more complete in neighboring Illinois, where the Illinoian Glacial gets its name. The age range of 130,000 to 300,000 years (fig. 10.6) is for the entire Illinoian Glacial Stage. Iowa's record for this glacial interval is restricted to the oldest part of the Illinoian glaciation.

In western Iowa, the Loveland Loess, approximately 140,000 to 160,000 years old, accumulated during the latter part of Illinoian Glacial time. The Loveland Loess has a limited distribution in the state and does not extend across southern Iowa to the area of Illinoian drift in southeastern Iowa. The Loveland Loess is named for the village of Loveland in northwestern Pottawattamie County. Its distribution in Iowa is restricted to a belt 20 to 40 miles wide paralleling the Missouri River.

SANGAMONIAN INTERGLACIAL STAGE

Paleosols that developed in the upper part of the Illinoian drift in southeastern Iowa and in the Loveland Loess in western Iowa provide the main evidence for the Sangamonian Interglacial Stage in Iowa. Some sand-and-gravel deposits also accumulated during the Sangamon Interglacial.

Illinoian Glacial deposits are absent over most of southern Iowa, and there the paleosol that developed in the Pre-Illinoian drift formed during both the Yarmouthian and the Sangamonian interglacial stages. As a result, this ancient soil interval is often referred to as a composite Yarmouth-Sangamon paleosol. Weathering and landscape development in southern Iowa continued during the Yarmouthian Interglacial through the Illinoian Glacial and the Sangamonian Interglacial. The Yarmouth-Sangamon paleosol is almost always covered by Wisconsinan loess deposits (fig. 10.8). The duration of the Sangamon Interglacial in Iowa is estimated to be from 125,000 to approximately 55,000 years ago. Figure 10.8, based on the work of George R. Hallberg (Kemmis et al. 1992), shows current correlations between the Pleistocene of eastern and western Iowa.

WISCONSINAN GLACIAL STAGE

The Wisconsinan Glacial Stage in Iowa includes drift deposits of the Sheldon Creek and Dows formations and eolian deposits of the Pisgah and Peoria formations (fig. 10.6). The Wisconsinan ice advances entered Iowa from the northwest along paths similar to some of the Pre-Illinoian advances (fig. 10.7). At one time, geologists believed that Iowa was affected by four separate advances and retreats of Wisconsinan ice sheets, and four glacial substages were proposed (from oldest to youngest): Iowan, Tazewell, Cary, and Mankato. At present, only two Wisconsinan drift sheets (fig. 10.6) are recognized in the state, the Sheldon Creek Formation (older) and the Dows Formation (younger).

An area of 8,800 square miles in northeastern Iowa that was formerly mapped as the deposits of the Iowan Glacial Substage is now recognized as the Iowan Surface landform region — a complex erosion surface that developed on Pre-Illinoian drift during Late Wisconsinan time, between 16,500 and 21,000 years ago. The Iowan Surface landform region formed during the coldest interval of the Wisconsinan Glacial Stage. During that time, topographic relief in northern Iowa was reduced by a variety of processes, including freeze-thaw action, dislodgement and downslope movement of loosened material, sheetwash on slopes, and wind erosion.

The Sheldon Creek Formation, named for a creek in Clay County, occurs at the surface in a portion of northwestern Iowa (fig. 10.6) and is buried by younger Dows Formation deposits in the Des Moines Lobe landform region of northern Iowa. Radiocarbon dates suggest that the Sheldon Creek Formation is approximately 25,000 to 38,000 years old. The Wisconsinan glacier that deposited the Sheldon Creek Formation apparently entered the state from the northwest, where it derived Cretaceous rocks and fossils from the bedrock of Minnesota and the Dakotas. Cretaceous-age Pierre Shale fragments are common in the Sheldon Creek drift of northwestern Iowa.

Deposition of Wisconsinan loess (12,500–31,000 years ago) took place during

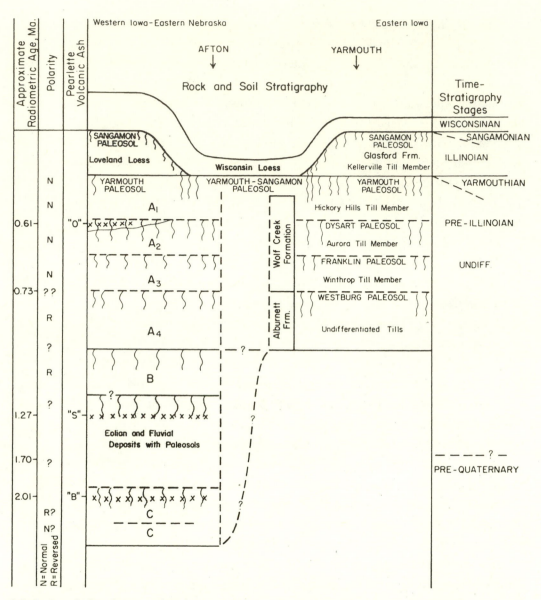

10.8 Correlations between the Pleistocene sections in eastern and western Iowa. The Wisconsinan drift sheets of northern Iowa are not shown. The generalized cross section shows the approximate locations of the type sections of the Afton and Yarmouth interglacials and the stratigraphic positions of Pearlette volcanic ashes O, S, and B. The wavy vertical lines portray paleosols and buried weathered zones, indicators of interglacial conditions. Radiometric ages and normal and reversed polarity events are indicated. From Kemmis et al. 1992.

and after deposition of the Sheldon Creek Formation. Loess deposits include the Pisgah Loess (older) and Peoria Loess (younger); the two deposits are separated by the Farmdale Geosol. For the most part, the Iowan Surface is younger than the deposition of the Sheldon Creek drift.

The Dows Formation (fig. 10.6), named for a town in Franklin County, includes all of the upland glacial deposits on the Des Moines Lobe landform region

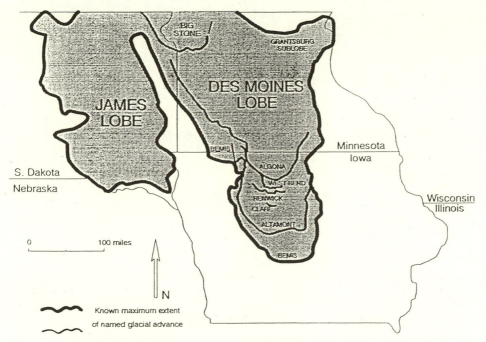

10.9 The location of the Des Moines and James lobes of Wisconsinan Glacial time. The Bemis moraine marks the southernmost advance of the ice into central Iowa. Readvances of the ice later produced the Altamont, Clare, Renwick, West Bend, and Algona moraines. From Bettis et al. 1996.

(fig. 10.9). The Dows drift marks the last interval of glaciation in Iowa, when ice of the Des Moines Lobe advanced into the state about 15,000 years ago. The Des Moines Lobe extended from central Canada through the Dakotas and Minnesota into Iowa. Ice of the Des Moines Lobe reached as far south as Iowa's capital city and retreated from the state about 12,000 years ago. A similar lobe, the James Lobe, extended into South Dakota at about the same time (fig. 10.9).

Since the late 1970s, the deposits of the Des Moines Lobe have been remapped and reevaluated. In particular, the efforts of Timothy J. Kemmis, Deborah J. Quade, and E. Arthur Bettis III (Bettis et al. 1996) have contributed to a better understanding of the Des Moines Lobe landforms and materials. According to Kemmis (Bettis et al. 1996), the Des Moines Lobe formed by a special type of glacial action known as surging. Surging is characterized by a rapid out-of-equilibrium glacial advance, followed by subsequent stagnation of the glacial ice.

The age of the Des Moines Lobe is known from numerous radiocarbon dates. Glacial ice of the lobe first surged into Iowa about 15,000 years ago and reached its terminal position in Des Moines approximately 13,800 years ago. The glacier then stagnated but readvanced to the position of the Altamont ice margin (fig. 10.9) near Ames and Boone about 13,500 years ago. The Altamont advance was followed by another interval of ice stagnation. Between 13,500 and 12,600 years ago, three minor readvances occurred and are marked by the Clare, Renwick, and West Bend moraines (fig. 10.9). The topography associated with these minor advances consists of discontinuous moraines. The final advance of the Des Moines Lobe ice took place 12,300 years ago and produced the Algona moraine (fig. 10.9).

The upper Des Moines River was established on the Des Moines Lobe during the Algona advance. By 12,000 years ago, the glacial ice had all melted from the state, and the subsequent history of the Des Moines Lobe area involved the development of a drainage network, infilling of depressions, and landscape modification.

One can only speculate as to the thickness of the ice sheet in Iowa. Based on topographic considerations, the James Lobe ice in South Dakota is estimated to have been 1,000 to 1,300 feet thick. The thickness of the Des Moines Lobe ice in Iowa may have been similar.

Three major end moraines (Bemis, Altamont, and Algona) are associated with the Des Moines Lobe (fig. 10.9). These topographic features formed along the margin of the Des Moines Lobe ice. Today, Iowa's landscape in the area of the moraines is characterized by hummocky topography. Topography of this type, formed by the irregular deposition of glacial drift from melting ice, consists of alternating rounded or knobby hills and poorly drained depressions. Swaledale in Cerro Gordo County, Pilot Mound in Boone County, and Pilot Knob State Park in Hancock County all derive their names from the irregular glacial-drift topography on which they are located.

Formation of the Iowan Surface had ceased prior to deposition of the Dows Formation of the Des Moines Lobe, but loess deposition was occurring during the advance of the Des Moines Lobe ice. The till deposits (Dows Formation) are silty where the ice advanced over the loess. Eolian sands overlie the Dows Formation in places. Several fens (peat-forming wetlands) developed on the Des Moines Lobe during postglacial time, and they contain a paleobotanical record that is helpful in deciphering the state's Holocene history.

In addition to the Dows Formation (glacial drift), three additional formations (fig. 10.6) are mapped within the area of the Des Moines Lobe. The formations are the Noah Creek Formation (glaciofluvial deposits), the Peoria Formation (wind-transported sediments), and the DeForest Formation (Holocene alluvium, slope deposits, and pond and lake sediments).

The Noah Creek Formation is composed of glaciofluvial sediments, primarily sand-and-gravel deposits. Named for a locality in Boone County, the Noah Creek Formation occurs in present and abandoned stream valleys and on outwash plains. There are few major rivers within the Des Moines Lobe, but those present are located at the margins of former glacial advances. All of these rivers accumulated Noah Creek sediments at various times during late Wisconsinan time. The Noah Creek ranges in age from about 11,000 to 14,000 years old.

The Peoria Formation is named for Peoria, Illinois; it includes wind-transported materials throughout most of Iowa. Two facies are recognized: a silt facies (loess) and a sand facies (eolian sand). On the Des Moines Lobe, the silt facies is buried by glacial deposits of the Dows Formation. However, at some localities on the Des Moines Lobe, the sand facies of the Peoria Formation unconformably overlies the Dows Formation. The silt facies accumulated between 22,000 and 12,500 years ago, whereas the sand facies formed contemporaneously with the silt facies and again during the Holocene Epoch.

HOLOCENE EPOCH

The Holocene (Recent) Epoch represents the last 10,000 to 10,500 years of geologic time. Holocene deposits in Iowa include local deposits of eolian sand, sometimes mapped as the Peoria Formation and deposits of the DeForest Formation (fig. 10.6). The DeForest Formation is a Holocene unit named for a locality in Harrison County, Iowa. Varied in composition, the formation is represented primarily by fine-grained alluvium, slope colluvium, and pond sediment. The DeForest Formation abruptly and unconformably overlies the Dows and Noah Creek formations, as well as any older Quaternary and bedrock units into which it is incised. The DeForest deposits are time transgressive. On the Des Moines Lobe, the DeForest deposits are younger than 11,000 years, but elsewhere in the state, the formation is as old as 14,000 years. Deposition of the DeForest Formation (fig. 10.6) continues today in the form of modern alluvial deposits.

The Landform Regions of Iowa

In 1991, Jean C. Prior described and defined seven landform regions of Iowa (fig. 10.10) in a beautifully illustrated and highly informative book entitled *Landforms of Iowa*. Anyone who is interested in Iowa's landscape should study Prior's book. Two of the landform regions, the Iowan Surface and the Des Moines Lobe, were mentioned previously because of their connection with Iowa's Wisconsinan glacial history. The Iowan Surface was once attributed to deposition by glacial ice; it is now recognized as a complex erosional feature. The Des Moines Lobe is underlain primarily by deposits of the Wisconsinan Glacial Stage.

The relative ages of Iowa's landform regions can be determined to some extent by the ages of loess deposits. Loess deposition occurred in Iowa during the advance of the Des Moines Lobe ice and up to the time of deglaciation, 12,500 years ago. However, any loess deposited on the surface of the glacial ice was disturbed when the ice melted. The surface of the Des Moines Lobe is essentially loess-free. The Alluvial Plains (fig. 10.10) are also loess-free. They are a young landscape area, carpeted primarily with Holocene alluvium. The other five landform regions of the state (Loess Hills, Southern Iowa Drift Plain, Iowan Surface,

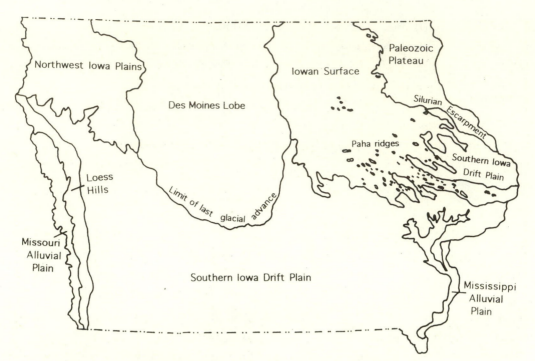

10.10 The seven landform regions of Iowa. Note the location of paha ridges within the Iowan Surface landform area. Adapted from Prior 1991.

Northwest Iowa Plains, and Paleozoic Plateau) are all blanketed to some extent by Wisconsinan loess.

DES MOINES LOBE

Although all of Iowa was glaciated, only the Des Moines Lobe preserves a topography directly produced by glacial action (fig. 10.11). The ice of the Des Moines Lobe surged into Iowa about 15,000 years ago and reached the area where the city of Des Moines is now located about 13,800 years ago. The present course of the Middle Raccoon River marks the approximate southern boundary of the glacier's excursion into central Iowa. Iowa's glacial lakes all lie within the borders of the Des Moines Lobe. This includes such well-known Iowa lakes and vacation spots as Clear Lake, Spirit Lake, Okoboji Lake, and Storm Lake. The highest elevation in the state, 1,670 feet above sea level, is on a protuberance of glacial drift at the northwestern edge of the Des Moines Lobe, near Sibley in Osceola County. This high point of Iowa's topography was once occupied, without fanfare, by a farmer's feedlot. It has recently been designated Hawkeye Point.

The advance of the Des Moines Lobe ice into the state was long after the bitter cold climatic regime of Late Wisconsinan time (16,500 to 21,000 years ago), when the Iowan Surface formed. The climate in Iowa 15,000 years ago was starting to moderate, as indicated by a varied fauna and flora preserved in glacial and ice-marginal deposits of the Des Moines Lobe.

10.11 The poorly drained landscape of the Des Moines Lobe in Pocahontas County, Iowa. Note the presence of lakes and ponds (dark areas) and the absence of a well-developed drainage system. From Faxlanger et al. 1973.

The most notable landforms on the Des Moines Lobe are the arcuate belts of hummocky topography that outline the margins of ice advances and the prolonged location of stagnant, debris-ridden glacial ice. The border of three major glacial advances mark the positions of the Bemis, Altamont, and Algona end moraines (fig. 10.9). Minor ice advances produced the Clare, Renwick, and West Bend moraines (fig. 10.9). The Bemis and Altamont moraines derive their names from towns in South Dakota, while the other moraines of the Des Moines Lobe are named for Iowa communities.

Iowa's end moraines, which accumulated along the borders of the stagnant ice, are variable in composition and appearance because of the diverse depositional settings in which they formed. These moraines do not conform with the classic recessional moraines described in states to the north and east of Iowa. The moraines in these states were formed by repeated episodes of glacial advance and retreat. Similar processes were once thought to be responsible for the moraines of the Des Moines Lobe. However, the moraines of north-central Iowa are now attributed to rapid surges of glacial ice, followed by regional stagnation of the ice. Previously, the Des Moines Lobe deposits were mapped primarily as either ground moraines or as end moraines. Geologists now realize that such terms do not adequately describe the variety and complexity of these deposits.

As the glacial ice stagnated after the various episodes of surging, tunnels formed within the ice. These ice-bordered conduits served as pathways for meltwater and sediment, and some of these corridors became linked. Later, these linked pathways for glacial meltwaters became the sites of the present-day rivers of the Des Moines Lobe.

Distinctive landform features called kames are produced where openings within glacial ice accumulate deposits of sand and gravel from meltwater. As the glacial ice melts away, these sediment fillings are let down on the underlying landscape as conical-shaped mounds. Ocheyedan Mound State Preserve in Osceola County is a classic example of a kame (fig. 10.12). Other prominent kames in Iowa include Pilot Knob in Hancock County and Pilot Mound in Boone County.

If an isolated block of ice melts within glacial drift, the result is often a bowl-shaped depression called a kettle. Freda Haffner Kettlehole in Dickinson County (fig. 10.13) is a noteworthy example of such a feature.

The landscape of the Des Moines Lobe is poorly drained (fig. 10.11), and the area is only moderately dissected by a network of sluggish drainages. There are few major stream valleys on the Des Moines Lobe. The Des Moines River valley is the most prominent of these, and it has cut deeply into Pennsylvanian bedrock at several localities. For example, Pennsylvanian rocks are well exposed at Dolliver Memorial and Ledges state parks and at the Saylorville Reservoir area in central Iowa. Meltwater derived from the Des Moines Lobe ice helped define the courses of the Cedar, Shell Rock, Iowa, Skunk, Raccoon, and Boyer rivers. The Little Sioux River basin in northwestern Iowa was transformed into a tributary system of the Missouri River when the ice of the Des Moines Lobe blocked and diverted streams that previously flowed toward the Mississippi River.

The names of certain towns on the Des Moines Lobe reflect the wetland heritage of the area. For example, the communities of Curlew, Plover, and Mallard were named for waterfowl. Although the natural wetlands of the Des Moines Lobe serve as great habitat for birds and other wildlife, they often impede the development of modern agriculture. Consequently, many of the original wetlands of the area have been drained.

ALLUVIAL PLAINS

Although all of the state's landform regions have been sculpted to some extent by running water, nowhere is the action of streams more evident than on the wide alluvial plains of the Missouri and Mississippi rivers (fig. 10.10). The wide, sandy lowlands along Iowa's western and eastern borders are large enough to plot on the map shown in figure 10.10. The alluvial plains shown in southeastern Iowa also include segments of the Iowa and Cedar river valleys, an area once attributed in origin to Lake Calvin. Based on the extensive nationwide media coverage of Iowa's floods in 1993, many people from outside the state envisioned

10.12 A drawing of Ocheyedan Mound in Osceola County, Iowa. The prominent mound, once thought to be Iowa's highest point, is a glacial kame and lies along the western edge of the Bemis moraine. Kames are conical hills or knobs where sand and gravel filled a large cavity once present within stagnant glacial ice. From Faxlanger et al. 1973.

10.13 Geologist Jean C. Prior, on the left, explains the origin of Freda Haffner Kettlehole in Dickinson County. The large depression (kettle) formed on the Altamont moraine, probably from an isolated mass of ice that melted within glacial drift. Photo by the author.

Iowa as one huge alluvial plain, swamped by floodwater. Actually, the Alluvial Plains represent one of the smallest of Iowa's landform regions in terms of area.

Typical landscape features produced by the action of running water are revealed along the valleys of the Mississippi and Missouri rivers, as well as along the interior streams of the state. Lewis and Clark State Park on the Missouri alluvial plain in Monona County and George Wyth State Park adjacent to the Cedar River in Black Hawk County illustrate the varied recreational uses of alluvial lands and floodplains. Floodplains are occasionally submerged when rivers carry excess water, as after heavy rains or following rapid melting of snow. Meandering rivers often leave segments of their previous channels abandoned on floodplains. Former channel loops are seen today as oxbow lakes. The broad alluvial flats and associated features of the modern Mississippi and Missouri river valleys provide important habitats for migratory waterfowl and other wildlife.

Evidence of change is often recorded in a stream valley by terraces and benches. Terraces are nearly flat features standing above the level of the current floodplain. They are remnants of former floodplains that were stranded on the landscape after a stream completed a new episode of erosion and downcutting. Terraces are underlain by floodplain deposits — sand, gravel, silt, and clay. In contrast, benches are covered with only a thin layer of alluvial sediment; they are

10.14 The flat Missouri alluvial plain (left) contrasts with the rugged Loess Hills (right), Harrison County, Iowa. From Faxlanger et al. 1973.

floored by more resistant materials such as bedrock or glacial drift. Multiple levels of terraces or benches in a stream valley provide the clues to interpret the past history of the steam valley. The history of the Des Moines River valley between Fort Dodge and Des Moines has been deciphered through such evidence.

Many of Iowa's rivers appear too small to have carved the broad valleys they currently occupy. The oversized valleys of these underfit streams provide evidence of former times, when huge volumes of meltwater flowed through the area.

Most of the landforms and deposits associated with alluvial plains owe their origin to flowing water. Occasionally, the work of wind is recorded. For example, wind-deposited sand dunes occur along the valleys of the Mississippi, Des Moines, Skunk, and Cedar rivers in eastern Iowa. Dune fields are also present at Sand Cove on a terrace of the Upper Iowa River in Allamakee County and at Big Sand Mound on a terrace remnant in the Mississippi Valley south of Muscatine. Both of these dune fields are currently active. Many of the eolian sands in Iowa trace their origin to Late Wisconsinan time, however, and most are currently stabilized.

The alluvial plains of the Missouri (fig. 10.14) and Mississippi rivers stand out because of their size. Both valleys were occupied by great volumes of glacial meltwater approximately 9,500 to 30,000 years ago. The floodplain deposits of the two rivers were major sources of the widespread loess of the Midwest. Both rivers have complex histories.

An ancestral Missouri River system once flowed eastward across western Iowa. The headwater region of the river system was in the Rocky Mountains. Advances by Pre-Illinoian glaciers probably diverted the ancient river system southward in Iowa. The current position of the Missouri Valley was established by or during Illinoian time.

The pathway of the Mississippi River in northeastern Iowa traces to ice-marginal drainages associated with Pre-Illinoian glacial events over 500,000 years ago. The course of the Mississippi from the northeastern corner of the state to near Clinton has occupied its present position since early in the Pleistocene. The valley of the modern Mississippi from near Clinton to the St. Louis area, however, is much younger. The path of the Mississippi once flowed southeastward near Clinton through a course in Illinois called the Princeton Channel. The Princeton Channel is named for a town in Illinois 60 miles east of the Quad Cities. The Princeton Channel merged with a bedrock valley 40 miles northnortheast of Peoria. From there, the ancestral Mississippi flowed southwestward to the St. Louis area, roughly parallel to the modern Illinois River valley.

During Wisconsinan Glacial time, between 21,000 and 25,000 years ago, a large lobe of glacial ice advanced southwestward out of the Lake Michigan Basin and blocked the Princeton Channel, diverting the course of the Mississippi to the west through the Port Byron Gorge near the Quad Cities. When the Wisconsinan ice finally melted, the Mississippi established its modern course in southeastern Iowa. Figure 10.15 shows the modern and ancestral courses of the Mississippi drainage system.

LOESS HILLS

The Loess Hills landform region (fig. 10.16) of western Iowa is distinctive. The range of elevation within a region defines its relief, and the change in relief is striking when you ascend the Loess Hills from the adjacent flat terrain of the Missouri River alluvial plain (fig. 10.14). The contrast in topography between the subdued landscape of the Great Plains and the adjacent towering Colorado Rockies is also impressive. The difference in elevation between these two regions is dramatic and nearly an order of magnitude greater than the total relief (1,190 feet) in the entire state of Iowa. Although no one would mistake the Loess Hill for the Rockies, Thaddeus Culbertson's observation in 1850 captures a common first impression of the loess-capped landscape of western Iowa. According to Mutel (1989), Culbertson, a nineteenth-century missionary, described the Loess Hills as "Mountains in Miniature."

The Loess Hills of western Iowa occupy a narrow band of varying width along the eastern side of the Missouri River valley (fig. 10.14). They border the entire length of the Missouri River valley in Iowa, a distance of approximately 200 miles. Deeply mantled loess topography is confined to a belt 2 to 10 miles wide adjacent

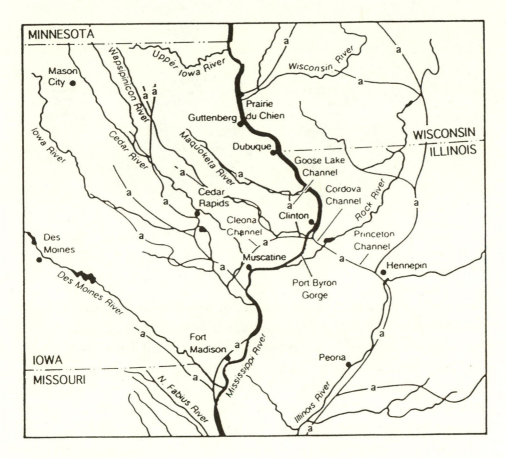

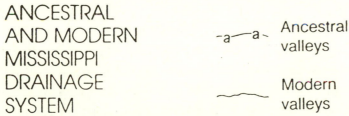

ANCESTRAL AND MODERN MISSISSIPPI DRAINAGE SYSTEM

-a—a- Ancestral valleys

——— Modern valleys

10.15 The location of the modern and ancient Mississippi drainage system. Adapted from Bettis 1987.

to the Missouri alluvial plain. Streams and gullies have dissected the area, and the region is characterized by narrow ridge crests and steep slopes (fig. 10.16). The vistas are magnificent from the hilltops, and you can see for tens of miles to the west across the vast alluvial plain below. Stone, Preparation Canyon, and Waubonsie state parks afford the visitor excellent views of Loess Hills landforms. The views are equally impressive from public lands north of Turin in Monona County. Two tracts within the Loess Hills have been designated as National Natural Landmarks.

The thickest deposits of loess in Iowa are in the Loess Hills, close to their source on the Missouri alluvial plain. The thickness of loess in the Loess Hills

10.16 The landscape of the Loess Hills in Pottawattamie County, north of Council Bluffs, Iowa. Drawing by geologist-artist Orestes St. John in 1868, looking southeast. From White 1870.

is usually more than 60 feet, and total thicknesses of 150 to 200 feet have been documented from well logs and field studies. At least three episodes of loess deposition are recorded in the Loess Hills area. The oldest deposit is the Loveland Loess, laid down in Illinoian time approximately 140,000 to 160,000 years ago. The Loveland Loess is capped by the Pisgah Formation. The Pisgah Loess formed in Wisconsinan time, 26,000 to 55,000 years ago. The Peoria Loess overlies the Pisgah deposits and includes Wisconsinan loess deposits from 12,500 to 25,000 years old. The Peoria Loess is the thickest and most widespread of Iowa's loess deposits.

Loess deposits are thicker on the eastern side of the Missouri River valley because the winds were predominantly from the west. Variations in wind direction are responsible for loess deposits on the western side of the Missouri River in Kansas, Nebraska, and South Dakota. Loess deposits are known elsewhere, too. Extensive deposits occur in the lower Mississippi Valley, along the Platte River valley in the Great Plains, in the Palouse area of eastern Washington, in Germany and elsewhere in Europe, and in northern China. The landscapes developed on the thick loess deposits along the Yellow River in China are similar in appearance to those of Iowa's Loess Hills landform region.

The wind-laid loess deposits of western Iowa have been modified subsequently by erosional agents such as running water and mass movement. The dissected topography of the Loess Hills bears the distinctive signature of running water. Rates of soil erosion are extremely high in the area, and streams draining the region carry huge amounts of eroded sediment. Gully erosion is pronounced in loess and often extends upslope into cropland. Stream channels and drainage ditches in the Loess Hills frequently become clogged with sediment after rainstorms and periods of heavy runoff.

Catsteps (terracelike features) are common in the Loess Hills. Scientists offer two hypotheses for the origin of catsteps: the trampling action of livestock or other animals and the periodic slumping and downslope movement of loess under the influence of gravity. Catsteps indicate that the slopes of the Loess Hills are unstable.

When loess is dry, it maintains nearly vertical slopes (fig. 10.4) because of its cohesiveness. The style of roadcuts in the Loess Hills illustrates this. The excavations often have only a single steep face, or they may be stepped back in a series of steep walls and alternating flat surfaces. As long as the loess remains dry, these nearly vertical faces remain stable. However, during wet conditions the slopes often become unstable, and troublesome landslides may occur.

Pre-Illinoian drift occurs below the loess in western Iowa. Included within these deposits are sand-and-gravel layers that yielded the so-called Aftonian fauna that was described by Samuel Calvin (1909a) and Bohumil Shimek (1909). This fauna provided important insights into Iowa's Pleistocene life and environments.

Buried within the Pre-Illinoian sequence are ash falls of the Pearlette Family of ash deposits. One such layer of ash near the Harrison-Monona county line exceeds 1 foot in thickness and yields a date of 620,000 years old. Pennsylvanian and Cretaceous bedrock are revealed below the surficial materials in a number of areas in the Loess Hills. For example, Cretaceous rocks are well exposed near Stone State Park in the Sioux City area, and Pennsylvanian strata occur in the vicinity of Council Bluffs and elsewhere. Locally, the Tertiary salt-and-pepper sands serve as an important aquifer.

SOUTHERN IOWA DRIFT PLAIN

The Southern Iowa Drift Plain borders the Loess Hills on the east. As figure 10.10 shows, the Southern Iowa Drift Plain is by far the largest of Iowa's landform regions. A traveler crossing Iowa from west to east on Interstate 80 passes over the Missouri Alluvial Plain and the Loess Hills landforms regions rather quickly. Both regions are situated in the western third of Pottawattamie County. The remainder of the route across western Iowa into central Iowa lies within the Southern Iowa Drift Plain. Across Polk County (the Des Moines area), the I-80 corridor encounters the southern margin of the Des Moines Lobe for a stretch of approximately 30 miles. The rest of the traverse across the state on I-80 is entirely within the Southern Iowa Drift Plain. It is no wonder that many out-of-state travelers remember Iowa's topography as uniform.

Grant Wood, the well-known Iowa artist, drew attention to the landscape of southern Iowa. His lithographs *March*, *July Fifteenth*, and *In the Spring* and paintings such as *Fall Plowing*, *Stone City*, and *Young Corn* capture the picturesque hills and valleys of what many people accept as the state's typical landscape — the Southern Iowa Drift Plain.

Subtle differences are observable in the landscape of southern Iowa. For ex-

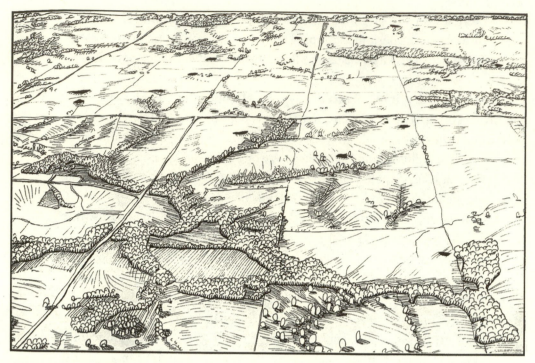

10.17 The undulating topography of the Southern Iowa Drift Plain in Lucas County, Iowa. Loess overlies Pre-Illinoian drift in this area. A multitude of farm ponds (dark areas) dot the landscape. From Faxlanger et al. 1973.

ample, the eastern part of the area is dominated by tabular uplands, whereas the western segment of the landform region is more rolling and has a thicker cover of loess. The central part of the region is also rolling (fig. 10.17), but it has a thinner cover of loess. At the extreme eastern edge of the Southern Iowa Drift Plain (Maquoketa Caves State Park, for example), the landscape is rugged, and the Maquoketa River is entrenched into bedrock. Talus slopes with cold-air drainage (algific slopes) have been reported from this segment of the Southern Iowa Drift Plain by Eilers and Roosa (1994). Algific slopes provide habitat for rare and endangered plant species that are more commonly associated with the Paleozoic Plateau of northeastern Iowa. The dissected terrain of southern Iowa is well suited for the location of dams, reservoirs, and farm ponds (fig. 10.17). Coralville, Rathbun, and Red Rock reservoirs help control flooding and provide Iowans with a variety of recreational opportunities.

The surface of the Southern Iowa Drift Plain is underlain primarily by Pre-Illinoian glacial drift, mantled with various thicknesses of Wisconsinan loess. In southern Iowa, erosion through time has removed the telltale features of glaciation that are so common in the northern part of the state. Evidence of morainal topography, glacial lakes and bogs, or ice-contact features such as kames and kettles no longer exists. Whereas the topography of the Des Moines

Lobe is poorly drained and characterized by low relief, southern Iowa exhibits a well-established drainage system and noticeable relief. The Pre-Illinoian glacial deposits of southern Iowa have been subjected to erosional processes for over 500,000 years, so it is not surprising that all evidence of primary glacial-depositional landforms has been erased.

A small area of southeastern Iowa (fig. 10.7) reveals Illinoian glacial deposits overlying the Pre-Illinoian sequence. Rivers of the Southern Iowa Drift Plain have carved deeply into the Pre-Illinoian drift, and Paleozoic bedrock is exposed in several places. In areas where Pennsylvanian bedrock is present, the groundwater quality and quantity may be poor in wells attempted in shale-dominated strata. Consequently, rural Iowans in these areas often depend on water from shallow seepage wells completed at the interface of permeable loess deposits with the underlying impermeable Pre-Illinoian till.

The modern geologist who specializes in landscape development sees the Southern Iowa Drift Plain as more steplike than typically portrayed by Grant Wood. The appreciation of the stepped character of the Southern Iowa Drift Plain and its evolution over time takes a trained eye. Erosion in the Southern Iowa Drift Plain was not uniform and continuous. Instead, episodes of rapid erosion and landscape development alternated with intervals of greater stability. Because of the intermittent nature of landscape formation, the slopes of southern Iowa are not one continuous feature, nor are they everywhere the same age. The landscape of southern Iowa has a somewhat stepped appearance, and the remnants of four generations of erosion surfaces have been recognized (figs. 10.18 and 10.19). The oldest of the surfaces is the Yarmouth-Sangamon Surface. It occurs at higher elevations and apparently represents a remnant of the Pre-Illinoian till plain that has endured for more than 500,000 years.

In a stack of sedimentary rocks in a quarry or canyon, the older units are at the bottom and are overlain by progressively younger units. Just the opposite is true for landscape surfaces. The older landscape surfaces are preserved at higher elevations, and the floodplains of the major rivers represent the youngest landscape surface in an area (fig. 10.19).

The Yarmouth-Sangamon Surface is marked by an ancient soil called the Yarmouth-Sangamon paleosol (fig. 10.18), which is often represented by a sticky, gray clay. Referred to as gumbotil in the early literature, this clay-rich interval impedes the downward movement of groundwater. Consequently, seeps and springs occur commonly on hillslopes at the interface of the Yarmouth-Sangamon paleosol and the overlying Wisconsinan loess.

A younger erosion surface, the Late Sangamon Surface (fig. 10.18), cuts into Pre-Illinoian drift below the level of the Yarmouth-Sangamon Surface. The erosion that produced this surface removed the Yarmouth-Sangamon paleosol. A distinctive reddish brown paleosol marks the Late Sangamon Surface in some

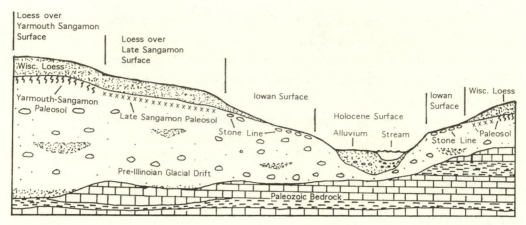

10.18 A schematic cross section of the Southern Iowa Drift Plain showing four surfaces: Yarmouth-Sangamon, Late Sangamon, Wisconsin (Iowan), and Holocene. The diagram is not to true scale. Adapted from Ruhe 1969 and Prior 1991.

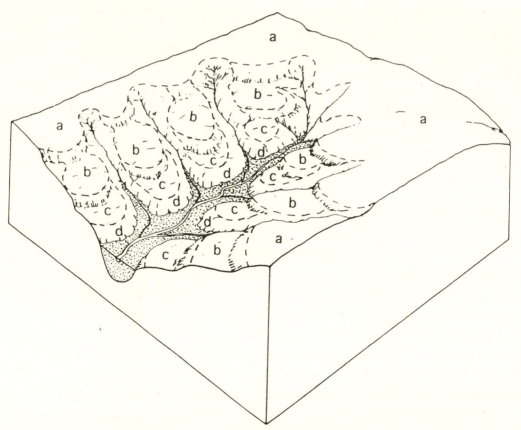

10.19 A diagram of the four surfaces of the Southern Iowa Drift Plain: (a) loess-mantled Yarmouth-Sangamon surface and paleosol, (b) loess-mantled Late Sangamon erosion surface and paleosol, (c) loess-mantled Wisconsinan-age erosion surface (Iowan surface), and (d) Holocene erosion surface. Holocene alluvium occurs on the floodplains and valley floors. Adapted from Kemmis et al. 1992.

areas. Thick loess, 15 to 16 feet deep, blankets both the Yarmouth-Sangamon and Late Sangamon surfaces.

Lower on the landscape of southern Iowa, a third surface occurs. According to Kemmis et al. (1992), this surface has been referred to by various terms ("Iowan surface," "Iowan Erosion Surface," and "Early Wisconsin pediment"). Labeled "Wisconsin (Iowan) Surface" (fig. 10.18), it is cut below the level of the Late Sangamon Surface and paleosol. The maximum thickness of loess on this surface is about 7.5 feet. The erosion surface apparently formed concurrently with loess deposition from 16,500 to 21,000 years ago. The surface contains no paleosol. There was either insufficient time to form a soil on the Wisconsin Surface or conditions were not right to preserve it. Stone lines or pebble bands are common on the Wisconsin Surface and are found on the Late Sangamon Surface as well. These concentrations of coarse gravels and cobbles represent lag deposits. They were concentrated on Iowa's ancient landscapes during erosion intervals, when finer-grained materials were removed by erosion by wind and water. The Wisconsin Surface occurs on the lower slopes of drainage basins in southern Iowa, but the surface is not very conspicuous. However, in the adjacent landform region to the north, this surface is a dominant landscape feature.

The flatlands along the modern valley floors (fig. 10.18) mark the youngest erosion surface in the Southern Iowa Drift Plain. This surface is of Holocene (Recent) age and is covered with postglacial alluvium.

IOWAN SURFACE

The Iowan Surface landform region (fig. 10.10) has much in common stratigraphically with its neighbor to the south, the Southern Iowa Drift Plain. Both are underlain by Pre-Illinoian drift. The loess cover on the Iowan Surface region is less thick than the loess of the Southern Iowa Drift Plain, however. There are obvious topographic differences between the regions, too. Whereas the Southern Iowa Drift Plain has noticeable relief and distinctive hills and valleys, the Iowan Surface region displays low relief, with long slopes and poorly defined drainage divides. Prior to the late 1960s, geologists explained the topography and surface deposits of the Iowan area by a Wisconsinan ice advance. The paha (elongated hills) of the Iowan Surface were then considered as either smooth landforms of molded Wisconsinan till or as topographic highs that were bypassed by thin Wisconsinan ice. The proposed Wisconsinan ice advance was considered to predate the Des Moines Lobe.

Paha is a Dakota Sioux word for hill or ridge. W. J. McGee (1891) used the term to describe the streamlined "dolphin-backed hills" of northeastern Iowa. According to our current understanding, paha are erosional remnants of Pre-Illinoian drift preserved by thick, wind-blown deposits. As Prior (1991) notes, 80 percent of the 116 paha mapped in the Iowan Surface region occur in the

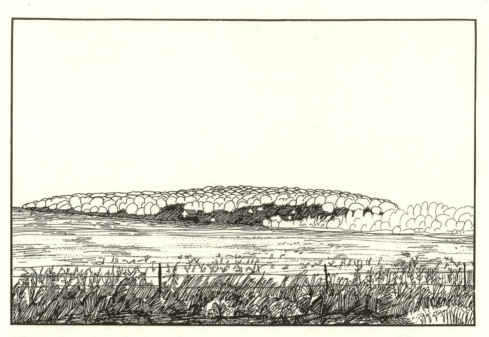

10.20 This paha in Benton County stands above the low-relief Iowan Surface landform region in the foreground. Paha are the highest part of the landscape in the Iowan Surface. From Faxlanger et al. 1973.

southern part of the landform area (fig. 10.10). The paha are oriented in a north-west-to-southeast direction; they are parallel to and in close proximity to river valleys: and they typically are mantled with silt and fine sand. According to Prior, the location and orientation of these paha suggest that they are wind-aligned, dunelike features deposited by strong northwest winds. Other paha occupy positions along interstream divides. These paha have a glacial stratigraphy identical to the upland areas of the Pre-Illinoian drift plain of southern Iowa. Core samples taken from such paha and core transects across the Iowan Surface to the Southern Iowa Drift Plain by Ruhe et al. (1968) demonstrate the similarity of the glacial stratigraphy throughout the area. The Iowan Surface in northeastern Iowa is cut deeply into Pre-Illinoian drift, the same material that underlies southern Iowa. It is now clear that a separate Iowan glacial drift of Wisconsinan age does not exist.

Paha (fig. 10.20) are often forested, and they represent essentially the only upland woods in the Iowan Surface area. The soils of the larger paha developed under the influence of forest settings rather than prairies. This suggests the paha have been wooded for some time.

Casey's Paha, mostly included in Hickory Hills Park in northern Tama County, is a classic example of a paha. A roadcut along Highway 21 at the west end of the feature reveals Wisconsinan loess overlying a thick, clay-rich Yarmouth-Sangamon paleosol on Pre-Illinoian drift. Cornell College in Mount Vernon and

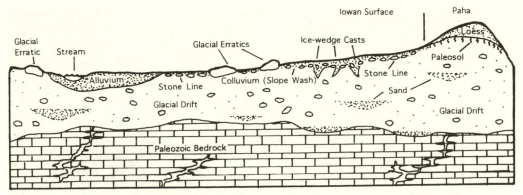

10.21 A schematic cross section of the Iowan Surface landform region showing a loess-capped paha, a stone line on Pre-Illinoian drift, and glacial erratics. Ice-wedge casts signify former periglacial conditions. The stratigraphy under paha matches that of the uplands of the Southern Iowa Drift Plain. Adapted from Prior 1991 and Walters 1994.

Kirkwood Community College in Cedar Rapids are located in paha country. Paha are usually the highest terrain on the Iowan Surface, so they are favored locations for radio and television transmitters in eastern Iowa. Commonly, paha rise 50 to 100 feet above the surrounding landscape of the Iowan Surface.

As we have seen, the Wisconsin (Iowan) Surface is one of four erosion surfaces recognized on the modern landscape of southern Iowa. However, in the Southern Iowa Drift Plain, the Winconsin Surface is much less conspicuous than the Yarmouth-Sangamon and the Late Sangamon surfaces. These older surfaces give the southern Iowa landscape its special character and appearance. Late Wisconsinan erosion and landscape development were much more pronounced in northern Iowa (fig. 10.21).

Fossils reveal that northern Iowa was a tundra 17,000 years ago. The Wisconsin Surface formed as a stepped erosional feature between 16,500 and 21,000 years ago during the coldest part of the Wisconsinan Glacial. Although glaciers did not cover northern Iowa then, the climate was very severe. It was cold and wet, with lots of freeze-thaw action. Huge volumes of surficial material were dislodged and moved downslope by gravitational processes and sheetwash. Strong winds buffeted and abraded the landscape. Patterned ground and ice-wedge casts are detected in some areas, indicating the presence of a permafrost. For nearly forty-five centuries, the northern Iowa landscape endured the big chill of a harsh periglacial. The end result was a landscape of low relief and subdued features. This is the landscape visible today in the Iowan Surface area of northeastern Iowa.

Stone lines under the stepped surfaces of the Iowan Surface landform region preserve resistant pebbles and cobbles, coarse materials that were lagged behind on the eroded slopes (fig. 10.21). Loess is thin to absent on the Iowan Surface because erosional scouring was occurring during the time of loess deposition.

10.22 An aerial photograph illustrates the topographic contrast of the low-relief terrain of the Iowan Surface landform region (left) and the rolling landscape of the Southern Iowa Drift Plain (right) near Blairstown in Benton County, Iowa. From Prior 1976.

Toward the end of the periglacial climatic regime, surface erosion became less intense. At that time, some loess accumulated and was preserved, particularly at the southern margin of the landform region.

The low-relief Iowan Surface contrasts with the typical rolling landscape of the Southern Iowa Drift Plain. Because of this, the juncture of these two landform regions often presents a sharp contrast in topography. This is particularly well displayed near Blairstown (fig. 10.22), north of Toledo, and west of Marshalltown. The Iowan Surface landform region has a distinctive look and a mappable boundary, especially along its southern border.

In north-central Iowa, the Wisconsin Surface that produced the subdued topography of the Iowan Surface landform region area is buried by the deposits of the Des Moines Lobe. In northwestern Iowa, the Wisconsin Surface is somewhat obscured by a covering of loess. The Paleozoic Plateau of northeastern Iowa (fig. 10.10) is dominated by bedrock, and only patches of Pre-Illinoian drift occur. Consequently, the Wisconsin Surface is not well expressed there.

The surficial material under the Iowan Surface is very thin in Floyd and Mitchell counties in northern Iowa. In this area of the state, sinkholes and other karst features are conspicuous in the Devonian bedrock (fig. 10.23). Elsewhere

10.23 Sinkholes are prominent in this aerial photograph of the Iowan Surface landform region near Floyd in Floyd County, Iowa. Photo by the author.

in northern Iowa, Paleozoic bedrock is revealed in numerous quarries. Streams have eroded through the Pre-Illinoian drift in several areas to expose bedrock, too. Near the eastern edge of the Iowan Surface, Backbone State Park displays bold cliffs of dolomite where the Maquoketa River has incised into Silurian bedrock.

Glacial erratics are common in the Iowan Surface region. Because of their size and resistance, many erratics survived the severe weathering and erosion that sculpted the Iowan Surface. Accordingly, some of the largest and most prominent glacial erratics in the state are found in the Iowan Surface (fig. 10.1). Erratics, of course, also occur in Pre-Illinoian deposits throughout the state and in the Wisconsinan glacial drifts. However, the combination of deep erosion of drift and thin cover of loess in northern Iowa explains why erratics are so noticeable in the Iowan Surface landform region (fig. 10.24).

NORTHWEST IOWA PLAINS

The gently rolling land of the Northwest Iowa Plains (fig. 10.10) has much in common with the Iowan Surface region of northeastern Iowa. The Northwest Iowa Plains was altered significantly by the erosional and weathering processes that produced the Iowan Surface. Both areas exhibit stones lines. Whereas the Iowan Surface is underlain by thin or no loess and Pre-Illinois drift, the Northwest Iowa Plains is covered by thicker loess over Pre-Illinoian drift or Wiscon-

10.24 Glacial erratics, like these in Butler County, Iowa, are characteristic features on the Iowan Surface landform region. Photo by the author.

sinan drift of the Sheldon Creek Formation (fig. 10.25). Even though the Sheldon Creek Formation is relatively young geologically speaking (26,000 – 40,000 years old), it preserves none of the telltale features of glaciation that are apparent on the Des Moines Lobe. The severe weathering and erosion associated with the Late Wisconsinan periglacial climate (16,500 – 21,000 years ago) modified the surface of the Sheldon Creek drift and obliterated primary depositional features such as moraines and kames. The erosional processes that formed the Iowan Surface also altered the upper surface of the Sheldon Creek or Pre-Illinoian deposits in north-central Iowa — an area that now lies buried under the younger deposits of the Des Moines Lobe.

The highest point in Iowa is a knob of glacial drift on the northwestern edge of the Des Moines Lobe, but overall the Northwest Iowa Plains region is uniformly higher than any other landform region of the state. Northwest Iowa is also the driest region in the state, averaging only 25 inches of precipitation per year. The sparse rainfall explains why northwestern Iowa has relatively few woodlands.

It is rare to see exposed bedrock in northwestern Iowa because the surficial cover of loess and glacial drift is quite thick. Cretaceous bedrock occurs in the area but is rarely exposed. Iowa's oldest exposed bedrock, the Sioux Quartzite, crops out in the northwestern corner of Lyon County — which is also the northwestern corner of the state. These significant exposures are protected at Gitchie Manitou State Preserve, affording people the opportunity to view 1.6-billion-

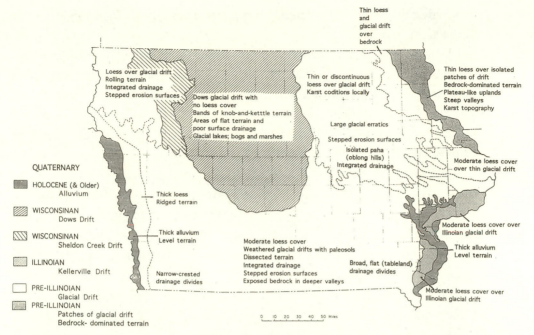

10.25 The landform materials and terrain characteristics of Iowa. Note the area of Wisconsinan-age Sheldon Creek drift to the west of the Des Moines Lobe. Adapted from Prior 1976, 1991.

year-old bedrock. A large chunk of the Sioux Quartzite is visible along the Little Sioux River valley, 3 miles south of Cherokee. Called Pilot Rock, the quartzite boulder is a glacial erratic that has long served as a local landmark. According to a sign sponsored by the Pilot Rock Chapter of the Daughters of the American Revolution, Pilot Rock was "a landmark on the prairie for the guidance of the indians, early settlers, and pioneers during the settlement and development of northwestern Iowa." Pilot Rock is 14 feet in height, and its perimeter measures 180 feet.

PALEOZOIC PLATEAU

The Paleozoic Plateau (fig. 10.10) is the most rugged of Iowa's seven landform regions. The area is often called Little Switzerland, but that is a bit of an exaggeration. Switzerland has three main land regions: the Jura Mountains, the Swiss Plateau, and the Swiss Alps. Certainly, no one would confuse northeast Iowa with the Jura Mountains or the Swiss Alps. The Swiss Plateau is a hilly region between the two Swiss mountain ranges. The topography of the Swiss Plateau has some similarity to the Paleozoic Plateau, but there are differences, too. The plateau province of Switzerland has numerous glacial lakes, such as Lake Geneva and Lake Constance. Even though comparisons with Switzerland are not altogether accurate, the Paleozoic Plateau region is still very scenic. It stands apart from the rest of Iowa with its rocky appearance and high relief. Bedrock imparts a special character to the land in northeastern Iowa and contributes significantly to the area's beauty.

The landscapes of northeastern Iowa, southwestern Wisconsin, southeastern Minnesota, and northwestern Illinois have much in common. The region was once described as the Driftless Area because of the apparent absence of glacial drift. Patches of Pre-Illinoian drift are now known in Iowa and Minnesota, but southwestern Wisconsin and the tip of northwestern Illinois are still mapped as driftless (fig. 10.7).

Glacial drift is rare in the Paleozoic Plateau, and bedrock exerts a strong influence on the area's landscape. Vertical breaks (joints) in the bedrock control some of the pathways of streams, and they impart a blocky character to cliffs. Resistant formations of limestone and dolomite cap the hills and bluffs, sandstone exposures add variety to the landscape, and less-resistant shales weather to form slopes.

The Silurian Escarpment is a prominent landscape feature in northeastern Iowa; it traces the western edge of the Paleozoic Plateau from Fayette County to southwestern Jackson County (fig. 10.10). Resistant Silurian dolomite underlies the prominent ridgelike feature across northeastern Iowa, although Devonian carbonates cap the escarpment locally at Goeken Park in Fayette County. Brush Creek Canyon, Mossy Glenn, Bixby, and White Pine Hollow state preserves and Bellevue State Park are located along the escarpment. Shales of the Ordovician Maquoketa Formation occur below the Silurian dolomites, and this juncture is marked by a change of slope. Large blocks of Silurian dolomite often slide and slump downslope on the underlying shale beds.

To the east of the Silurian dolomites and the shales of the Maquoketa Formation, carbonates of the Ordovician System contribute to the scenery of northeastern Iowa. Examples include the palisades along the Mississippi River at Dubuque, spectacular roadcuts near Guttenberg, bold bluffs in Clayton and Allamakee counties, and rocky cliffs along the Upper Iowa River in Winneshiek County. Sandstones of the Cambrian System crop out in Clayton and Allamakee counties and produce picturesque ledges and crags.

Slightly acidic groundwater has dissolved and etched the carbonate rocks of the Paleozoic Plateau in several areas. Sinkholes are particularly common in western Allamakee County and southern Clayton County. Caves such as Coldwater Cave and Spook Cave also attest to the action of groundwater in northeastern Iowa.

Ice caves occur in the Paleozoic Plateau region and include fine examples such as Decorah Ice Cave in Winneshiek County and the ice cave at Bixby State Preserve in Clayton County. In settings such as these, cold-air drainage through caves and talus at the base of north-facing slopes provides a special habitat for plants and animals. Northeast Iowa's algific (cold-air drainage) slopes are home to some of the state's rarest plants and a species of snail that traces to Ice Age time.

Although the present landscape surface of northeastern Iowa is strongly influenced by ancient Paleozoic bedrock, the landscape is relatively young geologically. The modern landscape developed during the Pleistocene and Holocene (Recent) epochs. Patches of Pre-Illinoian glacial drift cap the bedrock in several places, indicating that the Pre-Illinoian drift was once widespread in the Paleozoic Plateau. Loess-covered, reddish brown, Late Sangamon paleosol occurs on some of the remnants of glacial drift. The Late Sangamon interval also left its mark on the bedrock with a weathering product of reddish clay. Part of the modern landscape is inset below the level of the Late Sangamon Surface and its paleosol, so the landscape below the level of truncation is younger than the Late Sangamon Interglacial. Therefore, some of the relief in the Paleozoic Plateau developed after Late Sangamon time. However, a great deal of the relief of the Paleozoic Plateau probably predates the Late Sangamon Interglacial.

At other locations in northeastern Iowa, loess lies directly on eroded bedrock with no intervening paleosol. This bedrock surface corresponds to the Wisconsin Surface so prominently developed in the Iowan Surface region. The harsh periglacial climate under which this surface formed produced massive deposits of slope wash (colluvium) and talus blocks along the entrenched drainages of northeastern Iowa.

Quaternary Life

A general pattern of vegetation in the central and eastern United States during Wisconsinan glaciation (about 15,000 years ago) is shown in figure 10.26. As the glacial ice melted back to the north, the vegetation and climatic zones shifted to the north as well. Coniferous forests gave way to deciduous forests, and the deciduous forests in turn gave way to prairie grasslands.

Similar shifts in vegetation patterns probably took place during the earlier glacial episodes, too. The older record is not as well preserved as the Wisconsinan and postglacial records, however. Also, the pre-Wisconsinan materials are too old to date by radiocarbon techniques, so they are not as easily placed in a meaningful time sequence.

Paleobotany (the study of fossil plants) and palynology (the study of fossil pollen) have played important roles in the reconstruction of glacial and interglacial environments in Iowa. Fossil wood in Iowa occurs occasionally as trees rooted in place but more often as fragments of logs. Specific types of trees can often be identified on the basis of the structural characteristics of the wood. In addition, the wood can be dated by radiocarbon dating techniques if it is less

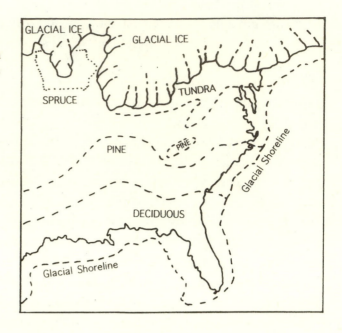

10.26 The generalized pattern of vegetation in the central and eastern United States during the Wisconsinan Glacial of the Pleistocene Epoch. This was the time of the surge of the Des Moines Lobe in Iowa, and climates were starting to moderate. Earlier in the Late Wisconsinan, 16,500 to 21,000 years ago, tundra was present in Iowa, and harsh periglacial conditions prevailed. Adapted from Matsch 1976.

than approximately 50,000 years old. Wood of greater age no longer contains enough radioactive carbon to measure.

Spruce wood is particularly common in Iowa's Pleistocene record. The basal Dows Formation and associated units of the Des Moines Lobe contain spruce, hemlock, and larch wood, which indicates that coniferous forests were present in Iowa at the time of the Des Moines Lobe, 13,500 to 15,000 years ago. Conifer wood is also common in the Sheldon Creek Formation and in the Pre-Illinoian drifts of the state. Additional information about the state's past vegetation and climate is provided by pollen samples from glacial, interglacial, and postglacial sediments. Studies of pollen from postglacial wetlands have been particularly helpful in deciphering the major changes in Iowa's vegetation and climate during the Holocene. Baker et al. (1992) provide a summary of midwestern vegetation history from the Late Wisconsinan to present (fig. 10.27).

Four pollen sequences along a transect from north-central Iowa to southeastern Wisconsin document the presence of a spruce-larch forest during the Late Wisconsinan. This conifer-dominated forest gave way to a conifer-hardwood forest with fir, birch, and ash. Between 10,000 and 9,000 years ago, a deciduous forest with associated oaks developed under mesic (moderately moist) conditions. Prairie was dominant in central Iowa, and the climate was drier than present from about 8,000 to 3,000 years ago. During the driest part of this period in central Iowa (6,500–5,500 years ago), a mesic forest flourished in eastern Iowa and southeastern Wisconsin. Apparently, conditions in eastern Iowa and southeastern Wisconsin then were wetter than at present. Prairie replaced the mesic

10.27 The major types of vegetation in Iowa during the Late Wisconsinan and Holocene. Adapted from Baker et al. 1992 and Eilers and Roosa 1994.

Years Before Present, Based on Radiocarbon Dates	Vegetation
30,000–21,500	Spruce-dominated forests
21,500–16,000	Open tundra and parkland during the last glacial maximum
16,000–10,000	Climate begins to warm and spruce forests return; the late glacial record suggests a spruce-larch forest with associated black ash and abundant sedges; this conifer-dominated forest gave way rapidly to a conifer-hardwood forest
10,000–9,000	Warming continues and deciduous forests with oaks are common, suggesting moderately moist conditions
9,000	Prairie enters western Iowa
8,500–6,500	Prairie extends into central Iowa and is dominant from about 8,000 to 3,000 years B.P. Mesic (requiring moderate moisture) forests prevail in eastern Iowa until about 5,400 years B.P.
6,500–5,500	Warmest and driest conditions in western and central Iowa, with abundant prairie
5,400	Mesic deciduous forests prevail in eastern Iowa until 5,400 years B.P., when they are replaced by prairie

forest in eastern Iowa about 5,400 years ago, but it did not extend much farther east. In southern Wisconsin and northern Illinois, the mesic forest gave way to a xeric (low or deficient in moisture) oak forest about 5,400 years ago.

Oak savanna was present when Euro-American settlers first arrived in Iowa. At the time of Euro-American settlement, about 85 percent of Iowa was covered by prairie (fig. 10.28). Much of the forested land in the state has since been cleared for farming, and nearly all of the original prairie has been converted to cropland. According to Eilers and Roosa (1994), the prairie today is only 0.01 percent of its extent at the time of Euro-American settlement. Wetlands have also declined significantly — from 1,500,000 acres to about 27,000 acres (excluding sedge meadows, fens, and wet prairies).

The major climatic and vegetational shifts in Pleistocene and postglacial times in Iowa caused shifts in the animal populations. The animal fossils range in size from tiny, air-breathing land snails that were buried in layers of loess to the large bones, teeth, and tusks of wooly elephants found in glacial outwash deposits. Iowa's Ice Age mammals include such diverse types as giant ground sloth, horse, peccary, camel, llama, deer, elk, caribou, reindeer, musk ox, bison, pantherlike cats, wolf, fox, beaver, and small rodents. The remains of small mammals such as hare, vole, shrew, and lemming are helpful in interpreting environmental settings. Even insects left a fossil record. Fossil beetles from Iowa's Pleistocene help

10.28 The forested areas (dark) and prairies (light) at the time of Euro-American settlement. Reconstructed from an original land survey conducted between 1832 and 1859. After Oschwald et al. 1965.

to document tundra conditions in the state 16,500 to 21,000 years ago, when the average July temperatures were about 11 to 13 degrees C cooler than today.

Mastodons and mammoths were elephant-like creatures whose remains are fairly common in Iowa's Pleistocene record. Mastodons had teeth with knobs or cusps that were well adapted for browsing on branches, twigs, and leaves (fig. 10.29). Mammoths had high-crowned teeth (fig. 10.29), adaptations for grinding abrasive grasses.

Other mammals that inhabited Iowa during the Pleistocene and early post-glacial times, such as the musk ox, caribou, and moose, are now at home in regions far to the north of Iowa. These animals apparently migrated north with the retreat of the Wisconsinan glacial ice. Their presence in Iowa during the Wisconsinan Glacial suggests that Iowa's climate then was much like that of Canada today. Figure 10.30 shows a postglacial woodland scene in the midcontinent region after the retreat of the Wisconsinan ice.

Many of the large Pleistocene mammals of North American became extinct only recently in geologic terms. Some of the animals, such as mammoths and mastodons, survived the Wisconsinan Glacial and became extinct shortly after the last ice sheets melted. The arrival of humans in North America appears to correlate with the extinction of some of the Ice Age fauna. The overkilling of Ice Age mammals by Paleo-Indians may have contributed to their extinction.

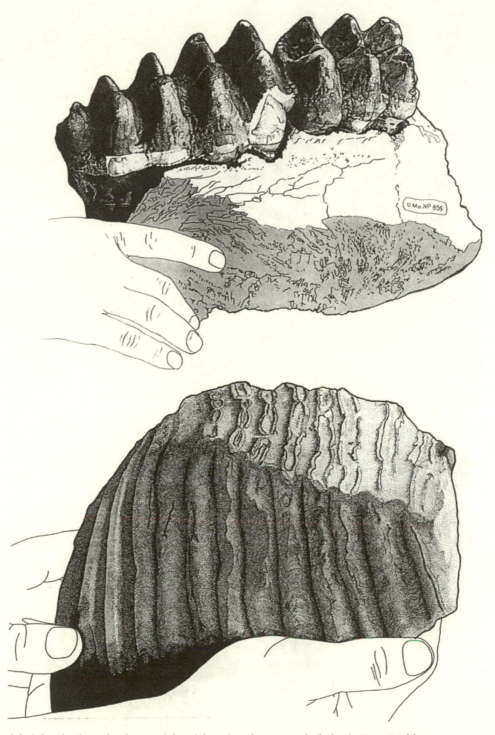

10.29 Cheek teeth of a mastodon (above) and mammoth (below). From Mehl 1962.

10.30 A mixed woodland and grassland scene in the midcontinent after the retreat of the Wisconsinan ice. Shown are, from top to bottom: giant ground sloth, mastodon, mammoth, bison, and saber-tooth cat. Drawing by Lori McNamee; based on an original drawing by Mary Tanner, University of Nebraska State Museum.

A Brief Culture History of Iowa

The Paleo-Indian interval in North America dates from approximately 9,500 to 7,500 B.C. Paleo-Indians in Iowa experienced a climate that was cooler and wetter than present averages for the state. In north-central Iowa, Paleo-Indians inhabited a recently deglaciated landscape covered by boreal and conifer-hardwood forests. These forests shifted though time to elm- and oak-dominated woodlands. Clovis and other fluted projectile points were made during the first two-thirds of the Paleo-Indian period; Dalton and unfluted point styles date to the latter third of the interval. In addition to these lanceolate (lance-shaped) points, the Paleo-Indian period is characterized by distinctive butchering tools, widespread use of exotic types of chert, and specialized lithic technologies. Both fluted and unfluted point forms have been recovered as surface finds from upland and valley locations throughout the state. Paleo-Indians were mobile peoples, and they hunted various animals, including the now-extinct mastodon, mammoth, and giant bison. Many of the Paleo-Indian sites documented in the United States are large-mammal kill sites. No Paleo-Indian base camps have yet been documented in Iowa.

The Early Archaic period (7,500–5,500 B.C.) was a transitional time between cultures dependent on big game animals for subsistence and those with a broader adaptation for foraging. Environments changed rapidly in the state during this period. Prairies and deciduous woodlands were present in western and central Iowa, whereas deciduous woodlands were common in the eastern part of the state. Peoples of the Early Archaic probably depended on bison for food in western Iowa and on deer and elk in eastern Iowa. In addition, they supplemented their diet with smaller game and by an increased use of plant foods. Somewhat permanent base camps were established, as well as seasonally occupied sites for hunting and resource procurement. Excavated sites, such as the Cherokee Sewer locality, suggest that local populations of Early Archaic peoples were small and were tied to seasonal hunting and resource exploitation. Typical artifacts include medium-to-large spear points, frequently with serrated and beveled blade edges.

The Middle Archaic period (5,500–2,500 B.C.) is poorly known in Iowa and is often lumped with the Early Archaic. This interval includes the warmest and driest postglacial climatic conditions (the so-called Hypsithermal). During this time, human populations throughout the Midwest became concentrated along river valleys because of the presence of water there. Such sites are now often deeply buried by alluvial deposits; thus, they are difficult to locate.

By the Late Archaic period (2,500–500 B.C.), human populations of the Midwest were on the rise, and evidence is present for encounters involving various groups. The dry Hypsithermal interval was followed by a more favorable climate, making previously unsuitable areas suitable for settlement. Populations were apparently becoming more sedentary, and multiple-interment cemeteries are known.

The Woodland tradition (500 B.C.–A.D. 1000) saw improved technologies, including ceramic production and horticulture. Woodland peoples made extensive use of fish and clams along the major river valleys of the state, and they hunted deer and bison. In addition, they utilized native plants, cultivating several varieties for their nutritious seeds. Woodland farmers developed domesticated varieties of selected grains, long before corn or beans became important. Woodland peoples are well known for their construction of burial mounds. Groups of linear, conical, and effigy mounds in northeastern Iowa record a distinctive aspect of the Effigy Mound Culture (A.D. 650–1000).

The Late Prehistoric (A.D. 1000–1650) includes the Plains Village pattern, an adaptation to the prairie/short-grass plains setting of South Dakota, Nebraska, southern Minnesota, and western Iowa. Bison meat was common in the diet of these peoples, and bison hides were processed for clothing and coverings for lodges and tepees. Bison bones were modified into tools such as scapula hoes.

The Mill Creek culture of northwestern Iowa falls within the Late Prehistoric; it is known from villages established on terraces above the Big and Little Sioux rivers and their tributaries. The Mill Creek villages were well planned and compact, and many of them were fortified with log palisades and encircling ditches. Mill Creek people grew corn, along with other crops. They probably participated in communal bison hunts and traded with other Late Prehistoric peoples.

The Late Prehistoric also includes the Central Plains tradition. Many Central Plains sites were farming communities, where the residents built impressive earthlodges. More than eighty earthlodges have been documented near Glenwood in Mills County. The Central Plains culture extended over Kansas, Nebraska, western Missouri, and southwestern Iowa.

During the Late Prehistoric period, the Oneota culture dominated much of eastern Iowa, along with substantial areas of central and northwestern Iowa. Oneota peoples inhabited the Midwest from approximately A.D. 1050 to 1700. Their villages were large and permanent or semipermanent. The Oneota had a subsistence economy based on hunting, fishing, plant collecting, and agriculture.

Several Oneota sites in northeastern and northwestern Iowa bridge or overlap the prehistoric to historic transition in the state. Early French trade goods such as glass beads, gunflints, and finger rings are found at some sites that otherwise contain mostly Oneota artifacts. In Iowa, the term "protohistoric" is given to this interval — when European goods were arriving but before European peoples began to make extensive written records of the area. Groups residing in Iowa or using portions of the state seasonally during the protohistoric interval included the Iowa, Oto, Omaha, and perhaps the Missouri and the Middle and Eastern Dakota. These peoples were mostly sedentary, but portions of the populations made seasonal forays for hunting and warfare.

From about A.D. 1650 to the present falls within historic time in Iowa. A discussion of that interval is best left to historians.

Interpreting Quaternary Environments

During the past decade, significant progress has been made in documenting the details of some of Iowa's Quaternary environments. Three examples will be discussed: the St. Charles site of south-central Iowa, with a mid-Wisconsinan record; the Conklin Quarry locality near Coralville in Johnson County, with a Late Wisconsinan record; and the Roberts Creek area of northeastern Iowa, with a nearly complete record of vegetational changes during the last 12,000 years.

The St. Charles site was investigated by R. G. Baker et al. (1991). Named for a village with the same name, the St. Charles site lies on the Southern Iowa Drift Plain, beyond the border of Wisconsinan glaciation. Two fluvial sequences are exposed. The alluvial fills date at about 34,000 years, and gravel in the basal fill contains seeds and mammoth bones. The seeds suggest a weedy floodplain setting. The younger alluvial fill yields plant macrofossils, well-preserved pollen, and fossil insects. The pollen indicates a prairie border or savanna environment. The insects match existing types that live in mixed conifer and hardwood forests at latitudes comparable to those from eastern North Dakota to New England. The St. Charles fossil assemblage suggests that July temperatures in south-central Iowa 34,000 years ago were probably 3 to 5 degrees C cooler than today.

The Conklin Quarry locality has a record of Late Wisconsinan sediments 16,710 to 18,090 years old. It is situated on the Southern Iowan Drift Plain, just south of the border with the Iowan Surface region. The Conklin site was investigated by R. G. Baker et al. (1986). Key discoveries include a diverse biota with pollen, bryophytes, vascular-plant macrofossils, small mammals, mollusks, and insects. The plant macrofossils include a number of tundra species, spruce and larch needles, and pieces of wood. The fossil insect fauna contains species now represented by their living counterparts in the forest-tundra transition zone of the northwestern Yukon Territory and Alaska. Collard lemmings and singing voles occur in the Conklin sediments, and they, too, suggest a tundra environment.

The sediments and fossils at the Conklin locality represent silty to sandy or gravelly uplands, wetlands, pond or stream-side deposits, and clear-water ponds. The plant fossils indicate a spruce-larch krummholz (stunted forest) with extensive tundra openings. The mean July temperature in the Conklin area during Late Wisconsinan time was likely 11 to 13 degrees C cooler than at present.

The Roberts Creek area is located near Elkader in the Paleozoic Plateau. Research by C. A. Chumbley et al. (1990) provides the background for this locality. A nearly complete record of vegetational changes has been documented for the last 12,500 years. By 9,100 years ago, deciduous forest had replaced spruce forest. Prairie was dominant from 5,400 to 3,500 years ago and oak savanna from 3,500 years ago to pre-Euro-American settlement time. A key finding was that the prairie invasion in northeastern Iowa was nearly 3,000 years later than at other sites in the state.

Soil Development

Iowa's soils owe their origin to the events of the Quaternary Period. A brief overview of soils is presented here because the rich, fertile soils of Iowa rank as the state's most important natural resource. A more extensive discussion of Iowa's soils can be found in Oschwald et al. (1965).

Soil is defined somewhat differently by engineers, geologists, and soil scientists. Soil, as used here, refers to the unconsolidated surface material of the earth that supports plant growth. The key factors that influence soil formation are parent material, climate, soil biota, topography, and time. These five factors work interdependently to produce a particular soil. Differences or similarities between soils are due to the differences or similarities of these key factors.

The principal parent materials of Iowa's soils are glacial drift, loess, and alluvium (fig. 10.25). Drift, loess, and alluvium are all materials that were previously weathered before they were transported and deposited. Consequently, 95 percent of Iowa's soils developed on materials that had already undergone some degree of breakdown or change. The loess-derived soils include some of the state's most productive soils.

About 5 percent of the soils in Iowa formed from bedrock such as limestone, dolomite, sandstone, and shale. These bedrock-derived soils occur mainly in the Paleozoic Plateau of northeastern Iowa and along deeper stream valleys, such as the Des Moines River valley. A few soils, chiefly in the Des Moines Lobe area, formed on organic-rich deposits such as peat. The distribution of the major parent materials on which Iowa's soils developed is shown in figure 10.25.

Climate is another key factor that influences soil development. The influence of climate on soil formation is particularly noticeable on a regional basis. The climate in Iowa is relatively uniform over the state at present, although a strong climatic gradient was apparently present between northern and southern Iowa 16,500 to 21,000 years ago. Currently, there are variations in annual rainfall and in the length of the growing season. Southeastern Iowa receives on average 6 to 8 inches more rain per year than does northwestern Iowa. Also, the growing season in southeastern Iowa is generally thirty to thirty-five days longer than the growing season in northwestern Iowa. Overall, the soils in northwestern Iowa are similar to those in southeastern Iowa except that they are less leached.

Organisms play a significant role in the soil-forming process. In addition to the mineral matter provided by the parent material, soils also contain organisms and organic matter. The organic material is provided by the remains of organisms that once lived in the area where the soil developed. Plant life, in particular, affects soil development. Many soils in Iowa developed under prairie landscapes that supported tall prairie grasses with deep roots. Forest soils formed in parts of eastern Iowa and along the state's principal stream valleys (fig. 10.28).

Topography also influences soil development. The topographic setting of an area affects drainage, runoff, and erosion rates. In turn, drainage, runoff, and erosion rates exert important controls over soil development. For example, soil formation is minimal in areas with steep slopes because erosion there proceeds at a faster rate than does soil development. Only thin, incomplete soils develop in such areas.

An additional factor in soil formation is time. It takes time to form a thick layer of topsoil. Generally, as soils continue to weather, the soil layers become thicker, and the soil develops a profile with well-developed A (upper) and B (middle) horizons. Prolonged weathering may not always be beneficial for soils, however. Soils that have undergone weathering for a long time often show a leaching of soluble materials, and this decreases soil productivity.

The characteristic features of Iowa's landform regions (fig. 10.10) were discussed previously. These are the landscapes where Iowa's modern soils formed. In addition, old soil horizons (paleosols, also called geosols) occur in many areas of the state and provide clues to help decipher Iowa's Quaternary history.

Although the soils of the state certainly rank as Iowa's prime natural resource, there are other significant natural resources. Mention has been made in preceding chapters of economic products associated with the state's bedrock record, and some of Iowa's bedrock formations serve as important aquifers for groundwater. In the final chapter, the relationship of the geologic resources of the state to the welfare of humans will be discussed in greater detail.

Summary

Ancient Iowa was subjected to weathering and erosion during most of the 65 million years of the Cenozoic. A few Tertiary deposits and fossils are known from the western part of the state. During the last 2.5 million years, Iowa experienced several episodes of glaciation. These continental ice sheets dumped poorly sorted glacial drift over all of Iowa, although subsequent erosion has since removed most of the drift from the northeastern part of the state. Meltwaters derived from the glaciers deposited sorted and stratified glacial outwash deposits. Silt and other fine-grained sediments on outwash plains were blown by the wind and deposited across the Iowa landscape as loess. Iowa's fertile soils have developed principally from weathered glacial drift, loess, and alluvium.

The events of the Quaternary Period are divided into the Pleistocene and Holocene (Recent) epochs. Studies made in Iowa during the late nineteenth and early twentieth centuries provided the basis for several of the original subdivisions of the Pleistocene. Iowa's extensive Pleistocene record includes several Pre-Illinoian drifts, deposits of the Illinoian Glacial Stage, and two Wisconsinan glacial drifts. In addition, an interglacial record is well documented, including the Yarmouthian and Sangamonian interglacials. Both Illinoian and Wisconsinan loess deposits are known. Ice of the Des Moines Lobe melted from the state about 12,000 years ago. The deposits of the last 10,500 years are assigned to the Holocene Stage and consist primarily of alluvium, slope deposits, and pond and lake sediments, with some local accumulations of wind-blown sand.

During Pleistocene time in Iowa, coniferous forests with spruce, larch, hemlock, fir, and yew flourished in areas beyond the continental glaciers. Iowa's climate then was much like that of southern Canada today. Tundra existed in the state from 16,500 to 21,000 years ago during a time of an intensely cold periglacial climate. As the glaciers melted from Iowa for the last time, coniferous forests gave way to deciduous forests. The deciduous forests, in turn, gave way to prairie grasslands.

Iowa's glacial and interglacial sediments have revealed the fossil remains of a wide variety of animals, including mastodon, mammoth, musk ox, bison, caribou, camel, deer, horse, moose, giant sloth, giant beaver, and wolf. The presence of humans in Iowa can be traced back to the Paleo-Indian interval, 9,500 to 7,500 years B.C.

The seven principal landform regions of the state are the Des Moines Lobe, Alluvial Plains, Loess Hills, Southern Iowa Drift Plain, Iowan Surface, Northwest Iowa Plains, and Paleozoic Plateau. Glacial-depositional features such as moraines, kames, and lakes occur within the Des Moines Lobe, but elsewhere running water has been the chief agent of landscape development.

Acknowledgments

For this chapter, I have relied primarily on the research and publications of Lynn M. Alex, Richard G. Baker, David W. Benn, E. Arthur Bettis III, Craig A. Chumbley, James M. Collins, Lawrence A. Eilers, William Green, George R. Hallberg, Timothy J. Kemmis, Greg A. Ludvigson, Louis J. Maher, Cornelia F. Mutel, Jean C. Prior, Deborah J. Quade, Richard S. Rhodes II, Dean M. Roosa, Robert V. Ruhe, Shirley J. Schermer, Donald P. Schwert, Holmes A. Semken Jr., Kent L. Van Zant, and Brian J. Witzke. Figures credited to Faxlanger et al. (1973) were originally published in *Land Patterns of Iowa* by David Faxlanger, James Sinatra, and C. John Uban. The figures used here are courtesy of the Department of Landscape Architecture, Iowa State University, Ames. Complete references are listed at the end of the book. Art Bettis and Jean Prior reviewed this chapter and provided helpful suggestions.

> Few Iowans are not too busily engaged with their own affairs to ponder for a little while that they may see how important a part the mineral resources have played in the material advancement of the State; how great will be the advantage of closer attention to this subject. Laborer, merchant, artisan, agriculturist, manufacturer, and capitalist, all are affected by a knowledge of the natural products of the community.
>
> Charles Keyes (1893a) on the importance of Iowa's geological resources

One of the meanings that is generally attributed to the Native American word "Iowa" is "the beautiful land." Although the translation may not be exact, the reference to beauty is certainly not misplaced. And it is logical to assume that Native Americans who inhabited the region we now call Iowa appreciated the beauty of their land.

Later, after Euro-Americans settled the state, Iowa's land was recognized for its productivity as well. The Iowa landscape has changed considerably since the days of the early North Americans and pioneers. Today, Iowa's landscape and agriculture are intimately related (fig. 11.1). Iowa poet James Hearst captured the essence of this interrelationship in "Landscape — Iowa":

No one who lives here
knows how to tell the stranger
what it's like, the land I mean,
farms all gently rolling,
squared off by roads and fences,
creased by streams, stubbled with groves,
a land not known by mountain's height
or tides of either ocean.
A land in its working clothes,
sweaty with dew, thick-skinned loam,
a match for the men who work it,
breathes dust and pollen, wears furrows
and meadows, endures drought and flood,
muscles swell and bulge in horizons
of corn, lakes of purple alfalfa,
a land drunk on spring promises,
half-crazed with growth — I can no more

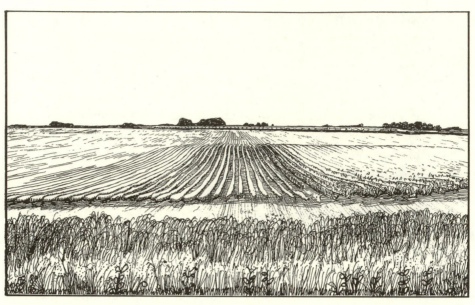

11.1 Iowa's landscape and agriculture are intimately linked today. The state's productive soils are largely products of the Quaternary Period, the last few paragraphs of Iowa's immense 3-billion-year-old chronicle of rocks, sediments, and shifting environments. Landscape sketch of nearly level cropland in Franklin County, Iowa. From Faxlanger et al. 1973.

> tell the secrets of its dark depths
> than I can count the banners in a
> farmer's eye at spring planting.

Although Iowa is famous for its fertile soils and agricultural wealth, the state is blessed with other important natural resources. Industrial materials such as limestone and dolomite, gypsum, clay and clay shale, and sand and gravel are plentiful in Iowa. Coal deposits are also abundant and were mined extensively at one time. Surface and underground water are important resources in Iowa. All of these natural resources are directly linked to Iowa's geologic history.

The geology of Iowa contributes to people's welfare and enjoyment in a number of ways. Many people derive pleasure from hobbies related to geology, such as collecting fossils, rocks, and minerals. Information about collecting and collecting localities is available in publications of the Geological Survey Bureau in Iowa City. Individuals who do collecting on private property should be certain to obtain permission from the owner.

Many geologic features occur in Iowa's state parks and recreation areas, and bedrock is dominant in such state parks and lands as Backbone, Bixby, Dolliver Memorial, Gitchie Manitou, Ledges, Maquoketa Caves, Pikes Peak, Stone, and Wildcat Den. Being informed about Iowa geology will benefit those who seek to appreciate and know more about the state's scenic and recreation areas. To that end, Robert Wolf's *Iowa's State Parks* and Jean Prior's *Landforms of Iowa* provide an introduction for those interested in exploring Iowa's state parks, forests, and recreation areas.

Geology is significant in fields such as construction, engineering, and land-use planning. However, a discussion of these fields is beyond the scope of this book. In this chapter, the focus will be on Iowa geology as it relates to mineral resources, water resources, and state parks and recreation areas.

Mineral Resources

French explorers observed lead in the Upper Mississippi Valley region in the 1600s. By 1788, Julien Dubuque had made a treaty with the Meskwaki Indians to work their lead mines in the vicinity of Dubuque. Lead mining in the Dubuque area grew steadily, and from the 1830s to 1870 the Upper Mississippi Valley mining district that centered around Dubuque was the principal metal-mining district in the United States. In 1852, the district accounted for 10 percent of the world's lead production and 87 percent of the lead mined in the United States. Zinc was also mined in the district. With the discovery of lead, silver, and other metals in the western United States in the 1860s, the importance of lead and zinc mining in the Upper Mississippi Valley district diminished. The last of the large lead mines in Iowa closed in 1910.

Lead mining in the Dubuque area reached its peak in 1848, shortly after Iowa became a state in 1846. Iowa's early history as a leading lead producer is depicted in the Great Seal of the State adopted by the First General Assembly in 1847. Several castings (pigs) of lead are shown stacked in the foreground of the seal, with a smelting furnace and the banks of the Mississippi in the background.

Iowa also produced iron at one time, although the total production was quite small. Limonite iron ore was mined from Cretaceous strata near Waukon in the early 1900s. However, no mining has taken place in the Waukon area since 1921. Iron and other metallic minerals occur in Iowa's Precambrian basement rocks, but such deposits are too deep to be mined with current technology.

A small number of oil and gas exploratory wells have been drilled in the state, but no commercial wells have yet been developed. A few hundred barrels of oil were recovered from a well near Keota in Washington County in 1963. The well did not prove to be commercially feasible, however. Of the 123 known exploration wells in Iowa, only 3 (all in Washington County) have yielded oil. The deepest oil test yet drilled in the state was to a depth of 17,851 feet in Carroll County in 1987. No petroleum was found, but the well provided valuable information about Iowa's Precambrian history. According to Anderson (1986), the best prospects for future oil finds in Iowa are in small structures in southeastern Iowa, in the southwestern corner of the state within the Forest City Basin, and along the flanks of the Precambrian Midcontinent Rift.

Gas storage has been developed in natural subsurface domal structures in south-central and southeastern Iowa (Dallas, Louisa, and Washington counties). Wells are drilled into these structures, and natural gas is injected into porous sandstones of the Cambrian and Ordovician systems. Natural gas is stored during the warm months and is used to supplement flow in gas pipelines during peak usage in the winter months. Iowa also has suitable geologic conditions for the underground storage of liquid petroleum gas, and two areas are currently in use (Johnson and Polk counties).

At present, Iowa's chief mineral production comes from the so-called industrial minerals: limestone, portland cement, gypsum, sand and gravel, and clay and clay shale. These products make important contributions to the state's economy; their value exceeded $483 million in 1995. A short description of each follows.

LIMESTONE

Limestone is used in this book in the commercial sense and includes both limestone and dolomite rocks. Technically, carbonate rocks that consist principally of the mineral calcite $(CaCO_3)$ are called limestones, and carbonate rocks that are composed mainly of the mineral dolomite $[CaMg(CO_3)_2]$ are called dolomites. Iowa's carbonate rocks are often composed of various proportions of these two carbonate minerals (calcite and dolomite), along with minor amounts of other constituents such as clay, quartz, and chert.

Limestone is the leading material mined in the state on a tonnage basis. Production of limestone achieved a multimillion-dollar level after World War II during the postwar boom in highway construction and road development. The value of limestone production in Iowa, including building stone and rock for cement, exceeded $230 million in 1995.

Limestone is the most important industrial mineral mined in Iowa — particularly when consideration is given to the amount of limestone used in the manufacture of portland cement. The state produced over 40,000,000 tons of limestone in 1995, which is nearly 15 tons (29,200 pounds) per Iowan. If the annual production of 15 tons of limestone per person seems excessive, consider the many uses of this material. If you have driven on a granular-surfaced road, paved road, driveway, or parking lot, you have benefited from one of the most common uses of crushed limestone. In addition, limestone aggregate is a key component in portland cement concrete, ready-mix, and asphalt, and the availability of crushed limestone is of vital importance to the road-building and construction industries of the state. A typical new home requires about 100 tons of limestone. Blocks of limestone are used for riprap and jetty stone. Coarse aggregate serves as railroad ballast. Other uses of limestone include the construction of athletic tracks, tennis courts, and bike paths. Limestone is also used in a variety of in-

11.2 The principal limestone-producing strata in Iowa. Adapted from Anderson 1983.

Geologic System	Group or Formation	Geologic System	Group or Formation
Pennsylvanian	Shawnee Group	Mississippian	McCraney Formation
Pennyslvanian	Lansing Group	Devonian	Cedar Valley Group
Pennsylvanian	Kansas City Group	Devonian	Wapsipinicon Group
Pennsylvanian	Bronson Group	Silurian	Gower Formation
Pennsylvanian	Marmaton Group	Silurian	Scotch Grove Formation
Mississippian	Pella Formation	Silurian	Hopkinton Formation
Mississippian	"St. Louis" Formation	Silurian	Blanding Formation
Mississippian	Keokuk Formation	Silurian	Tete des Morts Formation
Mississippian	Burlington Formation	Silurian	Mosalem Formation
Mississippian	Gilmore City Formation	Ordovician	Galena Group
Mississippian	Maynes Creek Formation	Ordovician	Platteville Formation
Mississippian	Starrs Cave Formation	Ordovician	Prairie du Chien Group

dustries, from plastics to sugar refining. Stomach-acid neutralizers contain powdered limestone, and finely pulverized limestone (ag lime) is spread on fields to treat soil acidity. Limestone even has agricultural uses as poultry grit and mineral food.

Iowa limestone deposits are generally mined by open-pit (quarry) methods, although a few underground operations have been developed. The principal limestone-producing strata in Iowa are shown in figure 11.2. Quarry locations are shown in figure 11.3.

Limestone is rarely transported a great distance because of freight costs. Most communities need limestone aggregate, so there are numerous local quarries, with over three hundred quarrying operations statewide. Most of these are in the eastern half of the state, where marine deposits of Paleozoic limestones and dolomites are abundant. Pennsylvanian strata constitute the bedrock in southwestern Iowa, and part of that record is nonmarine. Consequently, limestone quarries are less numerous in southwestern Iowa. Northwestern Iowa has only a few limestone quarries. There, the surficial materials are quite thick, and the bedrock is dominated by the Cretaceous System, which contains few limestones. The Paleozoic limestones are too deep to quarry throughout most of northwestern Iowa, although excellent quarries have been developed in Mississippian strata near Gilmore City and Humboldt.

PORTLAND CEMENT

Portland cement is made primarily from limestone and shale; it is a manufactured product rather than an industrial mineral in the strict sense of the term. Nevertheless, portland cement has traditionally been included as an industrial

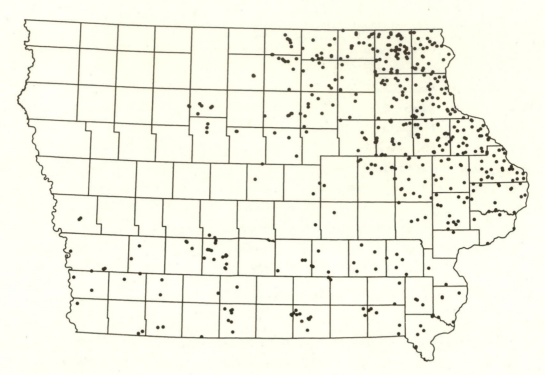

I I.3 The location of active limestone and dolomite quarries. Courtesy of Robert McKay, Geological Survey Bureau, Iowa City.

mineral in the listings of the United States Bureau of Mines, and for that reason portland cement is included among Iowa's important mineral resources.

Portland cement plants are located in Mason City, Des Moines, and Buffalo (fig. 11.4). These plants produced 3,113,000 tons of cement in 1995. Iowa generally ranks in the top ten states in the production of portland cement, and in 1995 the state placed seventh nationally.

SAND AND GRAVEL

Sand and gravel are composed principally of quartz, chert, and carbonate clasts. These deposits are predominantly of fluvial origin and generally occur along present-day streams and stream valleys. Some of the state's sand-and-gravel deposits formed as glacial outwash sediments. Extraction of sand-and-gravel deposits involves dredging, washing, and screening to sort the materials into size gradations.

Sand and gravel are used primarily as aggregate in concrete and road-surfacing material. In dollar value, sand-and-gravel production in Iowa in 1995 was valued at $53 million. Iowa usually ranks in the top twenty states in the production of sand and gravel.

Sand also occurs in Iowa's sedimentary rock column. One such deposit, the St. Peter Sandstone, was once mined near Clayton in northeastern Iowa. This

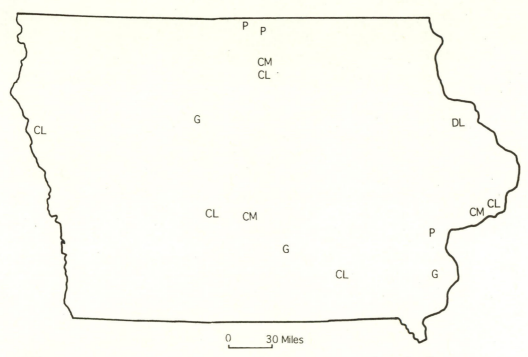

11.4 The location of clay production, cement plants, gypsum mines, dimension limestone production, and peat production in Iowa. Symbols: CL = clay production in Cerro Gordo, Dallas, Scott, Wapello, and Woodbury counties; CM = cement plants in Cerro Gordo, Polk, and Scott counties; G = gypsum mining in Des Moines, Marion, and Webster counties; DL = dimension limestone production in Dubuque County; P = peat production in Muscatine, Winnebago, and Worth counties. Adapted from Bureau of Mines 1995.

sand is nearly pure quartz sand; it found use as a foundry sand at John Deere plants in Dubuque and Waterloo.

GYPSUM

Gypsum is composed of hydrated calcium sulfate. Pure gypsum is $CaSO_4 \cdot 2H_2O$. When gypsum is heated to about 170 degrees C, part of its water is driven off, and gypsum is converted to plaster of paris ($CaSO_4 \cdot \frac{1}{2}H_2O$). Plaster of paris can be shaped, molded, and poured into a variety of products for the building industry. When plaster of paris is combined with water, it sets, or hardens, and converts to gypsum. The chief use of gypsum is in the manufacture of building materials like wallboard, sheathing, and plaster. Gypsum is also used in the manufacture of portland cement. Four companies currently mine gypsum and produce gypsum products in the Fort Dodge area.

The Fort Dodge gypsum deposits are exceptionally pure, as discussed in chapter 9. They are also located conveniently to major markets, such as Des Moines, Omaha, Kansas City, Minneapolis, St. Paul, and Chicago. The gypsum quarries at Fort Dodge have provided significant quantities of gypsum for more than a hundred years, and Iowa traditionally ranks either second or third nationally in the production of gypsum. In 1995, Iowa was second in the nation in crude gypsum production with 2,464,000 tons, valued at $12 million. Generally, Iowa gypsum production constitutes 12 to 13 percent of the nation's total production.

The Jurassic gypsum deposits in the Fort Dodge area are limited in extent. Based on current rates of extraction, these deposits may be depleted within the next forty to fifty years. However, there are extensive gypsum deposits in Mississippian strata in southern Iowa. Gypsum was formerly mined near Centerville, and it is currently quarried in Marion County (fig. 11.4). Gypsum is also mined in an underground operation near Sperry in Des Moines County. The Sperry Mine is the deepest underground mine in the state at a depth of more than 600 feet below the surface. The deposits at Sperry are within the Devonian Wapsipinicon Group. The Sperry Mine contributes approximately 25 to 30 percent of Iowa's total annual crude gypsum production.

CLAY AND CLAY SHALE

Clay is the mineral name for a group of aluminum-and-silicon-bearing minerals that have a layered structure. Clay shale is the term applied to a sedimentary rock composed chiefly of clay minerals. It is common practice to refer to unconsolidated deposits of either clay or clay shale as clay. Indurated accumulations of clay are referred to as clay shale.

The clay and clay shale mined in Cerro Gordo County (fig. 11.4) are used in the manufacture of portland cement. Much of the clay and clay shale for the cement plant in the Davenport area (Scott County) is currently supplied from out of state, although some clay-rich loess is obtained from local sources. Clay and clay shale are also mined in Dallas and Woodbury counties. Traditionally, these operations have utilized clay and clay shale for the production of brick, sewer pipe, drainage tile, and building tile. In recent years, plastic products have replaced clay tile for drainage purposes in many markets.

In the 1920s, more than three hundred brick and tile plants existed in Iowa, and nearly every region in the state was involved in the production of these products to serve local builders. Since then, Iowa's ceramic industry has gradually declined because of competition from other products. At present, clay and clay shales are mined from the Pennsylvanian System in central Iowa, from Devonian strata in northern and eastern Iowa, and from the Cretaceous System in western Iowa (fig. 11.4). The state produced 348,700 tons of clay and clay shale in 1995 with a value of $1.28 million.

COAL

Iowa has huge reserves of coal in the Pennsylvanian rocks of the southern part of the state. General figures on the state's coal reserves were given in chapter 8.

Production of coal in Iowa began in the 1840s and reached a peak between 1917 and 1926. Since then, coal production in Iowa gradually declined, and the last operating mine closed in 1994. The decrease and demise of the coal industry were partly due to competition from natural gas and fuel oil for home heating. Moreover, railroads converted their locomotives from coal to diesel fuel, and the railroads were at one time prime users of Iowa coal. Prior to the closing of the last coal mines in the 1990s, Iowa coal had limited use as an energy source at coal-burning electric power plants in the state.

Although Iowa coal has a fairly high energy rating in that it produces 9,000 to 11,000 BTU per pound, its sulfur content is very high. Sulfur compounds combine with oxygen to produce harmful oxides of sulfur when high-sulfur coals are burned. Unfortunately, the sulfur content of most of Iowa's coal exceeds the present restrictions by the federal Environmental Protection Agency on the use of high-sulfur coals.

Water Resources

Water is an important resource in Iowa. Without abundant supplies of good-quality water, Iowa's agricultural and industrial economy would surely decline. Current information indicates that adequate supplies of good-quality water are generally available throughout the state. Obviously, prolonged periods of drought could adversely affect supplies of surface water and shallow groundwater.

The basic relationships of groundwater occurrences are shown in figure 11.5. Groundwater occurs within the pores and fractures of subsurface rocks and sediments. The upper level of the rock or sediment that is saturated with groundwater is called the water table, and it separates the zone of saturation from the zone of aeration (fig. 11.5). Instead of occurring in underground pools or streams, the groundwater in rocks or sediments is much like water soaked up in a sponge. In the zone of aeration, the pores and fractures of the rocks and sediments are filled with air instead of water. Perched groundwater can occur where a body of porous and permeable material is saturated with water above the level of the main water table of the area.

Wherever the water table emerges on the land surface, wetlands, springs or seeps, or bodies of water such as streams, lakes, or ponds are found. Generally, the water table is fairly close to the surface in Iowa. In most cases, the water level of the principal streams represents the level of the local water table of the area. If it were not for the discharge of groundwater into the streams, many Iowa streams would dry up or be at very low levels within a few weeks following any major rainfall in the area. The slow, continual seepage of groundwater into streams accounts for the continuous flow of most of the state's permanent streams.

The state has a number of unusual wetlands called fens. These are wetlands with organic soils that receive water primarily from groundwater. Iowa's fens are

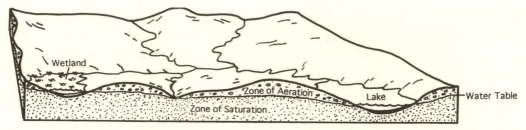

11.5 The occurrence of groundwater. The water table represents the upper level of the zone of saturation. Above the water table, the pores of sediments and rocks are filled with air (zone of aeration). Below the water table, the pores and fractures in sediments and rocks are saturated with water. Adapted from Anderson 1983; based on a drawing in Lemish 1969.

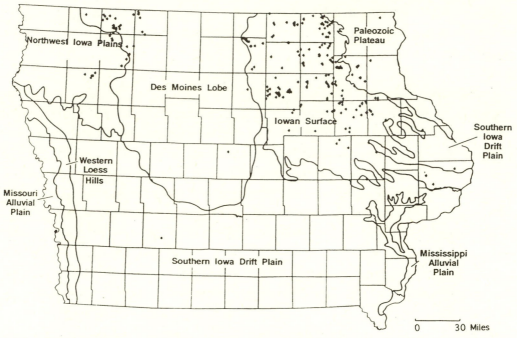

11.6 The location of fens in Iowa. From Thompson 1996.

dominated by herbaceous (nonwoody) vegetation. Peat deposits and rare plant communities are associated with fens. The fens of Iowa range in size from 25 acres to less than 1 acre. Most occur on sloping land and on upland positions on the landscape. According to Thompson (1996), Iowa has over three hundred fens located in twenty-seven counties. Fens occur in all seven of Iowa's landform regions, but they are most common in the Iowan Surface and the Des Moines Lobe (fig. 11.6). Thompson et al. (1992) recognized six types of fens based on landscape position, stratigraphic setting, and hydrology of the water system that recharges the fen. The six categories are fens along valley walls where sand-and-gravel deposits occur between glacial tills; fens in hummocky settings along the margins of the Des Moines Lobe where sand-and-gravel bodies occur within glacial till; fens on the lower slopes of exhumed sand-and-gravel deposits on the Iowan Surface; fens located at the base of alluvial terraces or outwash deposits; fens recharged by bedrock aquifers that are exposed on the landscape; and fens occupying sand-and-gravel deposits in abandoned channels.

The quality of groundwater in Iowa, in terms of dissolved solids, is variable. Some of the water is described as hard because it contains ions such as calcium and magnesium. Hardness is expressed in terms of parts per million of calcium carbonate. Calcium, magnesium, and other ions are often present in groundwater and are usually derived from the rock formation or rock body in which the groundwater is stored. Some of Iowa's groundwater is characterized by a high sulfate content because of the presence of sulfate-bearing evaporites in the subsurface. The quality of the state's groundwater is generally good, but some surficial aquifers and shallow bedrock aquifers show signs of contamination, particularly from agricultural chemicals.

Agricultural Chemicals and Groundwater Quality in Iowa

In 1870, State Geologist Charles A. White (1870) described Iowa's groundwater as "pure, limpid, and wholesome, the greatest objection to its use being the difficulty experienced in using it for washing purposes." To that, we would now add the concern about agricultural chemicals (ag-chemicals) in shallow wells throughout the state.

Groundwater is the principal source of drinking water for most Iowans. Surface-water impoundments also serve as important supplies of water, especially in southern Iowa. Of the state's thirty-seven rural water systems (RWS), fourteen use surface impoundments as their main source of water. It is important to guard both groundwater and surface water from contamination.

During the past two decades, scientists at Iowa's natural resource agencies and universities have investigated the occurrence of ag-chemicals in groundwater throughout the state. Much has been learned. George Hallberg (1986a, 1992) reported on ag-chemicals and groundwater quality in 1986 and 1992. Robert Libra (1988, 1995a) described the status of ag-drainage wells and groundwater quality in 1988 and provided an update on the Big Spring Demonstration Project in 1995.

Ag-chemicals are routinely applied to croplands over a vast area of the state, and applications of this type contribute chemical pollutants to Iowa's groundwater. This is described as nonpoint-source pollution because the chemical contaminants are derived from a broad area rather than from one specific point. The most effective way to deal with such pervasive and diffuse contamination is to reduce or eliminate the application of the chemicals. This is easier said than

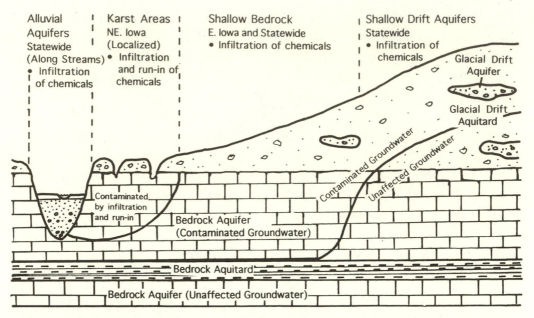

Alluvial Aquifers Statewide (Along Streams)
• Infiltration of chemicals

Karst Areas NE. Iowa (Localized)
• Infiltration and run-in of chemicals

Shallow Bedrock E. Iowa and Statewide
• Infiltration of chemicals

Shallow Drift Aquifers Statewide
• Infiltration of chemicals

Glacial Drift Aquifer

Glacial Drift Aquitard

Contaminated Groundwater

Unaffected Groundwater

Contaminated by infiltration and run-in

Bedrock Aquifer (Contaminated Groundwater)

Bedrock Aquitard

Bedrock Aquifer (Unaffected Groundwater)

11.7 A schematic cross section showing geologic conditions in Iowa and the pattern of ground-water contamination. Adapted from Hallberg 1986a.

done, however, because the use of chemicals contributes to increased agricultural productivity.

Studies in Iowa document that nitrates and pesticides leach through the soil and reach the underlying groundwater. Shallow aquifers are the most affected, and many farm wells are particularly susceptible to contamination from ag-chemicals. At present, the contaminants are confined to aquifers that are relatively shallow or to the recharge areas of deeper aquifers (fig. 11.7). Bedrock aquifers that are particularly prone to contamination include those in areas with abundant sinkholes or those in areas with relatively thin surficial cover (fig. 11.7). Alluvial aquifers along river valleys and shallow glacial-drift aquifers are also susceptible. The deeper bedrock aquifers are protected to some degree from infiltrating contaminants by overlying impermeable rock units known as aquicludes (fig. 11.7). Aquicludes consist of relatively impervious strata such as shale. Still, the contamination of the deeper bedrock aquifers of the state may be only a matter of time. The widespread use of ag-chemicals in Iowa dates to the early 1960s (fig. 11.8), so there has not been enough time for pollutants to reach the deeper aquifers. Note the correspondence between the increased use of nitrogen fertilizers in the state and the increased concentrations of nitrate in Iowa's groundwater between 1960 and 1980 (fig. 11.8).

Data for the nitrate concentrations in groundwater shown in figure 11.8 came from the Little Sioux River alluvial system in west-central Iowa, the Rock River alluvial system in northwestern Iowa, and the Big Spring Basin in northeastern

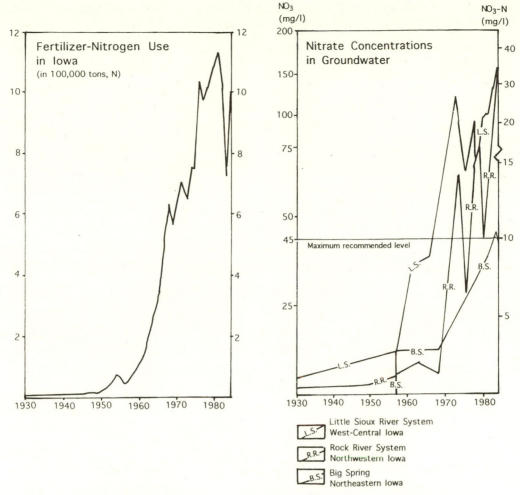

11.8 Use of nitrogen fertilizer in Iowa, 1930 to 1983 (left), and nitrate concentrations in groundwater in the Little Sioux River System, the Rock River System, and Big Spring (right). Adapted from Hallberg 1986a.

Iowa. The Big Spring Basin, near Elkader in Clayton County, serves as a natural laboratory for studying groundwater. Nearly 90 percent of the groundwater discharging from the basin emanates at Big Spring. This allows for quantitative measurements of both the volume of groundwater and the ag-chemicals in the groundwater. Data from the Big Spring Basin document that nitrate levels in the groundwater at Big Spring have increased in direct proportion to the increased use of nitrogen fertilizers within the 100-square-mile groundwater basin. The same pattern shows up throughout the state. Nitrate concentrations in groundwater are particularly high in western Iowa (fig. 11.8).

The initial findings at Big Spring spurred the development of the Big Spring Demonstration Project in 1987. This cooperative effort involved area farmers within the Big Spring drainage basin and several local, state, and federal agencies. Improved nitrogen management was undertaken, and the input of nitrogen fertilizers in the basin decreased by a third from 1981 to 1993. Applications of fertilizers were reduced from 174 pounds per acre to 115 pounds per acre with no noticeable reduction in yields. This represented a 2 million–pound reduction in nitrogen use and a savings to local farmers of $360,000 annually.

Have these reductions resulted in lower nitrate concentrations in the discharge at Big Spring? That question is difficult to answer. The volume of precipitation falling in the Big Spring Basin affects the nitrate concentrations, and nitrate concentrations at Big Spring rise and fall with the volume of water being discharged. Nitrate concentrations declined from 1982 to 1989, but so did the total discharge from Big Spring. Extreme drought conditions prevailed in the basin during 1988 and 1989. Although some of the decline in nitrate concentrations at Big Spring probably reflects the decrease in nitrogen applications in the basin, it is difficult to evaluate this because of the overall decrease in the volume of water discharged.

After the drought, wetter than normal conditions occurred in the early 1990s. In particular, 1993 was a very wet year. The volume of water discharged at Big Spring in 1993 was four times greater than in 1989, and the nitrogen concentrations increased significantly. Any improvements in water quality brought about by decreased applications of nitrogen fertilizers in the Big Spring Basin were masked by the effects of the increased discharge associated with greater than normal precipitation.

The research at Big Spring provides a detailed record of the relationships between agricultural practices and groundwater quality, and it is important to continue this type of research. Improvements in groundwater quality are difficult to detect at Big Spring, even after significant reductions in the amount of fertilizer applied since 1981. Climatic variations influence both the volume of groundwater discharged and the nitrate concentrations at Big Spring.

Although short-term improvements in water quality may be difficult to document, the long-term goal of improving Iowa's groundwater is important for the public health of all Iowans. Reductions in use of ag-chemicals may have positive economic results as well. As Hallberg (1992) points out, Iowa farmers reduced nitrogen application rates on corn by about 15 percent between 1985 and 1991, reducing statewide use of nitrogen by over 1 billion pounds. These reductions saved Iowa farmers over $200 million without significantly affecting crop yields.

An additional connection between groundwater contamination and agricultural practices involves ag-drainage wells. Such wells funnel excess water from farm fields directly into porous limestone formations. Often, these formations are sources of drinking water. This practice would be of little concern if only pure, clean water was involved. However, drainage water from farm fields often contains bacteria, nitrates, pesticides, and sediment. These are undesirable additions to Iowa's bedrock aquifers. Ag-drainage wells vary in depth from approximately 30 to 300 feet. Although no accurate count exists, the number of such wells is estimated to be between six hundred and seven hundred. The greatest concentration of ag-drainage wells is in north-central Iowa, including Floyd, Humboldt, Pocahontas, and Wright counties.

Aquifers of Iowa

Groundwater is recovered most readily from rock bodies called aquifers. Aquifers have good permeability (ability to transmit water) and high porosity (percent of void space). Two general types of aquifers occur in Iowa — surficial (unconsolidated) aquifers and bedrock (consolidated) aquifers.

The surficial aquifers consist mainly of sand-and-gravel bodies of fluvial or glaciofluvial origin. Bedrock aquifers are primarily porous and permeable sandstone or carbonate formations. Iowa's rock record also includes a number of formations that are essentially impermeable. These aquicludes, or aquitards, yield little or no groundwater.

Aquicludes serve as confining beds that hold the water in aquifers under pressure. In addition, aquicludes retard the vertical movement of water downward into underlying aquifers, protecting them to some extent from contamination. Aquicludes yield minor amounts of water in some localities, such as sandstone beds within the Pennsylvanian System in southern Iowa and porous carbonates of the Fort Atkinson Member of the Maquoketa Formation in northeastern Iowa. Brecciated Precambrian rocks produce groundwater in the Manson area of north-central Iowa, but elsewhere the Precambrian is generally impermeable. The principal aquicludes in Iowa are, from top to bottom: the shales, impure limestones, and chalks of the Carlile, Greenhorn, and Graneros formations (Cretaceous); the entire Pennsylvanian System; shales and impure carbonates of the Upper Devonian; shales and dense dolomites of the Ordovician Maquoketa Formation; shales and impure carbonates of the Ordovician Decorah Formation; dense dolomites, argillaceous sandstones, and shales of the Cambrian St. Lawrence and Lone Rock formations; and rocks of the Precambrian basement.

A hydrogeologic map (fig. 11.9) shows the distribution of the principal bedrock aquifers and aquicludes of the state. A hydrogeologic cross section (fig. 11.10) portrays the subsurface arrangement of these units. Six major aquifers are recognized in Iowa: surficial aquifers of the Quaternary System, the Dakota aquifer (Cretaceous), the Mississippian aquifer, the Silurian-Devonian aquifer, the Cambrian-Ordovician aquifer (the Jordan Formation, the Prairie du Chien Group, and the St. Peter Sandstone), and the Dresbach aquifer (the Wonewoc, Eau Claire, and Mt. Simon formations of the Cambrian System). In addition, the salt-and-pepper sands, an intradrift or subdrift unit, serve as an aquifer in western Iowa. A brief discussion of each of the six major aquifer systems follows.

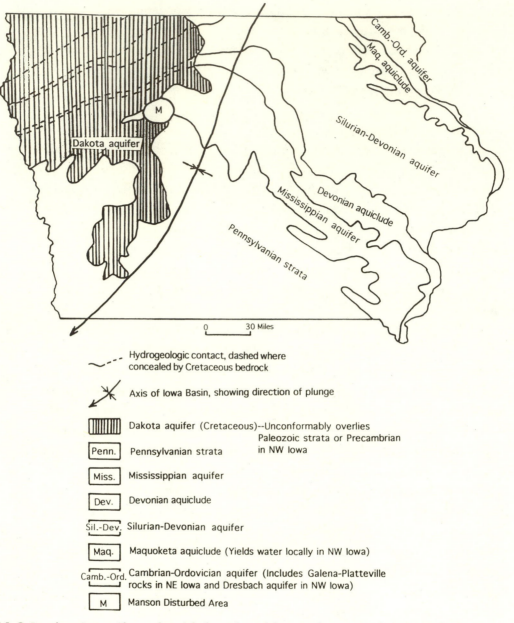

11.9 Iowa's major aquifers and aquicludes. Adapted from Anderson 1983; based on information in Steinhilber and Horick 1970.

SURFICIAL AQUIFERS

Iowa's surficial aquifers are of three principal types: alluvial aquifers in present-day stream valleys; buried-channel aquifers in ancient valleys, now covered by younger deposits; and drift aquifers, where discontinuous layers such as sand and gravel occur within impermeable glacial deposits (fig. 11.11). Surficial aquifers are usually economical to develop because of their shallow nature.

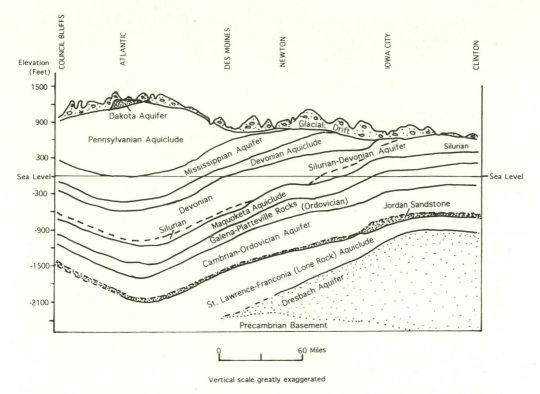

11.10 A generalized cross section (hydrogeologic section) showing the principal aquifers and aquicludes of Iowa. Vertical scale greatly exaggerated. Adapted from Steinhilber and Horick 1970.

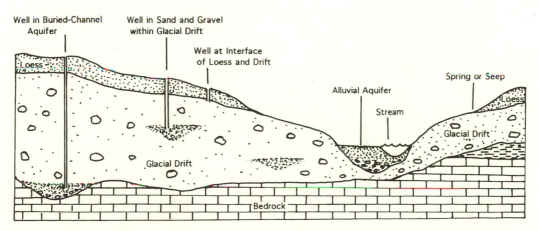

11.11 Three common types of Quaternary aquifers: alluvial aquifers, buried-channel aquifers, and porous and permeable zones within glacial drift. Note the location of springs or seeps, where loess overlies glacial drift. Adapted from Thompson and Kemmis 1990 and Prior 1991.

Groundwater quality can be satisfactory in such aquifers, but they are vulnerable to contamination from surface sources.

Thick deposits of alluvium occur along Iowa's major stream valleys, and these alluvial deposits contain large quantities of groundwater. Yields of more than 2,000 gallons per minute are common in irrigation wells in the Missouri River alluvium. Wells in the Mississippi River alluvium commonly produce at the rate of 1,000 gallons per minute. The alluvial aquifers of the state's interior streams are also excellent sources of groundwater. Alluvial water sources are easily developed. Consequently, they are often the favored water supplies for municipalities, rural water systems, industries, and irrigators.

Buried-channel aquifers (fig. 11.11) are similar to alluvial aquifers overall, except that they are buried by younger deposits. Aquifers of this type occur throughout the state and include some large systems, such as the Fremont Channel in southwestern Iowa and the Belle Plaine and Cleona channels in southeastern Iowa. Buried-channel aquifers are recharged either by percolation through overlying sediments or by groundwater moving laterally from bedrock or other materials. Some buried-channel aquifers are under artesian conditions, and water will rise in wells developed in such pressurized systems. The Jumbo well (fig. 11.12) at Belle Plaine is a classic example of a well that tapped a buried-channel aquifer with high artesian pressure.

According to Robert Libra (1995b), the city of Belle Plaine contracted for a well in 1886 to provide a source of water for fire protection. The driller underestimated the artesian pressure conditions and lost control of the well, which flowed unchecked for thirteen months. Estimates of the initial flow rates were from 30,000 to 50,000 gallons per minute. Large stones and chunks of fossil wood were ejected from the well, along with huge amounts of sand. It required a great effort to keep ditches open to carry off the flow from the runaway well. The Jumbo well was described in newspapers nationally and internationally, often in exaggerated fashion. European newspaper accounts told of water spouting hundreds of feet high and Belle Plaine residents being rescued from the third or fourth stories of their homes by boats.

Drift aquifers exhibit great variety, but a common occurrence is sand-and-gravel bodies within glacial drift (fig. 11.11). Wells completed in the permeable sands and gravels often provide small but adequate water supplies for rural homesteads. Another common drift well is one completed at the interface of slightly permeable loess deposits and the underlying less-permeable glacial drift (fig. 11.11). Typically, these are large-diameter seepage wells with low yields. However, such wells often supply adequate amounts of water for a farm or rural residence. Recharge to drift aquifers is by precipitation and infiltration; a prolonged drought can place the water supplies in jeopardy.

11.12 Jumbo, a runaway artesian well, drilled at Belle Plaine, Iowa, in 1886. The well encountered strong artesian pressure in a buried-channel aquifer and flowed out of control for thirteen months. Courtesy of the Calvin Photographic Collection, Department of Geology, University of Iowa.

THE DAKOTA AQUIFER

The Dakota Formation is the chief bedrock aquifer of western Iowa; it consists of a lower member (Nishnabotna Member) and an upper member (Woodbury Member). The Nishnabotna is composed of medium-to-coarse-grained, poorly consolidated sandstones and conglomerates and thin lenticular mudstones. The member ranges in thickness from 50 to 300 feet. It is a highly productive aquifer and commonly yields water at the rate of a few hundred gallons per minute to 1,000 gallons per minute or more. In contrast, the Woodbury Member generally yields only a few tens of gallons of water per minute. The Woodbury consists of shales, mudstones, lignites, and lenticular sandstones of local extent.

The regional groundwater flow in the Dakota aquifer is from north to south. The groundwater discharges into the Missouri River and its tributaries and into the Des Moines River and its tributaries. The static head of the aquifer is low, generally 200 to 300 feet below the elevation of the uplands in northwestern Iowa. Concentrations of dissolved solids are high enough to restrict use of the water for drinking purposes in certain localities in O'Brien, Osceola, and Plymouth counties. The Dakota aquifer is hydrologically connected with the underlying Paleozoic aquifers in northwestern Iowa.

THE MISSISSIPPIAN AQUIFER

The Mississippian System, discussed in chapter 7, consists primarily of limestone and dolomite. The system covers 60 percent of southeastern, central, and southwestern Iowa. Thin sandstone units and some shale deposits occur in the upper part of the system. Gypsum and anhydrite are known in the subsurface in an area between Des Moines and Centerville. Mississippian strata in Iowa reach their maximum thickness of approximately 600 feet in the subsurface of Appanoose and Davis counties in southeastern Iowa. In the Mississippian outcrop belt from southeastern Iowa to north-central Iowa, the system ranges between 100 and 300 feet in thickness.

Groundwater in the Mississippian carbonate rocks moves through secondary openings enlarged by solution. This includes movement along bedding planes, crevices, and joints, in addition to flow through vugs and solution cavities. Because these pathways are irregular in occurrence and distribution, the yields of groundwater from the Mississippian aquifer vary considerably throughout the state. In north-central Iowa where creviced and cavernous Mississippian strata occur, the storage and transmission characteristics of the aquifer are quite high.

The aquifer is recharged in its outcrop area through precipitation that infiltrates through loess and glacial drift. In the subsurface of northwestern Iowa, the Mississippian aquifer is recharged to some extent by infiltration from the overlying Dakota aquifer. Regional flow within the Mississippian aquifer is southward into neighboring Missouri and southeastward into the Des Moines and Skunk rivers and their tributaries.

Nearly 70 percent of the withdrawals from the aquifer are in north-central Iowa, where the highly creviced and cavernous Mississippian carbonates produce water at the rate of several hundred gallons per minute. Cities that obtain their water supplies from the aquifer include Story City, Roland, Eagle Grove, Belmond, Iowa Falls, and Marshalltown. The quality of the water is excellent in north-central Iowa. However, high concentrations of sodium, fluoride, and sulfate often limit the acceptability of the water for domestic use elsewhere. In western and southern Iowa, the dissolved solids in the water generally make it unsuitable for human use.

THE SILURIAN-DEVONIAN AQUIFER

The Silurian-Devonian aquifer is represented by a thick succession of limestones and dolomites of the Silurian and Devonian systems. The various stratigraphic units of the aquifer comprise fourteen formations and underlie about

89 percent of the state. The Silurian-Devonian aquifer is the uppermost bedrock over approximately 21 percent of Iowa. Across most of eastern and northern Iowa, the aquifer ranges between 200 and 400 feet in thickness; it reaches a maximum thickness of 700 feet in the subsurface of southwestern Iowa.

The Cedar Valley and Wapsipinicon groups (Devonian) and the Gower, Scotch Grove, Hopkinton, and Blanding formations (Silurian) are the principal water-yielding units of the aquifer. In northern Iowa, the Devonian portion of the aquifer is a tri-part system, where shales and argillaceous dolomites within the Cedar Valley and Wapsipinicon groups serve as local aquicludes.

The Silurian-Devonian aquifer is recharged by precipitation that infiltrates through Quaternary surficial materials and bedrock. Groundwater flow through the aquifer is along fractures, crevices, joints, bedding planes, caverns, and various solution openings. In addition, intercrystalline porosity occurs in dolomites, particularly those of the Silurian System. Moldic porosity is common in many dolomites, where fossils and fossil debris have dissolved. Furthermore, the carbonate formations are often highly broken or brecciated. Sinkholes and associated karst features occur within both the Silurian and Devonian systems in northern Iowa.

In the Silurian-Devonian outcrop area of eastern Iowa, the natural discharge for groundwater is to the principal streams, including the Cedar, Iowa, Maquoketa, Shell Rock, and Winnebago rivers and their tributaries. Groundwater discharge contributes 40 to 75 percent of the total stream flow of these eastern Iowa streams. This explains why most of the streams continue to flow even during prolonged periods of drought. The regional flow of groundwater in the Silurian-Devonian aquifer is to the southeast across Iowa.

The water in the aquifer is usually confined and under artesian pressure. In areas where the cover of glacial drift is thin, the aquifer is unconfined and is under water-table conditions. The aquifer is widely used in eastern Iowa, where at least 178 municipal water systems are supplied totally or partially by wells completed in Silurian or Devonian strata. These municipal systems withdraw approximately 12.5 billion gallons of water per year. In addition, rural wells utilize about 10 billion gallons of water from the aquifer annually; industrial wells extract 3.5 billion gallons; and irrigation wells use approximately 2 billion gallons. The municipal and rural wells provide water for some 490,000 Iowans. Yields from the aquifer are particularly high in the Waterloo–Cedar Falls area, where wells in fractured and cavernous Devonian carbonates test from 2,000 to 4,000 gallons per minute, with only a few feet of drawdown. Statewide, at least 27 to 28 billion gallons of water are withdrawn from the Silurian-Devonian aquifer annually, and about 17 percent of the population of Iowa is served by the aquifer.

Although it is widely utilized in eastern Iowa, the aquifer finds little use in the subsurface of western and southwestern Iowa, where it is deeply buried and its water is highly mineralized. In the outcrop area of the Silurian-Devonian aquifer, the water is generally of good quality and acceptable for practically every use. In a few areas, where the glacial drift is thin, the underlying aquifer is susceptible to infiltration of surface contaminants. Evaporites occur within the Cedar Valley and Wapsipinicon groups in the subsurface of southern Iowa. In these areas, the aquifer carries excessive amounts of sulfate, and the water is generally unsuited for human consumption.

THE CAMBRIAN-ORDOVICIAN AQUIFER

Although often referred to as the Jordan aquifer, the Cambrian-Ordovician aquifer actually consists of three stratigraphic units: the Cambrian Jordan Formation, the Ordovician Prairie du Chien Group, and the Ordovician St. Peter Sandstone. The porous sandstones of the Jordan Formation comprise the basal part of the aquifer and the principal water-yielding interval of the system. The aquifer is confined by impermeable beds of the overlying Glenwood, Platteville, and Decorah formations and by the underlying St. Lawrence and Lone Rock formations.

The Cambrian-Ordovician aquifer is a highly dependable source of water for large-capacity wells in Iowa. Yields of several hundred gallons per minute to 1,000 gallons per minute are common. Groundwater movement in the aquifer is southeasterly from its recharge area in southern Minnesota and northern Iowa. Discharge is into the alluvial fill of the Mississippi River valley in Dubuque and Jackson counties of northeastern Iowa and into the Illinois Basin from subsurface flow in southeastern Iowa and northeastern Missouri. The potentiometric surface of the aquifer (level to which water rises in a well) has declined 50 to 150 feet regionally. The decline is even greater in the major pumping centers such as Des Moines, Cedar Rapids–Marion, and Mason City.

The quality of the water in the aquifer is best in northeastern Iowa, where the dissolved solids concentration is reasonably low. Mineralization of the water increases significantly to the southwest. In northwestern Iowa where the Cambrian-Ordovician aquifer is recharged through overlying Cretaceous strata, high sulfate concentrations occur. Offensive concentrations of radium are present in the water over a wide area in southeastern, central, and western Iowa. This poses a potential health problem unless the radium is removed by treatment methods, such as water-softening procedures. Water from the aquifer often requires treatment to remove iron. In spite of these water-quality problems, the Cambrian-Ordovician aquifer is one of the best sources for large-capacity municipal and industrial wells over nearly two-thirds of the state.

Even though large withdrawals from the aquifer are expected to continue for the foreseeable future, there is little chance that the source will be totally exhausted. According to one estimate, the Cambrian-Ordovician aquifer stores

80 trillion gallons of water. About 16 billion gallons of water move into the aquifer annually. However, declining pumping levels are likely at major pumping centers unless new wells in the aquifer are spaced judiciously.

THE DRESBACH AQUIFER

The Dresbach aquifer system consists of three Cambrian formations, from top to bottom: the Wonewoc, Eau Claire, and Mt. Simon. The aquifer includes coarse to fine-grained sandstones with interbedded siltstones, shales, and a few dolomites. It is overlain by the relatively impermeable Lone Rock and St. Lawrence formations of the Cambrian System; below the Dresbach lies undifferentiated Precambrian crystalline rocks or the Red Clastic series.

The depth of the aquifer and the mineralization of its water limit use outside of eastern Iowa. High-capacity wells with satisfactory water quality have been completed at New Albin, Lansing, Marquette, Dubuque, Bellevue, Maquoketa, Preston, De Witt, Clinton, and New London. Wells at Burlington and Decorah tap the Dresbach, but they also receive flow from the Cambrian-Ordovician aquifer. Consequently, it is difficult to determine the exact contribution of the Dresbach aquifer to these wells. The Dresbach is a prolific source of water in the Clinton area, where industrial wells yield 2,500 gallons per minute or more. However, huge withdrawals have produced the largest drawdown cone anywhere in the state. Water levels have dropped more than 250 feet at the major pumping sites and more than 125 feet throughout the city.

The water quality of the aquifer is generally good in the easternmost part of the state, from New Albin to Clinton. However, the McGregor area is an exception. There, a deep well encountered salty water from the Mt. Simon portion of the Dresbach aquifer. Salty water also occurs in Dresbach wells in the Quad Cities area. Radium concentrations are at or near the maximum allowable contaminant levels in some eastern Iowa Dresbach wells. Accordingly, water from these wells warrants treatment (softening) for removal of radium or mixing with better-quality water to dilute the concentration of radium.

Geology of Selected Public Lands

Although Iowa's geology is not the spectacular and awesome geology of the western mountains, canyons, and plateaus, it is nevertheless interesting and scenic. Geology has exerted an important influence on some of Iowa's major parks, recreation areas, and public lands. For example, the terrain of Iowa's Great Lakes region owes its origin to continental glaciers, and the topography in such state parks and preserves as Backbone, Bellevue, Bixby, Brush Creek Canyon, Dolliver Memorial, Echo Valley, Ledges, Pikes Peak, and Wildcat Den was produced by stream erosion of Paleozoic bedrock.

Other recreation areas in the state have some connection to geology as well. Yellow River State Forest in northeastern Iowa has exposures of Cambrian and Ordovician rocks. Red Rock Reservoir is named for red Pennsylvanian strata that are exposed in the vicinity. Geode State Park takes its name from Iowa's state rock, which occurs in Mississippian strata in southeastern Iowa and adjacent Illinois and Missouri. Gitchie Manitou State Preserve in the northwestern corner of the state reveals Iowa's oldest exposed bedrock — the Precambrian Sioux Quartzite, 1.6 billion years old. Palisades-Kepler State Park in Linn County provides a glimpse of the core of an ancient Silurian carbonate mound. Gull Point, McIntosh Woods, and Clear Lake state parks are located adjacent to glacial lakes. Fort Atkinson State Park was constructed from local Ordovician bedrock, and the type section of the Fort Atkinson Member of the Maquoketa Formation is located nearby. Lake Macbride State Park and the spillway below Coralville Reservoir (Devonian Fossil Gorge) in Johnson County expose fossiliferous Devonian bedrock. Stainbrook Geological Preserve, also in Johnson County, reveals glacial grooves on Devonian bedrock. Maquoketa Caves State Park displays solution features in Silurian dolomites, including thirteen caves. Preparation Canyon State Park in western Iowa shows stream-dissected topography on Pleistocene loess, and Pilot Knob State Park in northern Iowa is a good example of topographic features produced by stagnant glacial ice. Figure 11.13 provides a list of geological features found in selected state parks, preserves, recreation areas, forests, and other public lands.

Summary

Iowa's rock record contains a number of valuable resources, including limestone, gypsum, sand and gravel, and clay and clay shale. Limestone is used for aggregate in concrete and asphalt and for crushed-rock roads and parking lots, building stone, agricultural limestone, and various other uses. In addition, limestone is a key ingredient in the manufacture of portland cement. Iowa consistently ranks high nationally in the production of gypsum, which is used in the manufacture of wallboard and plaster. The sand-and-gravel deposits of the state furnish aggregate for cement. Iowa's clay and clay shales serve as raw materials in the manufacture of building brick, drainage tile, and sewer pipe. Iowa has abundant reserves of coal; if methods are perfected to remove sulfur from the coal economically, the coal industry may experience a resurgence. No coal is mined in the state at present, however.

Water is an important and abundant natural resource in Iowa. The water resources of the state consist of both surface water and groundwater supplies. The bulk of Iowa's domestic and industrial water needs are obtained from groundwater, and special care is needed to protect these supplies from contamination.

11.13 Geologic features in selected state parks, preserves, recreation areas, forests, and other public lands in Iowa. Adapted from Anderson 1983; Gwynne 1951–1960; Prior 1991; and Wolf 1991.

Geologic Features (Arranged in Order of Age, Precambrian–Quaternary)	State Park, Preserve, Recreation Area, Forest, or Other Public Lands	County
Precambrian bedrock; exposures of 1.6 billion-year-old Sioux Quartzite	Gitchie Manitou State Preserve	Lyon
Cambrian and Ordovician bedrock	Yellow River State Forest	Allamakee
Steep bluffs underlain by Cambrian sandstone and Ordovician dolomite	Effigy Mounds National Monument	Allamakee
Magnificent view of the Mississippi River; bluffs of Ordovician strata with Cambrian at the base; colorful exposure of Ordovician St. Peter Sandstone at Sand Cave	Pikes Peak State Park	Clayton
Scenic pinnacles of Ordovician carbonates (Galena Group); 46 burial mounds	Turkey River Mound State Preserve	Clayton
Ice cave in Ordovician bedrock	Decorah Ice Cave State Preserve	Winneshiek
Bold bluffs of Ordovician carbonates along the Upper Iowa River	Cold Water Spring State Preserve	Winneshiek
Ordovician carbonates dominate the scenery along the Upper Iowa River	Bluffton Fir Stand State Preserve	Winneshiek
Ordovician limestone was used to construct historic Fort Atkinson; the type section of the Fort Atkinson Member of the Maquoketa Formation is nearby	Fort Atkinson State Preserve	Winneshiek
Lead ore was once mined in the area from the Galena Group carbonates (Ordovician)	Mines of Spain State Recreation Area	Dubuque
Ordovician dolomites outcrop prominently; Indian Mounds of the Woodland Period	Little Maquoketa Mounds State Preserve	Dubuque
Silurian dolomites exposed; ice cave and cold-air drainage slope (algific slope); edge of Silurian Escarpment	Bixby State Preserve	Clayton
Steep rocky exposures along the Silurian Escarpment; underlying shales of Maquoketa Formation	Mossy Glen State Preserve	Clayton County
Exposures of Silurian dolomite along the Silurian Escarpment; tilted slump blocks	Brush Creek Canyon State Preserve	Fayette
Narrow valleys cut into the strata underlying the Silurian Escarpment	Echo Valley State Park	Fayette
Outcrops of Silurian dolomite where the Volga River cuts through the Silurian Escarpment	Volga River State Recreation Area	Fayette

(*continued*)

Geologic Features (Arranged in Order of Age, Precambrian–Quaternary)	State Park, Preserve, Recreation Area, Forest, or Other Public Lands	County
Erosion by the Maquoketa River has isolated a narrow ridge of Silurian dolomite (Hopkinton Formation) within the Iowan Surface; one of Iowa's largest springs, Richmond Springs, flows from Silurian bedrock	Backbone State Park	Delaware
Sinkholes, caves, and rock exposures of Silurian bedrock along the forested Silurian Escarpment	White Pine Hollow State Preserve	Clayton
Exposures of Silurian strata of a carbonate mound facies; Scotch Grove and Gower formations exposed in the area	Palisades-Kepler State Park	Linn
Bluffs along the Wapsipinicon River and its tributaries reveal Silurian dolomite	Wapsipinicon State Park	Jones
Rugged topography on stream-carved Silurian dolomite	Indian Bluffs Primitive Area	Jones
Karst processes have produced sinkholes, caves, and a natural bridge in Silurian dolomite; park has at least 13 caves, more than any other Iowa state park	Maquoketa Caves State Park	Jackson
The banks of the Winnebago River (once known as Lime Creek) reveal fossiliferous strata of the Devonian Lime Creek Formation; this is the historic Hackberry Grove locality	Claybanks Forest State Preserve	Cerro Gordo
A diverse assortment of invertebrate fossils occur in the Devonian Lime Creek Formation	Bird Hill State Preserve	Cerro Gordo
An abandoned clay pit exposes the richly fossiliferous Lime Creek Formation (Devonian); hordes of fossils, and collecting is permitted	Rockford Fossil and Prairie Area	Floyd
Impounded water flows over a spillway of fossiliferous Devonian bedrock (Little Cedar Formation)	Lake Macbride State Park	Johnson
Devonian bedrock of the Cedar Valley Group preserves fossils and glacial grooves	Stainbrook Geological Preserve	Johnson
Fossiliferous Devonian strata are revealed; the type area of the State Quarry Formation, which was used in construction of the Old Capitol in Iowa City	Old State Quarry State Preserve	Johnson
Fossiliferous rocks of the Devonian Little Cedar Formation	Devonian Fossil Gorge	Johnson

Geologic Features (Arranged in Order of Age, Precambrian–Quaternary)	State Park, Preserve, Recreation Area, Forest, or Other Public Lands	County
An artificial lake along the eastern edge of the Des Moines Lobe, from the ponded water of Spring Creek (a former meltwater channel); the rock wall of the dam is of the Mississippian Maynes Creek Formation	Beeds Lake State Park	Franklin
Mississippian limestone underlying Pre-Illinoian drift	Oakland Mills State Park	Henry
Named for Iowa's state rock, the geode; an artificial lake within the Illinoian glacial drift; Mississippian rocks in the region contain crystal-lined geodes	Geode State Park	Henry
The valley of Flint Creek exposes the type section of the Mississippian Starrs Cave Formation and the McCraney, Prospect Hill, Wassonville, and Burlington formations (all of Mississippian age)	Starrs Cave State Preserve	Des Moines
A great meander bend in the Des Moines River; Mississippian bedrock is exposed	Lacey-Keosauqua State Park	Van Buren
Bluffs of Pennsylvanian sandstone along a valley that was excavated by glacial meltwater from the Des Moines Lobe; exposures of the so-called cooperas beds	Dolliver Memorial State Park	Webster
Pennsylvanian sandstone exposed in a steep ravine that opens into the Des Moines River	Woodman Hollow State Preserve	Webster
Ledges of Pennsylvanian sandstone exposed in a valley carved by meltwaters from the Des Moines Lobe glacier	Ledges State Park	Boone
Pennsylvanian bedrock exposed in the spillway below the dam at Saylorville Lake	Saylorville Canyon, below Saylorville Dam	Polk
Pennsylvanian limestones and shales exposed under Pre-Illinoian drift	Pammel State Park	Madison
Floodplain features, including a dried-up oxbow lake of the Nishnabotna River; Pennsylvanian strata exposed in the area	Cold Springs State Park	Cass
An artificial lake; exposures of Pennsylvanian sandstone	Lake Ahquabi State Park	Warren
Pennsylvanian sandstones exposed along the Iowa River Greenbelt, including the exposures known as Steamboat Rock and Tower Rock	Steamboat Rock Area of the Iowa River Greenbelt	Hardin

(*continued*)

Geologic Features (Arranged in Order of Age, Precambrian–Quaternary)	State Park, Preserve, Recreation Area, Forest, or Other Public Lands	County
Pennsylvanian sandstone outcrops along the bluffs of the Iowa River valley; near the interface of the Iowan Surface and the Des Moines Lobe	Fallen Rock State Preserve	Hardin
Cross-bedded sandstone of the Pennsylvanian System is visible below Pre-Illinoian drift	Pine Lake State Park	Hardin
Pennsylvanian sandstone is exposed where Pine Creek has cut through Illinoian glacial drift; large slump blocks of sandstone slide downslope on underlying shale	Wildcat Den State Park	Muscatine
Outcrops of Cretaceous sandstone, shale, and limestone (Dakota, Graneros, and Greenhorn formations); stream-dissected landscape of the Loess Hills	Stone State Park and Mount Talbot State Preserve	Woodbury
A wetland supplied with groundwater from Cretaceous sandstone; located on a floodplain	Cold Springs (Crystal Lake) State Park	Cass
Cretaceous sandstone and conglomerate exposed in the area; stream-dissected landscape with loess over Pre-Illinoian drift	Springbrook State Park	Guthrie
Stream-sculpted landscape on thick loess	Five-Ridge Prairie State Preserve	Plymouth
Stream-eroded landscape on loess; great views of the Loess Hills and the adjacent Missouri Alluvial Plain	Turin Loess Hills State Preserve	Monona
Stream-carved landscape on thick loess	Preparation Canyon State Park	Monona
Stream-sculpted terrain on thick loess deposits	Waubonsie State Park	Fremont
A 6-ton glacial erratic marks the entrance; shelter house constructed from local glacial erratics	Pioneer State Park	Mitchell
A conical hill of sand and gravel produced by meltwater deposition within stagnant ice; along the western margin of the Bemis end moraine; a classic example of a kame	Ocheyedan Mound State Preserve	Osceola
A glacial lake on the Des Moines Lobe	Okamanpedan State Park	Emmet
A glacial lake along the southern edge of the Des Moines Lobe	Swan Lake State Park	Carroll

Geologic Features (Arranged in Order of Age, Precambrian–Quaternary)	State Park, Preserve, Recreation Area, Forest, or Other Public Lands	County
Stream-dissected terrain on upland deposits of the Bemis end moraine	Mann Wilderness State Preserve	Hardin
Glacial lake on the Bemis end moraine	Clear Lake	Cerro Gordo
Northwestern shoreline of Clear Lake in the irregular terrain of the Bemis end moraine	McIntosh Woods State Park	Cerro Gordo
Bemis end moraine; hummocky topography	Black Hawk Lake State Park	Sac
Silver Lake is a kettle lake within the Bemis end moraine; water-saturated peat deposits form where ground-water upwells to produce Silver Lake Fen	Silver Lake Fen State Preserve	Dickinson
A peninsula along the lake with numerous glacial erratics; West Okoboji is the deepest of Iowa's glacial lakes	Gull Point State Park	Dickinson
A kettle depression on the Altamont moraine	Freda Haffner Kettlehole State Preserve	Dickinson
Swell-and-swale topography on the Altamont end moraine	Cayler Prairie State Preserve	Dickinson
Glaciated terrain of the Des Moines Lobe, along the northern shore of Iowa's largest glacial lake, Spirit Lake	Mini-Wakan State Park	Dickinson
A low-relief glacial plain of the Des Moines Lobe; buried structure of Manson Crater is below the surface but left no visible surface evidence	Kalsow Prairie State Preserve	Pocahontas
Pilot Mound is a glacial kame within the Altamont end moraine	Pilot Mound State Forest	Boone
Altamont moraine; upland deposits	Barkley State Forest	Boone
Prairie potholes in subtly linked depression left by the Altamont ice advance	Doolittle Prairie State Preserve	Story
Irregular knobs and kettles on the Altamont end moraine; Pilot Knob is a glacial kame; a wide assortment of glacial erratics were used in construction of the observation tower	Pilot Knob State Park and Preserve	Hancock
Elongated glacial lakes within the deposits of the Altamont glacial advance	Twin Lakes State Park	Calhoun
A glacial lake and wetland along the eastern edge of the Altamont end moraine	Rice Lake State Park	Winnebago

(*continued*)

Geologic Features (Arranged in Order of Age, Precambrian–Quaternary)	State Park, Preserve, Recreation Area, Forest, or Other Public Lands	County
Gravelly knobs and depressions on the Altamont end moraine	Anderson Prairie State Preserve	Emmet
Pothole marsh located within an outwash channel through the Altamont end moraine	Cheever Lake State Preserve	Emmet
Knob-and-kettle topography of the Altamont end moraine	Fort Defiance State Park	Emmet
Embayments of West Okoboji Lake; bordered by the Altamont end moraine	Emerson Bay State Park	Dickinson
Knobby topography on deposits of the Altamont ice advance	Lake Cornelia State Park	Wright
The lateral margin of the Altamont end moraine	Hoffman Prairie State Preserve	Cerro Gordo
Algona end moraine	Ambrose A. Call State Park	Kossuth
Native prairie along the edge of the Algona end moraine	Stinson Prairie State Preserve	Kossuth
An outwash channel formed from meltwater from the Algona ice advance	Union Slough National Wildlife Refuge	Kossuth
A kettle lake within a glacial outwash channel; within the Algona end moraine	Crystal Lake State Park	Hancock
The steep-sided Little Sioux valley formed when the waters of glacial Lake Spencer were released, carving into deposits of the Sheldon Creek drift	Wanata State Park	Clay and Buena Vista
Terraces and benches produced by meltwater from the Des Moines Lobe ice	Brushy Creek State Recreation Area and Preserve	Webster
Eastern edge of the ancestral diversion valley of the Mississippi (the Goose Lake Channel)	Manikowski Prairie State Preserve	Clinton
Wind-blown sand deposits, derived from the nearby Iowa River valley	Marietta Sand Prairie State Preserve	Marshall
Wetland depressions within windblown sand and diverse prairie habitats; one of the fens of the Iowan Surface	Cedar Hills Sand Prairie State Preserve	Black Hawk
Windblown sand from the Cedar River valley; a diverse upland prairie	Rock Island State Preserve	Linn
Windblown loess and sand mantle Pre-Illinoian drift in this typical paha; the paha is oriented northwest to southeast	Casey's Paha State Preserve	Tama

Geologic Features (Arranged in Order of Age, Precambrian–Quaternary)	State Park, Preserve, Recreation Area, Forest, or Other Public Lands	County
A remnant of a terrace of the Mississippi River; high bluffs underlain by Cambrian sandstone	Fish Farm Mounds State Preserve	Allamakee
A deeply entrenched meander loop of the Upper Iowa River; Ordovician bedrock exposed in the area	Slinde Mounds State Preserve	Allamakee
The braided river channel of the Mississippi River is visible; gorgelike valley with steep bluffs	Upper Mississippi Wildlife Refuge	Clayton
Floodplain features of the Mississippi River	Mark Twain National Wildlife Refuge	Louisa
A stable island in the Cedar River valley	Rock Creek Island Preserve	Cedar
A low-relief landscape with floodplain features of the Cedar River	George Wyth Memorial State Park	Black Hawk
A bedrock bench covered by gravels	Hardin City Woodland State Preserve	Hardin
A funnel-shaped trap built by Native Americans from glacial erratics in the Iowa River	Indian Fish Trap	Iowa
Meander scars and floodplain features of the Raccoon River valley	Walnut Woods State Park	Polk
An oxbow lake in an abandoned channel of the Missouri River	Browns Lake State Park	Woodbury
An abandoned meander loop of the Missouri River	DeSoto Bend National Wildlife Refuge	Harrison
An oxbow lake; marks a former course of the Missouri River	Lewis and Clark State Park	Monona
An oxbow lake, representing a change in course of the Missouri River	Lake Manawa State Park	Pottawattamie

Geologic processes played a key role in the formation of many of Iowa's state parks, preserves, recreation areas, and other public lands. Thus, a knowledge of geology is of value to anyone interested in understanding the origin of the state's scenic areas and landscape features.

Acknowledgments

Robert M. McKay provided information on Iowa's mineral production and reviewed this chapter. I acknowledge his helpful suggestions. Several articles from *Iowa Geology* were useful. In addition, I relied on the work of the following for portions of this chapter: Charles S. Gwynne, George R. Hallberg, Paul J.

Horick, Bernard E. Hoyer, Timothy J. Kemmis, John Lemish, Robert D. Libra, Greg A. Ludvigson, Jean C. Prior, Walter L. Steinhilber, Carol A. Thompson, Brian J. Witzke, and Robert C. Wolf. James Hearst's "Landscape — Iowa" appeared in *Shaken by Leaf-Fall* and *Snake in the Strawberries, Selected Poems by James Hearst*. It is reprinted here with the permission of Iowa State University Press.

References

Ager, D. V. 1961. "The Epifauna of a Devonian Spiriferid." *Quarterly Journal of the Geological Society of London* 117: 1–10.

———. 1967. *Principles of Paleontology*. McGraw-Hill, New York.

Agnew, A. F. 1955. "Facies of Middle and Upper Ordovician Rocks of Iowa." *American Association of Petroleum Geologists Bulletin* 39: 1703–1752.

———, A. V. Heyl, C. H. Behre, and E. J. Lyons. 1956. *Stratigraphy of Middle Ordovician Rocks in the Zinc-Lead District of Wisconsin, Illinois, and Iowa*. U.S. Geological Survey Professional Paper 274-K. Washington, D.C.

Alden, W. C., and M. M. Leighton. 1917. "The Iowa Drift: A Review of the Evidence of the Glacial Lobes of the Central Lowlands." *Geological Society of America Bulletin* 68: 1415–1449.

Alex, L. M. 1976. *Educational Series 1–7*. Office of the State Archaeologist, University of Iowa, Iowa City.

Alvarez, L. W., W. Alvarez, F. Asaro, and H. Michel. 1980. "Extraterrestrial Cause for the Cretaceous-Tertiary Extinction." *Science* 208: 1095–1108.

Anderson, D. C. 1975. *Western Iowa Prehistory*. Iowa State University Press, Ames.

———. 1981. *Eastern Iowa Prehistory*. Iowa State University Press, Ames.

———, and H. A. Semken Jr., eds. 1980. *The Cherokee Excavations: Holocene Ecology and Human Adaptations in Northwestern Iowa*. Academic Press, New York.

———, and P. M. Williams. 1974. "Western Iowa Proboscidians." *Proceedings of the Iowa Academy of Science* 81: 185–191.

Anderson, R. R. 1986. "Petroleum Test Drilling in Iowa." *Iowa Geology* 11: 10–11.

———. 1987a. "Precambrian Geology of Iowa." Iowa Geological Survey Open-File Map 87-1. Iowa City.

———. 1987b. "Precambrian Sioux Quartzite at Gitchie Manitou State Preserve, Iowa." In *Geological Society of America Centennial Field Guide — North-Central Section*, ed. D. L. Biggs. Geological Society of America, Boulder, Colo.

———. 1988. "Historic Oil Test Completed." *Iowa Geology* 13: 11.

———, ed. 1990a. *The Amoco M. G. Eischeid #1 Deep Petroleum Test, Carroll County, Iowa, Preliminary Investigation*. Geological Survey Bureau Special Report Series No. 2. Iowa City.

———. 1990b. "Iowa's Deepest Well." *Iowa Geology* 15: 26–27.

———. 1990c. "Review of Current Studies of Proterozoic Rocks in the Amoco M. G. Eischeid #1 Petroleum Test Well, Carroll County, Iowa." In *The Amoco M. G. Eischeid #1 Deep Petroleum Test, Carroll County, Iowa, Preliminary Investigations*, ed. R. R. Anderson. Geological Survey Bureau Special Report Series No. 2. Iowa City.

———. 1990d. "Review of the Precambrian Geological History of the Central United States and the Midcontinent Rift System." In *The Amoco M. G. Eischeid #1 Deep Petroleum Test, Carroll County, Iowa, Preliminary Investigations*, ed. R. R. Anderson. Geological Survey Bureau Special Report Series No. 2. Iowa City.

———. 1992. "Oil Exploration in Iowa." *Iowa Geology* 17: 19.

———. 1995. "Midcontinent Rift System in Iowa." Home Page. Geological Survey Bureau, Iowa City.

———, and G. A. Ludvigson. 1986. "Baraboo Interval Quartzite in Washington County, Iowa." *Geoscience Wisconsin* 10: 15–27.

———, R. M. McKay, and B. J. Witzke. 1979. *Field Trip Guidebook to the Cambrian Stratigraphy of Allamakee County*. Geological Society of Iowa Guidebook 32. Iowa City.

————, and P. E. VanDorpe. 1991. "Iowa Perspective on Midwestern Earthquakes." *Iowa Geology* 16: 25–27.

————, and B. J. Witzke. 1994. "The Terminal Cretaceous Manson Impact Structure in North-Central Iowa: A Window into the Late Cretaceous History of the Eastern Margin of the Western Cretaceous Seaway." In *Perspectives of the Eastern Margin of the Cretaceous Western Interior Basin*, eds. G. A. Shurr, G. A. Ludvigson, and R. H. Hammond. Geological Society of America Special Paper 287. Boulder, Colo.

————, B. J. Witzke, and D. J. Roddy. 1996. "The Drilling of the 1991–1992 Geological Survey Bureau and U.S. Geological Survey Manson Impact Structure Research Cores." In *The Manson Impact Structure, Iowa: Anatomy of an Impact Crater*, eds. C. Koeberl and R. R. Anderson. Geological Society of America Special Paper 302. Boulder, Colo.

Anderson, W. I. 1964. "Upper Devonian and Lower Mississippian Conodont Faunas, North-Central Iowa." *Proceedings of the Iowa Academy of Science* 71: 320–334.

————. 1966. "Upper Devonian Conodonts and the Devonian-Mississippian Boundary of North-Central Iowa." *Journal of Paleontology* 40: 395–415.

————. 1969. "Lower Mississippian Conodonts from Northern Iowa." *Journal of Paleontology* 43: 916–929.

————. 1973. "Mississippian Conodonts in Iowa." *Proceedings of the Iowa Academy of Science* 80: 34–38.

————. 1983. *Geology of Iowa: Over Two Billion Years of Change*. Iowa State University Press, Ames.

————. 1984. *General Geology of North-Central Iowa*. Guidebook for the Forty-eighth Annual Tri-State Geological Field Conference. Cedar Falls, Iowa.

————. 1989. "Iowa Geology: The Early Years." *Journal of the Iowa Academy of Science* 6: 81–91.

————. 1995a. "The Lime Creek Formation in the Rockford Area: A Geological Perspective." In *Geology and Hydrogeology of Floyd-Mitchell Counties*, ed. B. J. Bunker. Geological Society of Iowa Guidebook 62. Iowa City.

————. 1995b. "The Rockford Area: A Historical Perspective." In *Geology and Hydrogeology of Floyd-Mitchell Counties*, ed. B. J. Bunker. Geological Society of Iowa Guidebook 62. Iowa City.

————, and W. M. Furnish. 1983. "Iowa's Self-Trained Paleontologists." *Proceedings of the Iowa Academy of Science* 90: 1–12.

————, and W. M. Furnish. 1987. "The Lime Creek Formation of North-Central Iowa." In *Geological Society of America Centennial Field Guide — North-Central Section*, ed. D. L. Biggs. Geological Society of America, Boulder, Colo.

————, and K. Megivern. 1982. "Epibionts from the Cerro Gordo Member of the Lime Creek Formation (Upper Devonian), Rockford, Iowa." *Proceedings of the Iowa Academy of Science* 89: 71–82.

Andrews, G. W. 1958. "Windrow Formation of the Upper Mississippi Valley Region, a Sedimentary and Stratigraphic Study." *Journal of Geology* 66: 597–624.

Atherton, E., and J. E. Palmer. 1979. *The Mississippian and Pennsylvanian (Carboniferous) Systems in the United States — Illinois*. U.S. Geological Survey Professional Paper 1110-L. Washington, D.C.

Atkinson, W. 1989. *The Next New Madrid Earthquake: A Survival Guide for the Midwest*. Southern Illinois Press, Carbondale and Edwardsville.

Auge, T. 1976. "The Life and Times of Julien Dubuque." *Palimpsest* 57: 2–13.

Austin, G. S. 1972a. "Cretaceous Rocks." In *Geology of Minnesota: A Centennial Volume*, eds. P. K. Sims and G. B. Morey. Minnesota Geological Survey, St. Paul.

————. 1972b. "The Sioux Quartzite: Southwestern Minnesota." In *Geology of Minnesota: A*

Centennial Volume, eds. P. K. Sims and G. B. Morey. Minnesota Geological Survey, St. Paul.

Bain, H. F. 1895. "Geology of Woodbury County." *Iowa Geological Survey Annual Report* 5: 241–299.

———. 1896. "Geology of Guthrie County." *Iowa Geological Survey Annual Report* 7: 413–487.

Baker, C., B. F. Glenister, and C. O. Levorson. 1986. "Devonian Ammonoid *Manticoceras* from Iowa." *Proceedings of the Iowa Academy of Science* 93: 1–6.

Baker, R. G., T. J. Frest, and R. S. Rhodes II. 1984. "Paleoecology of Quaternary Sediments at Conklin Quarry." In *Underburden-Overburden: An Examination of Paleozoic and Quaternary Strata at the Conklin Quarry near Iowa City*, eds. B. J. Bunker and G. R. Hallberg. Geological Society of Iowa Guidebook 41. Iowa City.

———, L. J. Maher, C. A. Chumbley, and K. L. Van Zant. 1992. "Pattern of Holocene Environmental Change in the Midwestern United States." *Quaternary Research* 37: 379–389.

———, R. S. Rhodes II, D. P. Schwert, A. C. Ashworth, T. J. Frest, G. R. Hallberg, and J. A. Janssens. 1986. "A Full-Glacial Biota from Southeastern Iowa, USA." *Journal of Quaternary Science* 1: 91–107.

———, D. P. Schwert, E. A. Bettis III, T. J. Kemmis, D. G. Horton, and H. A. Semken. 1991. "Mid-Wisconsinan Stratigraphy and Paleoenvironments at the St. Charles Site, South-Central Iowa." *Geological Society of America Bulletin* 103: 210–220.

Baldwin, W. B. 1951. "Geology of the Sioux Formation." Ph.D. diss. Columbia University, New York.

Bandiozamani, K. 1973. "The Dorag Dolomitization: Application to the Middle Ordovician of Wisconsin." *Journal of Sedimentary Petrology* 43: 965–984.

Bankes, M. 1994. *The History of Big Spring*. Geological Survey Bureau Technical Information Series 30. Iowa City.

Bates, R. L., and J. A. Jackson, eds. 1980. *Glossary of Geology*. 2nd ed. American Geological Institute, Falls Church, Va.

Beinert, R. J. 1968. *Development Report of Coralville Reservoir Geological Preserve by Mehaffey Bridge, and State Quarry Geological Preserve*. Iowa State Advisory Board for Preserves, Development Series Report 1. Iowa City.

Beitz, R. S. 1962. "Swarthy King of the Palace Age." *Iowan* (summer): 42–45.

Belanski, C. H. 1927. "The Shell Rock Stage of the Devonian of Iowa, Part 1." *American Midland Naturalist* 10: 317–370.

———. 1928a. *Pentameracea of the Devonian of Northern Iowa*. State University of Iowa Studies 7. Iowa City.

———. 1928b. "The Shell Rock Stage of the Devonian of Iowa, Part 3." *American Midland Naturalist* 11: 165–215.

———. 1928c. *Terebratulacea of the Devonian of Northern Iowa*. State University of Iowa Studies 8. Iowa City.

Benn, D. W. 1990. "Depositional Stratigraphy, Site Context and Prehistoric Cultural Overview." In *Holocene Alluvial Stratigraphy and Selected Aspects of the Quaternary History of Western Iowa*, ed. E. A. Bettis III. Geological Survey Bureau Guidebook 9. Iowa City.

———. 1992. "Archaeological Overview of the Upper Mississippi River Valley, Rock Island District." In *Late Wisconsinan and Holocene Alluvial Stratigraphy, Paleoecology, and Archaeological Geology of East-Central Iowa*, eds. E. A. Bettis III, R. G. Baker, W. R. Green, M. K. Whelan, and D. W. Benn. Geological Survey Bureau Guidebook 12. Iowa City.

Bettis, E. A., III. 1986. "Saylorville Canyon." *Iowa Geology* 11: 18–19.

———. 1987. "History of the Upper Mississippi Valley." *Iowa Geology* 12: 12–15.

———, ed. 1990. *Holocene Alluvial Stratigraphy and Selected Aspects of the Quaternary History of Western Iowa*. Geological Survey Bureau Guidebook 9. Iowa City.

———, R. G. Baker, W. R. Green, M. K. Whelan, and D. W. Benn, eds. 1992. *Late Wisconsinan and Holocene Alluvial Stratigraphy, Paleoecology, and Archaeological Geology of East-Central Iowa*. Geological Survey Bureau Guidebook 12. Iowa City.

———, and D. W. Benn. 1984. "An Archaeological and Geomorphic Survey in the Central Des Moines River Valley, Iowa." *Plains Anthropologist* 29: 211–227.

———, and B. E. Hoyer. 1986. *Late Wisconsinan and Holocene Landscape Evolution and Alluvial Stratigraphy in the Saylorville Lake Area, Central Des Moines River Valley, Iowa*. Iowa Geological Survey Open-File Report 86-1. Iowa City.

———, T. J. Kemmis, and B. J. Witzke, eds. 1985. *After the Great Flood: Exposures in the Emergency Spillway, Saylorville Dam*. Geological Society of Iowa Guidebook 43. Iowa City.

———, J. Pearson, M. Edwards, D. M. Gradwohl, N. M. Osborn, T. J. Kemmis, and D. J. Quade. 1988. *Natural History of Ledges State Park and the Des Moines River Valley in Boone County, Iowa*. Iowa Natural History Association Field Trip Guidebook 6 and Geological Society of Iowa Guidebook 43. Iowa City.

———, J. C. Prior, G. R. Hallberg, and R. L. Handy. 1986. "Geology of the Loess Hills Region." *Proceedings of the Iowa Academy of Science* 93: 78–85.

———, D. J. Quade, and T. J. Kemmis. 1996. *Hogs, Bogs, and Logs: Quaternary Deposits and Environmental Geology of the Des Moines Lobe*. Geological Survey Bureau Guidebook 18. Iowa City.

———, B. J. Witzke, and B. J. Bunker. 1994. "Flood of '93 Uncovers Devonian Seafloor." *Iowa Geology* 19: 2–7.

Beyer, S. W. 1897. "Sioux Quartzite and Certain Associated Rocks." *Iowa Geological Survey Annual Report* 16: 67–112.

Bickford, M. E., W. R. Van Schmus, and I. Zietz. 1986. "Proterozoic History of the Midcontinent Region of North America." *Geology* 14: 492–496.

Bleifuss, R. L. 1972. "The Iron Ores of Southeastern Minnesota." In *Geology of Minnesota: A Centennial Volume*, eds. P. K. Sims and G. B. Morey. Minnesota Geological Survey, St. Paul.

Boellstorff, J. D. 1978. "North American Pleistocene Stages Reconsidered in Light of Probable Pliocene-Pleistocene Continental Glaciation." *Science* 202: 305–307.

Bolt, J. R., R. M. McKay, B. J. Witzke, and M. P. McAdams. 1988. "A New Lower Carboniferous Tetrapod Locality in Iowa." *Nature* 333: 768–770.

Bounk, M. J., and E. A. Bettis III. 1984. "Karst Development in Northeastern Iowa." *Proceedings of the Iowa Academy of Science* 91: 12–15.

Boyt, R. 1962. *Crinoid and Starfish Fossils from Le Grand, Iowa*. Iowa State Department of History and Archives Special Publication. Des Moines.

Brenner, R. L., R. F. Bretz, B. J. Bunker, D. L. Iles, G. A. Ludvigson, R. M. McKay, D. L. Whitley, and B. J. Witzke. 1981. *Cretaceous Stratigraphy and Sedimentation in Northwest Iowa, Northeast Nebraska, and Southeast South Dakota*. Iowa Geological Survey Guidebook 4. Iowa City.

Bretsky, P. W., and J. J. Bermingham. 1970. "Ecology of the Paleozoic Scaphopods Genus *Plagioglypta* with Special Reference to the Ordovician of Eastern Iowa." *Journal of Paleontology* 44: 908–924.

Briggs, D. E., E. N. Clarkson, and R. J. Aldridge. 1983. "The Conodont Animal." *Lethaia* 16: 1–14.

Briggs, J. 1920. "A Geologic Palimpsest." *Palimpsest* 1: 133.

Brower, J. C. 1995. "Dendrocrinid Crinoids from the Ordovician of Northern Iowa and Southern Minnesota." *Journal of Paleontology* 69: 939–960.

———, and H. L. Strimple. 1983. "Ordovician Calceocrinids from Northern Iowa and Southern Minnesota." *Journal of Paleontology* 57: 1261–1281.

Brown, D. E. 1974. "Phosphatic Zone in the Lower Part of the Maquoketa Shale in Northeastern Iowa." *Journal of Research, U.S. Geological Survey* 2 (2): 219–232.

Bucher, W. H. 1933. "Cryptovolcanic Structures in the United States." *Sixteenth International Geological Congress, United States*: 1055–1084.

Bunker, B. J. 1981. "Configuration of the Precambrian Surface in Iowa and Adjacent States." Iowa Geological Survey Open-File Map. Iowa City.

———. 1988. "Late Middle through Lower Upper Devonian Stratigraphy across Eastern Iowa." In *New Perspectives on the Paleozoic History of the Upper Mississippi Valley: An Examination of the Plum River Fault Zone*, eds. G. A. Ludvigson and B. J. Bunker. Geological Survey Bureau Guidebook 8. Iowa City.

———. 1991. "Lithograph City." *Iowa Geology* 16: 16–19.

———, ed. 1995. *Geology and Hydrogeology of Floyd-Mitchell Counties*. Geological Society of Iowa Guidebook 62. Iowa City.

———, G. Klapper, and B. J. Witzke. 1983. *New Stratigraphic Interpretations of the Middle Devonian Rocks of Winneshiek and Fayette Counties, Northeastern Iowa*. Geological Society of Iowa Guidebook 39. Iowa City.

———, G. A. Ludvigson, and B. J. Witzke, eds. 1985. *The Plum River Fault Zone and the Structural and Stratigraphic Framework of Eastern Iowa*. Iowa Geological Survey Technical Information Series 13. Iowa City.

———, and B. J. Witzke. 1987. "Cedar Valley Formation of the Coralville Lake Area, Iowa." In *Geological Society of America Centennial Field Guide — North-Central Section*, ed. D. L. Biggs. Geological Society of America, Boulder, Colo.

———, and B. J. Witzke. 1988. "Central Mid-Continent Region, United States, Plate 4." In *The Geology of North America*. Vol. D-2, *Sedimentary Cover — North American Craton*, ed. L. L. Sloss. Geological Society of America, Boulder, Colo.

———, and B. J. Witzke. 1989. *The Coralville and Lithograph City Formations of the Coralville Lake Area*. Geological Society of Iowa Guidebook 51. Iowa City.

———, and B. J. Witzke. 1992. "An Upper Middle through Lower Upper Devonian Lithostratigraphic and Conodont Biostratigraphic Framework of the Midcontinent Carbonate Shelf Area, Iowa." In *The Stratigraphy, Paleontology, Depositional and Diagenetic History of the Middle-Upper Devonian Cedar Valley Group of Central and Eastern Iowa*, eds. J. E. Day and B. J. Bunker. Geological Survey Bureau Guidebook 16. Iowa City.

———, and B. J. Witzke. 1995. "Lithograph City — Historical Overview." In *Geology and Hydrogeology of Floyd-Mitchell Counties*, ed. B. J. Bunker. Geological Society of Iowa Guidebook 62. Iowa City.

———, B. J. Witzke, and J. E. Day. 1986. *Upper Cedar Valley Stratigraphy, North-Central Iowa, Lithograph City Formation*. Geological Society of Iowa Guidebook 44. Iowa City.

———, B. J. Witzke, W. L. Watney, and G. A. Ludvigson. 1988. "Phanerozoic History of the Central Midcontinent, United States." In *The Geology of North America*. Vol. D-2, *Sedimentary Cover — North American Craton*, ed. L. L. Sloss. Geological Society of America, Boulder, Colo.

Bureau of Mines. 1995. "The Mineral Industry of Iowa." In *Minerals Yearbook, Area Reports: Domestic 1993–94*, Vol. 2. United States Department of the Interior, Washington, D.C.

Byers, C. W., and R. H. Dott Jr. 1995. "Sedimentology and Depositional Sequences of the Jordan Formation (Upper Cambrian), Northern Mississippi Valley." *Journal of Sedimentary Research* B65: 289–305.

Calvin, S. 1893a. "Cretaceous Deposits of Woodbury and Plymouth Counties, with Observations on Their Economic Uses." *Iowa Geological Survey Annual Report* 1: 145–161.

———.1893b. "Relations of Cretaceous Deposits of Iowa to Subdivisions of Cretaceous Proposed by Meek and Hayden." *Proceedings of the Iowa Academy of Science* 1: 7–12.

———. 1894. "Geology of Allamakee County." *Iowa Geological Survey Annual Report* 4: 35–120.

———. 1895. "Geology of Jones County." *Iowa Geological Survey Annual Report* 5: 35–112.

———. 1897. "Geology of Delaware County." *Iowa Geological Survey Annual Report* 8: 121–192.

———. 1902. "Geology of Chickasaw County." *Iowa Geological Survey Annual Report* 13: 255–292.

———. 1904. "Geology of Winneshiek County." *Iowa Geological Survey Annual Report* 16: 37–146.

———. 1906. "Fifteenth Annual Report of the State Geologist." *Iowa Geological Survey Annual Report* 17: 1–6.

———. 1909a. "Aftonian Mammalian Fauna." *Geological Society of America Bulletin* 20: 341–356.

———. 1909b. "Present Phase of the Pleistocene Problem in Iowa." *Geological Society of America Bulletin* 20: 133–152.

———. 1911. "The Iowan Drift." *Journal of Geology* 19: 577–602.

———, and H. F. Bain. 1899. "Geology of Dubuque County." *Iowa Geological Survey Annual Report* 10: 379–622.

Catlin, G. 1840. "Account of a Journey to the Coteau des Prairies, with a Description of the Red Pipe Stone Quarry and Granite Boulders Found There." *American Journal of Science* 38: 145.

Cheney, E. S. 1981. "The Hunt for Giant Uranium Deposits." *American Scientist* 69: 3–48.

Chowns, T. M., and J. E. Elkins. 1974. "The Origin of Quartz Geodes and Cauliflower Cherts through the Silicification of Anhydrite Nodules." *Journal of Sedimentary Petrology* 44: 885–903.

Christian, R. 1998. "John L. Lewis Memorial Museum of Mining and Labor." *Des Moines Sunday Register*, May 3.

Chumbley, C. A., R. G. Baker, and E. A. Bettis III. 1990. "Midwestern Holocene Paleoenvironments Revealed by Floodplain Deposits in Northeastern Iowa." *Science* 249: 272–274.

Clark, J. M. 1921. *James Hall of Albany, Geologist and Paleontologist.* N.p., Albany, N.Y.

Cody, R. D., R. R. Anderson, and R. M. McKay. 1996. *Geology of the Fort Dodge Formation (Upper Jurassic), Webster County, Iowa.* Geological Survey Bureau Guidebook 19. Iowa City.

Collinson, C. W., 1964. *Guide for Beginning Fossil Hunters.* Illinois Geological Survey Educational Series Publication 4. Urbana.

———, and R. Skartvedt. 1960. *Field Book: Pennsylvanian Plant Fossils of Illinois.* Illinois Geological Survey Educational Series 6. Urbana.

Cooper, T. C., ed. 1982. *Iowa Natural Heritage.* Iowa Natural Heritage Foundation, Des Moines, and Iowa Academy of Science, Cedar Falls.

Cowles, G. 1986. "Mike Cowles Looks Back; Mencken: Make Mate for Giant Fake." *Des Moines Register*, Sept. 9.

Cross, A. T. 1966. "Palynologic Evidence of Mid-Mesozoic Age of Fort Dodge (Iowa) Gypsum." In *Abstracts, Program of the 1966 Annual Meeting of the Geological Society of America* 101: 1.

Cummings, E. R., and R. R. Shrock. 1928. "Niagaran Coral Reefs of Indiana and Adjacent

States and Their Stratigraphic Relationships." *Geological Society of America Bulletin* 39: 597–620.

Dapples, E. C. 1955. "General Lithofacies Relationship of St. Peter Sandstone and the Simpson Group." *American Association of Petroleum Geologists Bulletin* 39: 444–467.

Davis, L. C., R. E. Eshelman, and J. C. Prior. 1972. "A Primary Mammoth Site with Associated Fauna in Pottawattamie County, Iowa." *Proceedings of the Iowa Academy of Science* 79: 62–65.

Day, J. E. 1988. "The Brachiopod Succession of the Late Givetian-Frasnian of Iowa." In *Devonian of the World*, eds. N. J. McMillan, A. F. Embry, and D. J. Glass. Canadian Society of Petroleum Geologists Memoir 14. Calgary.

———. 1990. "The Upper Devonian (Frasnian) Conodont Sequence in the Lime Creek Formation of North-Central Iowa and Comparison with Lime Creek Ammonoid, Brachiopod, Foraminifer, and Gastropod Sequences." *Journal of Paleontology* 64: 614–628.

———. 1992. "Middle-Upper Devonian (Late Givetian–Early Frasnian) Brachiopod Sequence in the Cedar Valley Group of Central and Eastern Iowa. In *The Stratigraphy, Paleontology, Depositional and Diagenetic History of the Middle-Upper Devonian Cedar Valley Group of Central and Eastern Iowa*, eds. J. E. Day and B. J. Bunker. Geological Survey Bureau Guidebook 16. Iowa City.

———. 1995. "The Brachiopod Fauna of the Upper Devonian (Late Frasnian) Lime Creek Formation of North-Central Iowa, and Related Deposits in Eastern Iowa." In *Geology and Hydrogeology of Floyd-Mitchell Counties*, ed. B. J. Bunker. Geological Society of Iowa Guidebook 62. Iowa City.

Delgado, D. J., ed. 1983. *Ordovician Galena Group of the Upper Mississippi Valley, Deposition, Diagenesis, and Paleoecology*. Guidebook for the Thirteenth Annual Field Conference, Great Lakes Section, Society of Economic Paleontologists and Mineralogists. Dubuque, Iowa.

Denison, R. H. 1985. "A New Ptyctodont Placoderm, *Ptyctodopsis*, from the Middle Devonian of Iowa." *Journal of Paleontology* 59: 511–522.

Dietz, R. S., and J. C. Holden. 1970. "The Breakup of Pangaea." *Scientific American* 223: 30–41.

Dirks, R. A., and C. R. Busch. 1969. "The Giant Boulders of the Iowan Drift and a Consideration of Their Origin." *Proceedings of the Iowa Academy of Science* 76: 282–295.

Dorale, J. A., L. A. Gonzalez, M. K. Reagan, D. A. Pickett, M. T. Murrell, and R. G. Baker. 1992. "A High-Resolution Record of Holocene Climate Change in Speleothem Calcite from Cold Water Cave, Northeast Iowa." *Science* 258: 1626–1630.

Dorheim, F. H., and R. B. Campbell. 1958. "Recent Gypsum Exploration in Iowa." *Proceedings of the Iowa Academy of Science* 65: 246–253.

———, and D. L. Koch. 1962. "Unusual Exposure of Silurian-Devonian Unconformity in Loomis Quarry near Denver, Iowa." *Proceedings of the Iowa Academy of Science* 69: 341–350.

———, D. L. Koch, and M. C. Parker. 1969. *The Yellow Spring Group of the Upper Devonian in Iowa*. Iowa Geological Survey Report of Investigations 9. Iowa City.

Dott, R. H., Jr. 1983. "The Proterozoic Red Quartzite Enigma in the North-Central United States: Resolved by Plated Collisions?" In *Early Proterozoic Geology of the Great Lakes Region*, ed. L. G. Medaris Jr. Geological Society of America Memoir 160. Boulder, Colo.

———. 1985. "James Hall's Discovery of the Craton." In *Geologists and Ideas: A History of North American Geology*, eds. E. T. Drake and W. M. Jordan. Geological Society of America Centennial Special Volume 1. Boulder, Colo.

———, and R. L. Batten. 1971. *Evolution of the Earth*. 1st ed. McGraw-Hill, New York.

———, C. W. Byers, G. W. Fielder, S. R. Stenzel, and K. E. Winfree. 1986. "Aeolian to

Marine Transition in Cambro-Ordovician Cratonic Sheet Sandstones of the Northern Mississippi Valley, U.S.A." *Sedimentology* 33: 345–367.

———, and D. R. Prothero. 1994. *Evolution of the Earth*. 5th ed. McGraw-Hill, New York.

Drake, L., and R. Neimann. 1987. "Strip Mine Reclamation." In *Environments of Deposition of the Carboniferous System along the Mississippi River from Burlington to East of Muscatine and Strip Mine Reclamation*, eds. G. G. McCormick and L. Johnson. Guidebook for the Fifty-first Annual Tri-State Geological Field Conference. Iowa City.

———, and T. Ririe. 1975. "Strip-Mine Reclamation in South-Central Iowa." In *Guidebook for the Thirty-ninth Annual Tri-State Geological Field Conference*. Iowa City.

Dryden, J. E. 1955. "A Study of a Well Core from Crystalline Rocks near Manson, Iowa." Master's thesis, University of Iowa, Iowa City.

DuMontelle, P. B., S. C. Bradford, R. A. Bauer, and M. M. Killey. 1981. *Mine Subsidence in Illinois: Facts for the Homeowner Considering Insurance*. Illinois Geological Survey Environmental Geology Notes 99. Urbana.

Dunham, R. J. 1962. "Classification of Carbonate Rocks According to Depositional Texture." In *Classification of Carbonate Rocks*, ed. W. E. Ham. American Association of Petroleum Geologists Memoir 1. Tulsa, Okla.

Dunn, J. T. 1948. "The Cardiff Giant." *New York History* (July). Reprint, *Iowan* 1960 (Aug.–Sept.): 10–13.

Eastman, C. R. 1907. "Devonian Fishes of Iowa." *Iowa Geological Survey Annual Report* 18: 29–386.

Ehlers, E., and H. Blatt. 1982. *Petrology, Igneous, Sedimentary, and Metamorphic*. W. H. Freeman, San Francisco.

Eilers, L. J., and D. M. Roosa. 1994. *The Vascular Plants of Iowa*. University of Iowa Press, Iowa City.

Esling, S. P. 1983. "Quaternary Stratigraphy of the Lower Iowa and Cedar River Valleys, Southeast Iowa." Ph.D. diss., University of Iowa, Iowa City.

Faxlanger, D., J. Sinatra, and C. J. Uban. 1973. *Land Patterns of Iowa*. Department of Landscape Architecture, Iowa State University, Ames.

Fenton, C. L. 1919. "The Hackberry Stage of the Upper Devonian." *American Journal of Science* 48: 355–376.

———. 1931. *Studies of the Evolution of the Genus* Spirifer. Vol. 2. Wagner Free Institute of Science, Philadelphia.

———, and M. A. Fenton. 1924. *The Stratigraphy and Fauna of the Hackberry Stage of the Upper Devonian*. Vol. 1. University of Michigan Contributions from the Museum of Geology. Ann Arbor.

———, and M. A. Fenton. 1958. *The Fossil Book*. Doubleday, Garden City, N.Y.

Fisher, D. W. 1978. "James Hall — Patriarch of American Paleontology, Geological Organizations, and State Geological Surveys." *Journal of Geological Education* 26: 146–152.

Folk, R. L., and L. S. Land. 1975. "Mg/Ca Ratio and Salinity: Two Controls Over Crystallization of Dolomite." *American Association of Petroleum Geologists Bulletin* 59: 60–68.

Foster, J. D., and R. C. Palmquist. 1969. "Possible Subglacial Origin for 'Minor Moraine' Topography." *Proceedings of the Iowa Academy of Science* 76: 296–310.

Frest, T. J., and J. R. Dickson. 1986. "Land Snails (Pleistocene–Recent) of the Loess Hills: A Preliminary Survey." *Proceedings of the Iowa Academy of Science* 93: 130–157.

Furnish, W. M. 1963. "Silurian-Devonian of Eastern Iowa." In *Guidebook for the Twenty-seventh Annual Tri-State Geological Field Conference*. Iowa City.

———, and W. L. Manger. 1973. "Type Kinderhook Ammonoids." *Proceedings of the Iowa Academy of Science* 80: 15–24.

Gallaher, R. A. 1921. "The Cardiff Giant." *Palimpsest* 2: 269–281.

Garvin, P. L. 1998. *Iowa's Minerals: Their Origins, Industries, and Lore.* University of Iowa Press, Iowa City.

———, N. Sammis, D. Berchenbriter, and O. J. Van Eck. 1975. *Strippable Coal Reserve Study in Selected Iowa Counties.* Iowa Geological Survey Miscellaneous Publication 10. Iowa City.

———, and O. J. Van Eck. 1976. *Strippable Coal Reserve Study in Selected Iowa Counties.* Iowa Geological Survey Miscellaneous Publication 11. Iowa City.

Giglierano, J. D. 1988. "Gravity and Magnetics: Tools for Exploring the Earth's Interior." *Iowa Geology* 13: 8–9.

Glenister, B. F. 1987a. "Limestones of the Type Mississippian, Burlington, Southeastern Iowa." In *Environments of Deposition of the Carboniferous Systems along the Mississippi River from Burlington to East of Muscatine and Strip Mine Reclamation,* eds. G. R. McCormick and L. Johnson. Guidebook for the Fifty-first Annual Tri-State Geological Field Conference. Iowa City.

———. 1987b. *Mississippian Carbonates of the Le Grand Area: Ancient Analogs of the Bahama Banks.* Geological Society of Iowa Guidebook 47. Iowa City.

———, A. C. Kendall, J. A. Person, and A. B. Shaw. 1987. "Starrs Cave Park, Burlington Area, Des Moines County, Southeastern Iowa." In *Geological Society of America Centennial Field Guide — North-Central Section,* ed. D. L. Biggs. Geological Society of America, Boulder, Colo.

———, and S. C. Sixt. 1982. *Mississippian Biofacies-Lithofacies Trends, North-Central Section.* Geological Society of Iowa Guidebook 37. Iowa City.

Goebel, K., E. A. Bettis III, and P. H. Heckel. 1989. "Upper Pennsylvanian Paleosol in Stranger Shale and Underlying Iatan Limestone, Southwestern Iowa." *Journal of Sedimentary Petrology* 59: 224–232.

Gordon, D. L. 1986. "Leasing Land for Oil and Gas Exploration." *Iowa Geology* 11: 8–9.

Gradwohl, D. M., and N. M. Osborn. 1984. *Exploring Buried Buxton: Archaeology of an Abandoned Iowa Coal Mining Town with a Large Black Population.* Iowa State University Press, Ames.

Griffiths, L. 1974. "Timing, Technology Killed Lithograph City." *Waterloo Courier,* June 16.

Grotzinger, J. P., S. A. Bowring, B. Z. Saylor, and A. J. Kaufman. 1995. "Biostratigraphic and Geochronologic Constraints on Early Animal Evolution." *Science* 270: 598–604.

Gwynne, C. S. 1942. "Swell and Swale Pattern of the Mankato Lobe of the Wisconsin Drift in Iowa." *Journal of Geology* 50: 200–208.

———. 1951–1960. *Iowa Conservationist,* Feb. 10, 1951–June 19, 1960 (various issues).

———. 1957. "Quarrying in Iowa." *Palimpsest* 38: 177–208.

———. 1961. "B. H. Beane and the Le Grand Crinoid Hunters." *Annals of Iowa* 35: 481–490.

Hake, H. V. 1968. *Iowa Inside Out.* Iowa State University Press, Ames.

Hale, W. E. 1955. *Geology and Groundwater Resources of Webster County, Iowa.* Iowa Geological Survey Water Supply Bulletin 4. Iowa City.

Hall, J. 1857. "Observations upon the Carboniferous Limestones of the Mississippi Valley." *American Journal of Science* 23: 187–201.

———, and J. D. Whitney. 1858. *Report on the Geological Survey of the State of Iowa, Embracing the Results of Investigations Made during 1855, 56, & 57.* Vol. 1. Part 1: *Geology,* Part 2: *Palaeontology.* State of Iowa.

Hallberg, G. R. 1979. "Wind-Aligned Drainage in Loess in Iowa." *Proceedings of the Iowa Academy of Science* 86: 4–9.

———, ed. 1980a. *Illinoian and Pre-Illinoian Stratigraphy of Southeast Iowa and Adjacent Illinois.* Iowa Geological Survey Technical Information Series 11. Iowa City.

———. 1980b. *Pleistocene Stratigraphy in East-Central Iowa*. Iowa Geological Survey Technical Information Series 10. Iowa City.

———. 1986a. "Ag-Chemicals and Groundwater Quality." *Iowa Geology* 11: 4–7.

———. 1986b. "From Hoes to Herbicides: Agriculture and Groundwater Quality." *Journal of Soil and Groundwater Quality* 41: 357–364.

———. 1986c. "Pre-Wisconsin Glacial Stratigraphy of the Central Plains Region in Iowa, Nebraska, Kansas, and Missouri." In *Quaternary Glaciations in the Northern Hemisphere*, ed. S. V. Sibrava, D. Q. Bowen, and G. M. Richmond. Quaternary Science Reviews 5. Pergamon, Elmsford, N.Y.

———. 1992. "Water Quality and Agriculture." *Iowa Geology* 17: 24–25.

———, and R. G. Baker. 1980. "Reevaluation of the Yarmouth Type Area." In *Illinoian and Pre-Illinoian Stratigraphy of Southeast Iowa and Adjacent Illinois*, ed. G. R. Hallberg. Iowa Geological Survey Technical Information Series 11. Iowa City.

———, E. A. Bettis III, and J. C. Prior. 1984. "Geologic Overview of the Paleozoic Plateau Region of Northeastern Iowa." *Proceedings of the Iowa Academy of Science* 91: 3–11.

———, and J. D. Boellstorff. 1978. "Stratigraphic 'Confusion' in the Region of the Type Areas of Kansan and Nebraskan Deposits." *Geological Society of America Abstracts with Programs* 10: 255.

———, C. K. Contant, C. A. Chase, G. A. Miller, M. D. Duffy, R. J. Killorn, R. D. Voss, A. M. Blackmer, S. C. Padgitt, J. R. DeWitt, J. B. Gulliford, D. A. Lindquist, L. W. Asell, D. R. Keeney, R. D. Libra, and K. D. Rex. 1991. *A Progress Review of Iowa's Agricultural-Energy-Environmental Initiatives: Nitrogen Management in Iowa*. Geological Survey Bureau Technical Information Series 30. Iowa City.

———, T. E. Fenton, T. J. Kemmis, and G. A. Miller. 1980. *Yarmouth Revisited*. Twenty-seventh Field Conference, Midwest Friends of the Pleistocene and Iowa Geological Survey Guidebook 3. Iowa City.

———, T. E. Fenton, G. A. Miller, and A. J. Lutenegger. 1978. "The Iowan Erosion Surface: An Old Story, an Important Lesson, and Some New Wrinkles." In *Geology of East-Central Iowa*, ed. R. R. Anderson. Guidebook for the Forty-second Annual Tri-State Geological Field Conference. Iowa City.

———, J. M. Harbough, and P. M. Witinok. 1979. *Changes in the Channel Area of the Missouri River in Iowa, 1879–1976*. Iowa Geological Survey Special Report Series 1. Iowa City.

———, and B. E. Hoyer. 1982. *Sinkholes, Hydrogeology, and Groundwater Quality in Northeast Iowa*. Iowa Geological Survey Open-File Report 82-3. Iowa City.

———, B. E. Hoyer, E. A. Bettis III, and R. D. Libra. 1983. *Hydrogeology, Water Quality, and Land Management in the Big Spring Basin, Clayton County, Iowa*. Iowa Geological Survey Open-File Report 83-3. Iowa City.

———, and T. J. Kemmis. 1986. "Stratigraphy and Correlation of the Glacial Deposits of the Des Moines and James Lobes and Adjacent Areas in North Dakota, South Dakota, Minnesota, and Iowa." In *Quaternary Glaciations in the Northern Hemisphere*, eds. V. Sibrava, D. Q. Bowen , and G. M. Richmond. Quaternary Science Reviews 5. Pergamon, Elmsford, N.Y.

———, T. J. Kemmis, N. C. Wollenhaupt, S. P. Esling, E. A. Bettis III, and T. J. Bicki. 1984. "The Overburden: Quaternary Stratigraphy of the Conklin Quarry." In *Underburden-Overburden: An Examination of Paleozoic and Quaternary Strata at the Conklin Quarry near Iowa City*, eds. B. J. Bunker and G. R. Hallberg. Geological Society of Iowa Guidebook 41. Iowa City.

———, R. D. Libra, D. J. Quade, J. Littke, and B. Nations. 1989. *Groundwater Monitoring in the Big Spring Basin, 1984–1987: A Summary Review*. Geological Survey Bureau Technical Information Series 16. Iowa City.

Handy, R. L. 1976. "Loess Distribution by Variable Winds." *Geological Society of America Bulletin* 87: 915–927.

Harris, S. E., and M. C. Parker. 1964. *Stratigraphy of the Osage Series in Southeastern Iowa.* Iowa Geological Survey Report of Investigations 1. Iowa City.

Hartung, J. B., and R. R. Anderson. 1988. *A Compilation of Information and Data on the Manson Impact Structure.* Lunar and Planetary Institute Report 88-08. Houston.

———, and R. R. Anderson. 1996. "A Brief History on Investigations of the Manson Impact Structure." In *The Manson Impact Structure, Iowa: Anatomy of an Impact Crater,* eds. C. Koeberl and R. R. Anderson. Geological Society of America Special Paper 302. Boulder, Colo.

Hatch, J. R., M. J. Avcin, and P. E. VanDorpe. 1984. *Element Geochemistry of Cherokee Group Coal from South-Central and Southeast Iowa.* Iowa Geological Survey Technical Paper 5. Iowa City.

Hay, O. P. 1914. "The Pleistocene Mammals of Iowa." *Iowa Geological Survey Annual Report* 23: 1–662.

Hayden, F. V. 1857. "Geological Structure of the Country Bordering on the Missouri River, from the Mouth of the Platte River to Fort Benton." *Proceedings of the Philadelphia Academy of Natural Science* 9: 109–116.

———. 1873a. *First Annual Report (for 1867) of the United States Geological Survey of the Territories, Embracing Nebraska.* Government Printing Office, Washington, D.C.

———. 1873b. *Third Annual Report (for 1869) of the United States Geological Survey of the Territories, Embracing Colorado and New Mexico.* Government Printing Office, Washington, D.C.

Hayes, J. B. 1963. "Kaolinite from Warsaw Geodes, Keokuk Region, Iowa." *Proceedings of the Iowa Academy of Science* 70: 261–272.

———. 1964. "Geodes and Concretions from the Mississippian Warsaw Formation, Keokuk Region, Iowa, Illinois, and Missouri." *Journal of Sedimentary Petrology* 34: 123–133.

Hearst, J. S. 1976. *Shaken by Leaf-Fall.* Kylix Press, Ann Arbor, Mich.

———. 1979. *Snake in the Strawberries, Selected Poems by James Hearst.* Iowa State University Press, Ames.

Heckel, P. H. 1974. "Carbonate Buildups in the Geologic Record." In *Reefs in Time and Space,* ed. L. F. Laporte. Society of Economic Paleontologists and Mineralogists Special Publication 18.

———. 1977. "Origin of Phosphatic Black Shale Facies in Pennsylvanian Cyclothems of Midcontinent North America." *American Association of Petroleum Geologists Bulletin* 61: 1045–1068.

———. 1980a. *Field Guide to Upper Pennsylvanian Cyclothems in South-Central Iowa.* Geological Society of Iowa Guidebook 33. Iowa City.

———. 1980b. "Paleogeography of Eustatic Model for Deposition of Midcontinent Upper Pennsylvanian Cyclothems." In *Paleozoic Paleogeography of West-Central United States,* eds T. D. Fouch and E. R. Magathan. Rocky Mountain Section, Society of Economic Paleontologists and Mineralogists, Denver.

———. 1983. "Diagenetic Model for Carbonate Rocks in Midcontinent Pennsylvanian Eustatic Cyclothems." *Journal of Sedimentary Petrology* 53: 733–759.

———. 1986. "Sea-level Curve for Pennsylvanian Eustatic Marine Transgressive-Regressive Depositional Cycles along Midcontinent Outcrop Belt, North America." *Geology* 14: 330–334.

———. 1987. "Pennsylvanian Cyclothem near Winterset, Iowa." In *Geological Society of America Centennial Field Guide — North-Central Section,* ed. D. L. Biggs. Geological Society of America, Boulder, Colo.

————. 1992. "Overview of Pennsylvanian Cyclothems in Southwestern Iowa." In *Stratigraphy and Cyclic Sedimentation of Middle and Upper Pennsylvanian Strata around Winterset, Iowa*, eds. P. H. Heckel and J. P. Pope. Geological Survey Bureau Guidebook 14. Iowa City.

————, and J. F. Baesemann. 1975. "Environmental Interpretation of Conodont Distribution in Upper Pennsylvanian (Missourian) Megacyclothems of Eastern Kansas." *American Association of Petroleum Geologists Bulletin* 59: 486–509.

————, and J. P. Pope. 1992. *Stratigraphy and Cyclic Sedimentation of Middle and Upper Pennsylvanian Strata around Winterset, Iowa*. Geological Survey Bureau Guidebook 14. Iowa City.

Heyl, A. V., A. F. Agnew, J. L. Erwin, and C. H. Behre. 1959. *The Geology of the Upper Mississippi Valley Zinc-Lead District*. U.S. Geological Survey Professional Paper 309. Washington, D.C.

————, A. F. Agnew, E. J. Lyons, and C. H. Behre. 1959. *The Geology of the Upper Mississippi Valley District*. U.S. Geological Survey Bulletin 1242 A. Washington, D.C.

————, E. J. Lyons, A. F. Agnew, and C. H. Behre. 1955. *Zinc-Lead-Copper Resources and General Geology of the Upper Mississippi Valley District*. U.S. Geological Survey Bulletin 1015-G. Washington, D.C.

Hickerson, W. 1992. "Trilobites from the Late Givetian Solon Member, Little Cedar Formation of Eastern Iowa and Northwestern Illinois." In *The Stratigraphy, Paleontology, Depositional and Diagenetic History of the Middle-Upper Devonian Cedar Valley Group of Central and Eastern Iowa*, eds. J. E. Day and B. J. Bunker. Geological Survey Bureau Guidebook 16. Iowa City.

————, and R. C. Anderson. 1994. *Paleozoic Stratigraphy of the Quad-Cities Region, East-Central Iowa, Northwestern Illinois*. Geological Society of Iowa Guidebook 59. Iowa City.

Hinman, E. E. 1968. *A Biohermal Facies in the Silurian of Eastern Iowa*. Iowa Geological Survey Report of Investigations 6. Iowa City.

Holtzman, A. F. 1970. "Gravity Study of the Manson Disturbed Area, Calhoun, Pocahontas, Humboldt, and Webster Counties, Iowa." Master's thesis, University of Iowa, Iowa City.

Hoppin, R. A., and J. E. Dryden. 1958. "An Unusual Occurrence of Precambrian Crystalline Rocks Beneath Glacial Drift near Manson, Iowa." *Journal of Geology* 66: 694–699.

Horberg, L. 1956. "Bedrock Topography and Pleistocene Glacial Lobes in Central United States." *Journal of Geology* 64: 101–116.

Horick, P. J., ed. 1970. *Water Resources of Iowa*. Iowa Academy of Science Special Publication. Iowa City.

————. 1973. "Some Hydrologic Aspects of the Mississippian of Iowa." *Proceedings of the Iowa Academy of Science* 80: 8–14.

————. 1974. *Minerals of Iowa*. Iowa Geological Survey Educational Series 2. Iowa City.

————. 1984. "Silurian-Devonian Aquifer of Iowa." Iowa Geological Survey Miscellaneous Map Series 10. Iowa City.

————. 1988. "Rock and Mineral Collecting." *Iowa Geology* 13: 16–17.

————. 1990a. "Hydrogeology of the Mississippian of Iowa." In *Iowa's Principal Aquifers: A Review of Iowa Geology and Hydrogeologic Units*. Iowa Groundwater Association, Iowa City.

————. 1990b. "Iowa's Principal Aquifers." In *Iowa's Principal Aquifers: A Review of Iowa Geology and Hydrogeologic Units*. Iowa Groundwater Association, Iowa City.

————. 1990c. "The Jordan (Lower Cambrian–Ordovician) Aquifer of Iowa." In *Iowa's Principal Aquifers: A Review of Iowa Geology and Hydrogeologic Units*. Iowa Groundwater Association, Iowa City.

————. 1990d. "The Silurian-Devonian Aquifer of Iowa." In *Iowa's Principal Aquifers: A*

Review of Iowa Geology and Hydrogeologic Units. Iowa Groundwater Association, Iowa City.

———, and R. M. McKay. 1990. "The Dresbach Aquifer of Iowa." In *Iowa's Principal Aquifers: A Review of Iowa Geology and Hydrogeologic Units.* Iowa Groundwater Association, Iowa City.

———, and W. L. Steinhilber. 1973. "Mississippian Aquifer of Iowa." Iowa Geological Survey Miscellaneous Map Series 3. Iowa City.

———, and W. L. Steinhilber. 1978. "Jordan Aquifer of Iowa." Iowa Geological Survey Miscellaneous Map Series 6. Iowa City.

Horne, J. L., J. C. Ferm, F. T. Caruccio, and B. P. Baganz. 1978. "Depositional Models in Coal Exploration and Mine Planning in the Appalachian Region." *American Association of Petroleum Geologists Bulletin* 62: 2379–2411.

Howes, M. R. 1990. "Iowa Coal: A Review of Geology, Resources, and Production." In *Coal Geology of the Interior Coal Province, Western Region*, eds. R. B. Finkelman, S. A. Friedman, and J. R. Hatch. Guidebook for the Coal Geology Division Field Trip, 1990 Geological Society of America Annual Meeting. Environmental and Coal Associates, Reston, Va.

———. 1993. "Upper Cherokee Group (Pennsylvanian) Exposures, Saylorville Emergency Spillway." In *Water, Water, Everywhere*, eds. W. W. Simpkins and R. R. Anderson. Geological Society of Iowa Guidebook 58. Iowa City.

———, M. A. Culp, and H. Greenberg. 1989. *Abandoned Underground Coal Mines of Des Moines, Iowa, and Vicinity.* Geological Survey Bureau Technical Paper 8. Iowa City.

———, M. A. Culp, H. Greenberg, and P. E. VanDorpe. 1986. *Underground Coal Mines of Centerville, Iowa, and Vicinity.* Geological Survey Bureau Open-File Report 86-2. Iowa City.

———, and L. L. Lambert. 1988. *Stratigraphy and Depositional History of the Cherokee Group, South-Central Iowa.* Geological Society of Iowa Guidebook 49. Iowa City.

Hoyer, B. E. 1980. *Geomorphic History of the Little Sioux River Valley.* Geological Society of Iowa Guidebook 34. Iowa City.

———. 1987. "Groundwater Policy and Geology." *Iowa Geology* 12: 22–23.

———. 1991. "Groundwater Vulnerability Map of Iowa." *Iowa Geology* 16: 13–15.

———, E. A. Bettis III, and B. J. Witzke. 1986. *Water Quality and the Galena Group in the Big Spring Area, Clayton County.* Geological Society of Iowa Guidebook 45. Iowa City.

Hudak, C. M. 1987. "Quaternary Landscape Evolution of the Turkey River Valley, Northeastern Iowa." Ph.D. diss., University of Iowa, Iowa City.

Iowa Geological Survey. 1969. "Bedrock Map of Iowa." Iowa Geological Survey, Iowa City.

Joeckel, R. M., G. A. Ludvigson, R. L. Brenner, B. J. Witzke, R. L. Ravn, E. P. Kvale, and J. B. Swinehart. 1996. "Stratigraphic Investigations of the Basal Cretaceous Section at Ash Grove Cement Quarry, Louisville, Nebraska." *Geological Society of America Abstracts with Programs* 28 (6): 47.

Johnson, C. J. 1975. "A New Life: The Iowa Coal Mines." *Palimpsest* 56: 56–64.

Johnson, D. B., and K. Swett. 1974. "Origin and Diagenesis of Calcitic and Hematitic Nodules in the Jordan Sandstone of Northeast Iowa." *Journal of Sedimentary Petrology* 44: 790–794.

Johnson, G. D., and C. Vondra. 1969. "Lithofacies of the Pella Formation (Mississippian), Iowa." *American Association of Petroleum Geologists Bulletin* 53: 1894–1908.

Johnson, J. G., G. Klapper, and C. A. Sandberg. 1985. "Devonian Eustatic Fluctuations in Euramerica." *Geological Society of America Bulletin* 96: 567–587.

Johnson, M. E. 1975. "Recurrent Community Patterns in Epeiric Seas: The Lower Silurian of Eastern Iowa." *Proceedings of the Iowa Academy of Science* 82: 130–139.

————. 1977a. "Community Succession and Replacement in Early Silurian Platform Seas: The Llandovery Series of Eastern Iowa." Ph.D. diss., University of Chicago, Chicago.

————. 1977b. "Early Geological Explorations of the Silurian System in Iowa." *Proceedings of the Iowa Academy of Science* 84: 150–156.

————. 1977c. "Succession and Replacement in the Development of Silurian Brachiopod Populations." *Lethaia* 10: 83–93.

————. 1979. "Evolutionary Brachiopod Lineages from the Llandovery Series of Eastern Iowa." *Palaeontology* 22: 549–567.

————. 1980. "Paleoecological Structure in Early Silurian Platform Seas of the North American Midcontinent." *Palaeogeography, Palaeoclimatology, Palaeoecology* 30: 191–216.

————. 1983. "New Member Names for the Lower Silurian Hopkinton Dolomite of Eastern Iowa." *Proceedings of the Iowa Academy of Science* 90: 13–18.

————. 1987. "Extent and Bathymetry of North American Platform Seas in the Early Silurian." *Paleoceanography* 2: 185–211.

————. 1988. "Early Silurian Carbonates from the Upper Mississippi Valley Area as a Key to Platform Development on a Cratonic Scale." In *New Perspectives on the Paleozoic History of the Upper Mississippi Valley: An Examination of the Plum River Fault Zone*, eds. G. A. Ludvigson and B. J. Bunker. Geological Survey Bureau Guidebook 8. Iowa City.

————, B. G. Baarli, H. Nestor, M. Rubel, and D. Worsley. 1991. "Eustatic Sea-level Patterns from the Lower Silurian (Landovery Series) of Southern Norway and Estonia." *Geological Society of America Bulletin* 103: 315–335.

————, L. R. M. Cocks, and P. Copper. 1981. "Late Ordovician–Early Silurian Fluctuations in Sea Level from Eastern Anticosti Island, Quebec." *Lethaia* 14: 73–82.

————, and H. L. Lescinsky. 1986. "Depositional Dynamics of Cyclic Carbonates from the Interlake Group (Lower Silurian) of the Williston Basin." *Palaios* 1: 111–121.

————, J. Y. Rong, and X. C. Yang. 1985. "Intercontinental Correlation by Sea-level Events in the Early Silurian of North America and China (Yangtze Platform)." *Geological Society of America Bulletin* 96: 1384–1397.

Johnson, W. H., A. K. Hansel, E. A. Bettis III, P. F. Karrow, G. J. Larson, T. V. Lowell, and A. F. Schneider. 1997. "Late Quaternary Temporal and Event Classifications, Great Lakes Region, North America." *Quaternary Research* 47: 1–12.

Jorgensen, D. G., and D. C. Signor. 1984. *Proceedings of the Geohydrology Dakota Aquifer Symposium*. Water Well Journal Publishing, Worthington, Ohio.

Kasler, D. 1990. "Quake Prediction Jiggles Insurers." *Des Moines Sunday Register*, Oct. 28.

Kay, G. F., and E. T. Apfel. 1929. "The Pre-Illinoian Pleistocene Geology of Iowa." *Iowa Geological Survey Annual Report* 34: 1–304.

————, and J. B. Graham. 1943. "The Illinoian and Post-Illinoian Pleistocene Geology of Iowa." *Iowa Geological Survey Annual Report* 38: 1–262.

Kay, M. 1929. "Stratigraphy of the Decorah Formation." *Journal of Geology* 37: 639–671.

Kemmis, T. J. 1981. *Glacial Sedimentation and the Algona Moraine in Iowa*. Geological Society of Iowa Guidebook 36. Iowa City.

————. 1991. "Glacial Landforms, Sedimentology, and Depositional Environments of the Des Moines Lobe, Northern Iowa." Ph.D. diss., University of Iowa, Iowa City.

————, E. A. Bettis III, and G. R. Hallberg. 1992. *Quaternary Geology of Conklin Quarry*. Geological Survey Bureau Guidebook 13. Iowa City.

————, G. R. Hallberg, and A. J. Lutenegger. 1981. *Depositional Environments of Glacial Sediments and Landforms on the Des Moines Lobe, Iowa*. Iowa Geological Survey Guidebook 6. Iowa City.

————, and D. J. Quade. 1988. "Sand and Gravel Resources." *Iowa Geology* 13: 18–21.

Kettenbrink, E. C. 1972a. "The Cedar Valley Formation at Pint's Quarry." In *General Geology in the Vicinity of Northern Iowa*, ed. W. I. Anderson. Guidebook for the Thirty-sixth Annual Tri-State Geological Field Conference. Cedar Falls, Iowa.

————. 1972b. "Depositional and Post-Depositional History of the Devonian Cedar Valley Formation, East-Central Iowa." Ph.D. diss., University of Iowa, Iowa City.

Keyes, C. R. 1893a. "The Geological Formations of Iowa." *Iowa Geological Survey Annual Report* 1: 13–472.

————. 1893b. "Geology of Des Moines County." *Iowa Geological Survey Annual Report* 3: 409–492.

————. 1894. "Coal Deposits of Iowa." *Iowa Geological Survey Annual Report* 2: 1–536.

————. 1895. "Gypsum Deposits of Iowa." *Iowa Geological Survey Annual Report* 3: 257–304.

————. 1896. "An Epoch in the History of American Science." *Annals of Iowa* 2: 345–364.

————. 1919. "A Century of Iowa Geology." *Proceedings of the Iowa Academy of Science* 26: 407–466.

Klapper, G., and J. E. Barrick. 1983. "Middle Devonian (Eifelian) Conodonts from the Spillville Formation of Northern Iowa and Southern Minnesota." *Journal of Paleontology* 57: 1212–1243.

————, and W. M. Furnish. 1962. "Conodont Zonation of the Early Upper Devonian in Eastern Iowa." *Proceedings of the Iowa Academy of Science* 69: 400–410.

Klug, C. R. 1992. "Distribution and Biostratigraphic Significance of Miospores in the Middle-Upper Devonian Cedar Valley Group." In *The Stratigraphy, Paleontology, Depositional and Diagenetic History of the Middle-Upper Devonian Cedar Valley Group of Central and Eastern Iowa*, eds. J. E. Day and B. J. Bunker. Geological Survey Bureau Guidebook 16. Iowa City.

Kneedler, H. S. 1890. "A Prairie Coal Palace." *Harper's Weekly*, Sept. 13: 717–718.

Knox, R., and D. Stewart. 1995. *The New Madrid Fault Finders Guide*. Gutenberg-Richter Publications, Marble Hill, Mo.

Koch, D. L. 1968. *Fort Atkinson Limestone Member*. Iowa State Advisory Board for Preserves, Development Series Report 2. Des Moines.

————. 1969. *The Sioux Quartzite Formation in Gitchie Manitou State Preserve*. Iowa State Advisory Board for Preserves, Development Series Report 8. Des Moines.

————. 1970. *Stratigraphy of the Upper Devonian Shell Rock Formation of North-Central Iowa*. Iowa Geological Survey Report of Investigations 10. Iowa City.

————. 1973. "Mississippian Stratigraphy of North-Central Iowa." *Proceedings of the Iowa Academy of Science* 80: 1–3.

————, and H. L. Strimple. 1968. *A New Upper Devonian Cystoid Attached to a Discontinuity Surface*. Iowa Geological Survey Report of Investigations 5. Iowa City.

Koeberl, C., and R. R. Anderson. 1996a. "Manson and Company: Impact Structures in the United States." In *The Manson Impact Structure, Iowa: Anatomy of an Impact Crater*, eds. C. Koeberl and R. R. Anderson. Geological Society of America Special Paper 302. Boulder, Colo.

————, and R. R. Anderson, eds. 1996b. *The Manson Impact Structure, Iowa: Anatomy of an Impact Crater*. Geological Society of America Special Paper 302. Boulder, Colo.

Kolata, D. R., H. R. Strimple, and C. O. Leverson. 1977. "Revision of Ordovician Carpoid Family *Iowacystidae*." *Paleontology* 20 (3): 529–557.

Kreiner, C. B. 1922. "The Ottumwa Coal Palace." *Palimpsest* 3: 336–342.

Kunk, M. J., G. A. Izett, R. A. Haurerud, and J. F. Sutter. 1989. "^{40}Ar-^{39}Ar Dating of the Manson Impact Structure: A Cretaceous-Tertiary Boundary Crater Candidate." *Science* 244: 1565–1568.

————, G. A. Izett, and J. F. Sutter. 1987. "^{40}Ar/^{39}Ar Age Spectra of Shocked K- Feldspar

Suggests K-T Boundary Age for Manson, Iowa, Impact Structure." *EOS Transactions of the American Geophysical Union* 68: 1514.

LaBerge, G. L. 1994. *Geology of the Lake Superior Region*. Geoscience Press, Phoenix, Ariz.

Ladd, H. S. 1929. "Stratigraphy and Paleontology of the Maquoketa Shale of Iowa, Part 1." *Iowa Geological Survey Annual Report* 34: 305–440.

Land, L. S. 1973. "Holocene Meteoric Dolomitization." *Journal of Sedimentary Petrology* 37: 914–930.

Landis, E. R., and O. J. Van Eck. 1965. *Coal Resources of Iowa*. Iowa Geological Survey Technical Paper 4. Iowa City.

Laudon, L. R. 1929. "The Stratigraphy and Paleontology of the Northward Extension of the Burlington Limestone." M.S. thesis, University of Iowa, Iowa City.

———. 1931. "Stratigraphy of the Kinderhook Series of Iowa." *Iowa Geological Survey Annual Report* 35: 335–451.

———. 1933. "Stratigraphy and Paleontology of the Gilmore City Formation." *State University of Iowa Studies in Natural History* 15 (2): 1–74.

———. 1935. "Supplemental Statement on the Mississippian System in Iowa." In *Kansas Geological Society Ninth Annual Field Conference Guidebook*.

———. 1937. "Stratigraphy of the Northern Extension of the Burlington Limestone in Missouri and Iowa." *American Association of Petroleum Geologists Bulletin* 21: 1158–1167.

———. 1948. "Osage-Meramec Contact." *Journal of Geology* 56: 288–302.

———. 1973. "Stratigraphic Crinoid Zonation in Iowa Mississippian Rocks." *Proceedings of the Iowa Academy of Science* 80: 25–33.

———, and B. H. Beane. 1937. "The Crinoid Fauna of the Hampton Formation at Le Grand, Iowa." *State University of Iowa Studies in Natural History* 17 (16): 227–272.

Lees, J. H. 1918. "Some Features of the Fort Dodge Gypsum." *Proceedings of the Iowa Academy of Science* 25: 587–598.

Lemish, J. 1969. *Mineral Deposits of Iowa*. Iowa Southern Utilities Company Special Publication. Centerville, Iowa.

———, D. R. Burggraf Jr., and H. J. White. 1981. *Cherokee Sandstones and Related Facies of Central Iowa, an Examination of Tectonic Setting and Depositional Environments*. Iowa Geological Survey Guidebook 5. Iowa City.

———, R. E. Chamberlain, and E. W. Mason. 1981. "Introduction and Regional Geology." In *Cherokee Sandstones and Related Facies of Central Iowa, an Examination of Tectonic Setting and Depositional Environments*, eds. J. Lemish, D. R. Burggraf Jr., and H. J. White. Iowa Geological Survey Guidebook 5. Iowa City.

Leonard, A. G. 1906. "Geology of Clayton County." *Iowa Geological Survey Annual Report* 15: 213–318.

Leopold, A. 1949. *A Sand County Almanac with Essays on Conservation from Round River*. Oxford University Press, Fair Lawn, N.J.

Leverett, F. 1898. "The Weathered Zone (Yarmouth) between the Illinoian and Kansan Till Sheets." *Journal of Geology* 6: 238–243.

———. 1921. "Outline of the Pleistocene History of the Mississippi Valley." *Journal of Geology* 29: 615–626.

Leverson, C. O., and A. J. Gerk. 1972. "A Preliminary Stratigraphic Study of the Galena Group of Winneshiek County, Iowa." *Proceedings of the Iowa Academy of Science* 79: 111–122.

———, A. J. Gerk, and T. W. Broadhead. 1979. "Stratigraphy of the Dubuque Formation (Upper Ordovician) in Iowa." *Proceedings of the Iowa Academy of Science* 86: 57–65.

Libra, R. D. 1988. "Agricultural Drainage Wells and Groundwater Quality." *Iowa Geology* 13: 22–23.

———. 1995a. "Agriculture and Groundwater: The View from Big Spring." *Iowa Geology* 20: 2–5.

———. 1995b. "Jumbo, a Runaway Artesian Well." *Iowa Geology* 20: 6–9.

———. 1996. "Natural Resource Mapping of Linn County." *Iowa Geology* 21: 2–7.

Lowenstam, H. A. 1950. "Niagaran Reefs of the Great Lakes Area." *Journal of Geology* 58: 430–487.

Ludvigson, G. A. 1989. "Plum River Fault Zone." *Iowa Geology* 14: 20–21.

———, and B. J. Bunker, eds. 1988. *New Perspectives on the Paleozoic History of the Upper Mississippi Valley: An Examination of the Plum River Fault Zone.* Geological Survey Bureau Guidebook 8. Iowa City.

———, and J. A. Dockal. 1984. "Lead and Zinc Mining in the Dubuque Area." *Iowa Geology* 9: 4–5.

———, and G. McCormick. 1975. "Ordovician Structure and Mineralization in Northeastern Iowa." In *Guidebook for the Thirty-ninth Annual Tri-State Geological Field Conference.* Iowa City.

———, R. M. McKay, and R. R. Anderson. 1990. "Petrology of Keweenawan Sedimentary Rocks in the M. G. Eischeid #1 Drillhole." In *The Amoco M. G. Eischeid #1 Deep Petroleum Test, Carroll County, Iowa, Preliminary Investigations*, ed. R. R. Anderson. Geological Survey Bureau Special Report Series No. 2. Iowa City.

———, and P. G. Spry. 1990. "Tectonic and Paleohydrologic Significance of Carbonate Veinlets in the Keweenawan Sedimentary Rocks of the Amoco M. G. Eischeid #1 Drillhole." In *The Amoco M. G. Eischeid #1 Deep Petroleum Test, Carroll County, Iowa, Preliminary Investigations*, ed. R. R. Anderson. Geological Survey Bureau Special Report Series No. 2. Iowa City.

———, and K. Swett. 1987. "Pennsylvanian Strata at Wyoming Hill and Wildcat Den State Park, Muscatine County, Iowa." In *Environments of Deposition of the Carboniferous System along the Mississippi River from Burlington to East of Muscatine and Strip Mine Reclamation*, eds. G. R. McCormick and L. Johnson. Guidebook for the Fifty-first Annual Tri-State Geological Field Conference. Iowa City.

———, and B. J. Witzke. 1990. "Cretaceous Geology and the Dakota Aquifer of Iowa." In *Iowa's Principal Aquifers: A Review of Iowa Geology and Hydrogeologic Units.* Iowa Groundwater Association, Iowa City.

———, B. J. Witzke, and L. A. Gonzalez. 1992. "Observations on the Diagenesis and Stable Isotopic Compositions of Silurian Carbonates in Iowa." In *Silurian Stratigraphy and Carbonate Mound Facies of Eastern Iowa.* Geological Survey Bureau Guidebook 11. Iowa City.

———, B. J. Witzke, L. A. Gonzalez, R. H. Hammond, and O. W. Plocher. 1994. "Sedimentology and Carbonate Geochemistry of Concretions from the Greenhorn Marine Cycle (Cenomanian-Turonian), Eastern Margin of the Western Interior Seaway." In *Perspectives on the Eastern Margin of the Cretaceous Western Interior Basin*, eds. G. W. Shurr, G. A. Ludvigson, and R. H. Hammond. Geological Society of America Special Paper 287. Boulder, Colo.

———, B. J. Witzke, P. H. Liu, and J. P. Pope. 1997. "Outlier of Cretaceous Niobrara Formation in Lyon County, Iowa." *Iowa Academy of Science Abstracts.*

Macurda, D. B., Jr. 1964. "The *Pentremites* (Blastoidea) of the Burlington Limestone (Mississippian)." *Journal of Paleontology* 49: 346–373.

Maliva, R. G. 1987. "Quartz Geodes: Early Diagenetic Silicified Anhydrite Nodules Related to Dolomitization." *Journal of Sedimentary Petrology* 57: 1054–1059.

Martin, P. S. 1973. "The Discovery of America." *Science* 179: 969–974.

Matsch, C. 1976. *North America and the Great Ice Age.* McGraw-Hill, New York.

McGee, W. J. 1891. "The Pleistocene History of Northeastern Iowa." *Eleventh Annual Report of the Director of the U.S. Geological Survey*. Part 1: *Geology*. Washington, D.C.

McKay, R. M. 1987. "Fossil Amphibian Site: A Significant North American Find." *Iowa Geology* 12: 27.

———. 1988. "Stratigraphy and Lithofacies of the Dresbachian (Upper Cambrian) Eau Claire Formation." In *New Perspectives on the Paleozoic History of the Upper Mississippi Valley: An Examination of the Plum River Fault Zone*, eds. G. A. Ludvigson and B. J. Bunker. Geological Survey Bureau Guidebook 8. Iowa City.

———. 1990. "Regional Aspects of the Mt. Simon Formation and the Placement of the Mt. Simon–Pre-Mt. Simon Sedimentary Contact in the Amoco M. G. Eischeid #1 Drillhole." In *The Amoco M. G. Eischeid #1 Deep Petroleum Test, Carroll County, Iowa, Preliminary Investigations*, ed. R. R. Anderson. Geological Survey Bureau Special Report Series No. 2. Iowa City.

———. 1992. "Mineral Production in Iowa." *Iowa Geology* 17: 13–14.

———. 1993. *Selected Aspects of Lower Ordovician and Upper Cambrian Geology in Allamakee and Northern Clayton Counties*. Geological Society of Iowa Guidebook 57. Iowa City.

———. 1994. "Ancient River Channels." *Iowa Geology* 19: 12–13.

———. 1995. "The Cambrian-Ordovician Aquifer and the Northeast Iowa Landscape." *Iowa Geology* 20: 10–11.

———, and M. J. Bounk. 1987. "Underground Limestone Mining." *Iowa Geology* 12: 24–26.

———, M. P. McAdams, B. J. Witzke, and J. R. Bolt. 1986. "Ancient Amphibians Discovered in Iowa." *Iowa Geology* 11: 20–23.

———, B. J. Witzke, and M. P. McAdams. 1987. *Early Tetrapods, Stratigraphy, and Paleoenvironments of the Upper St. Louis Formation, Western Keokuk County, Iowa*. Geological Society of Iowa Guidebook 46. Iowa City.

McKerrow, W. S., ed. 1978. *The Ecology of Fossils, an Illustrated Guide*. MIT Press, Cambridge.

Meek, F. B., and F. V. Hayden. 1862. "Descriptions of New Lower Silurian (Primordial), Jurassic, Cretaceous, and Tertiary Fossils, Collected in Nebraska by the Exploring Expedition under the Command of Capt. W. F. Reynolds, U.S. Topographical Engineers, with Some Remarks on the Rocks from Which They Were Obtained." *Philadelphia Academy of Natural Sciences Proceedings* 13: 415–457.

Mehl, M. G. 1962. *Missouri's Ice Age Animals*. Missouri Geological Survey and Water Resources Education Series 1. Rolla.

Menzel, M., and M. Pratt. 1968. "James Hall, Pioneer Geologist." *Iowan* (winter): 22–24, 50–51.

Miller, J. F., and J. H. Melby. 1971. "Trempealeauan Conodonts (and Comments)." In *Conodonts and Biostratigraphy of the Wisconsin Paleozoic*, ed. D. L. Clark. Wisconsin Geological and Natural History Survey Information Circular 19. Madison.

Miller, S. A., and W. F. E. Gurley. 1894. *New Genera and Species of Echinodermata*. Illinois State Museum Bulletin 5. Springfield.

Moore, R., and C. Teichert, eds. 1978. *Treatise on Invertebrate Paleontology*, Geological Society of America, Boulder, Colo., and University of Kansas Press, Lawrence.

Mossler, J. H., and J. D. Hayes. 1966. "Ordovician Potassium Bentonites of Iowa." *Journal of Sedimentary Petrology* 36 (2): 414–427.

Muhm, D. 1981. "Costly Restoration of Stripped Land." *Des Moines Sunday Register*, July 5.

———. 1986. "Sand Mine to Help House Harvest Bounty." *Des Moines Sunday Register*, Aug. 3.

Muller, K. J., and E. M. Muller. 1957. "Early Upper Devonian (Independence) Conodonts from Iowa, Part I." *Journal of Paleontology* 31: 1069–1108.

Munter, J. A., G. A. Ludvigson, and B. J. Bunker. 1983. *Hydrogeology and Stratigraphy of the Dakota Formation in Northwest Iowa*. Iowa Geological Survey Water Supply Bulletin 13. Iowa City.

Murray, R. A. 1968. *Pipes of the Plains*. Pipestone Indian Shrine Association, Omaha, Nebr.

Mutel, C. F. 1989. *Fragile Giants: A Natural History of the Loess Hills*. University of Iowa Press, Iowa City.

Nitecki, M. H. 1972. "North American Silurian Receptaculitid Algae." *Fieldiana: Geology* 28: 1–108.

———, and D. F. Toomey. 1979. "Nature and Classification of Receptaculitids." *Bulletin des Centres de Recherches Exploration-Production Elf-Aquitaine* 3 (2): 725–732.

Norton, W. H. 1894. "Geology of Linn County." *Iowa Geological Survey Annual Report* 4: 121–195.

———. 1928. "Deep Wells of Iowa." *Iowa Geological Survey Annual Report* 33: 246–254.

———. 1935. "Deep Wells of Iowa, 1928–1932." *Iowa Geological Survey Annual Report* 36: 311–364.

———, H. E. Simpson, W. S. Hendrixson, M. F. Arey, O. E. Meinzer, A. O. Thomas, W. J. Miller, and J. L. Tilton. 1912. "Underground Water Resources of Iowa." *Iowa Geological Survey Annual Report* 21: 29–1186.

Odom, I. E. 1978. "Mineralogy of Cambrian Sandstones, Upper Mississippi Valley." In *Lithostratigraphy, Petrology, and Sedimentology of Late Cambrian–Early Ordovician Rocks near Madison, Wisconsin*, ed. M. E. Ostrom. Wisconsin Geological and Natural History Survey Field Trip Guidebook 3. Madison.

———, and M. E. Ostrom. 1978. "Lithostratigraphy, Petrology, and Sedimentology of the Jordan Formation near Madison, Wisconsin." In *Lithostratigraphy, Petrology, and Sedimentology of Late Cambrian–Early Ordovician Rocks near Madison, Wisconsin*, ed. M. E. Ostrom. Wisconsin Geological and Natural History Survey Field Trip Guidebook 3. Madison.

Ojakangas, R. W., and R. E. Weber. 1984. "Petrography and Paleocurrents of the Early Proterozoic Sioux Quartzite, Minnesota and South Dakota." *Minnesota Geological Survey Report of Investigations* 32: 1–15.

Olin, H. L. 1965. *Coal Mining in Iowa*. Special Publication of the State Mining Board, Iowa Department of Mines and Minerals. Des Moines.

Oschwald, W. R., F. F. Riecken, R. I. Dideriksen, W. H. Scholtes, and F. W. Schaller. 1965. *Principal Soils of Iowa*. Iowa State University Department of Agronomy Special Report 42. Ames.

Osgood, E. S. 1964. *The Field Notes of Captain William Clark, 1803–1805*. Yale University Press, New Haven, Conn.

Ostrom, M. E. 1965. "Cambro-Ordovician Stratigraphy of Southwest Wisconsin." In *Guidebook for the Twenty-ninth Annual Tri-State Geological Field Conference*. Wisconsin Geological and Natural History Survey Information Circular 6. Madison.

———. 1966. *Cambrian Stratigraphy in Western Wisconsin*. Wisconsin Geological and Natural History Survey Information Circular 7. Madison.

———. 1970. "Sedimentation Cycles in the Lower Paleozoic Rocks of Western Wisconsin." In *Fieldtrip Guidebook for Cambrian-Ordovician Geology of Western Wisconsin*, eds. M. E. Ostrom, R. A. Davis, and L. M. Cline. North-Central Geological Society of America and Wisconsin Geological Natural History Survey Information Circular 11. Madison.

———, ed. 1978. *Lithostratigraphy, Petrology, and Sedimentology of Late Cambrian–Early Ordovician Rocks near Madison, Wisconsin*. Wisconsin Geological and Natural History Survey Field Trip Guidebook 3. Madison.

Owen, D. D. 1840. *Report of a Geological Exploration of Part of Iowa, Wisconsin, and Illinois*. 1st ed. U.S. 26th Congress, House of Representatives Executive Document 239.

———. 1844. *Report of a Geological Exploration of Part of Iowa, Wisconsin, and Illinois.* 2d ed. U.S. 28th Congress, Senate Executive Document 407.

———. 1852. *Report of a Geological Survey of Wisconsin, Iowa, and Minnesota.* Lippincott, Grambo, Philadelphia.

Palacas, J. G., J. W. Schmoker, T. A. Daws, M. J. Pawlewicz, and R. R. Anderson. 1990. "Petroleum Source-Rock Assessment of Middle Proterozoic (Keweenawan) Sedimentary Rocks, Eischeid #1 Well, Carroll County, Iowa." In *The Amoco M. G. Eischeid #1 Deep Petroleum Test, Carroll County, Iowa, Preliminary Investigations*, ed. R. R. Anderson. Geological Survey Bureau Special Report Series No. 2. Iowa City.

Palmer, T. J. 1978. "Burrows at Certain Omission Surfaces in the Middle Ordovician of the Upper Mississippi Valley." *Journal of Paleontology* 52: 109–117.

Palmquist, R. C. 1969. "The Configuration of the Prairie du Chien–St. Peter Contact in Southwestern Wisconsin: An Example of an Integrated Geological-Geophysical Study." *Journal of Geology* 77: 694–702.

Parker, M. C. 1967. *La Porte City Chert: A Devonian Subsurface Formation in Central Iowa.* Iowa Geological Survey Report of Investigations 4. Iowa City.

———. 1971. "The Maquoketa Formation (Upper Ordovician) in Iowa." Iowa Geological Survey Miscellaneous Map Series 1. Iowa City.

———. 1975. "Mississippian Stratigraphy in Southeastern Iowa." *Proceedings of the Iowa Academy of Science* 80: 4–7.

Paull, R. A. 1987. "Presentation of the Neil Miner Award to Lowell R. Laudon." *Journal of Geological Education* 35: 47–48.

Petersen, W. J. 1966a. "Earthquakes in Iowa." *Palimpsest* 47: 65–96.

———. 1966b. "Julien Dubuque." *Palimpsest* 47: 105–119.

———. 1966c. "Spanish Land Grants in Iowa: Iowa Under Spain." *Palimpsest* 47: 97–104.

Philcox, M. E. 1970a. "Coral Bioherms in the Hopkinton Formation (Silurian), Iowa." *Geological Society of America Bulletin* 81: 969–974.

———. 1970b. "Geometry and Evolution of the Palisades Reef Complex, Silurian of Iowa." *Journal of Sedimentary Petrology* 40: 177–183.

———. 1972. "Burial of Reefs by Shallow-water Carbonates, Silurian Gower Formation, Iowa, U.S.A." *Geologische Rundschau* 61: 686–708.

Pins, K. 1983. "Iowan Plans to Turn Mine into Big Storage Warehouse." *Des Moines Register*, Jan. 8.

Pitrat, C. W. 1962. "Devonian Corals from the Cedar Valley Limestone of Iowa." *Journal of Paleontology* 36: 158–159.

Plocher, O. W., ed. 1989. *Geologic Reconnaissance of the Coralville Lake Area.* Geological Society of Iowa Guidebook 51. Iowa City.

———, and B. J. Bunker. 1989. "A Geological Reconnaissance of the Coralville Lake Area." In *Geological Reconnaissance of the Coralville Lake Area*, ed. O. W. Plocher. Geological Society of Iowa Guidebook 51. Iowa City.

Prior, J. C. 1976. *A Regional Guide to Iowa Landforms.* Iowa Geological Survey Educational Series 3. Iowa City.

———, ed. 1984–1998. *Iowa Geology.* Geological Survey Bureau, Iowa City.

———. 1987. "Loess Hills: A National Natural Landmark." *Iowa Geology* 12: 16–19.

———. 1991. *Landforms of Iowa.* University of Iowa Press, Iowa City.

———. 1996. "Building with the Geologic Past." *Iowa Geology* 21: 28–29.

———, and C. R. Milligin. 1985. "The Iowa Landscapes of Orestes St. John." In *Geologists and Ideas: A History of North American Geology*, eds. E. T. Drake and W. M. Jordan. Geological Society of America Centennial Special Volume 1. Boulder, Colo.

———, D. M. Roosa, D. D. Smith, P. C. Christiansen, and L. J. Eilers. 1986. *Natural History of the Cedar and Wapsipinicon River Basins on the Iowan Erosion Surface.* Iowa Natural History Association Field Trip Guidebook 4. Iowa City.

Quade, D. J., and L. S. Seigley. 1994. "Hogs and the Environment." *Iowa Geology* 19: 20–21.

Ramsbottom, W. H. C. 1973. "Transgressions and Regressions in the Dinantian: A New Synthesis of British Dinantian Stratigraphy." *Proceedings of the Yorkshire Geological Society* 39: 567–607.

Ravn, R. L. 1986. *Palynostratigraphy of the Lower and Middle Pennsylvanian Coals of Iowa.* Iowa Geological Survey Technical Paper 7. Iowa City.

———, and D. J. Fitzgerald. 1982. "A Morrowan (Upper Carboniferous) Miospore Flora from Eastern Iowa, U.S.A." *Palaeontographica B* 183: 108–172.

———, J. W. Swade, M. R. Howes, J. L. Gregory, R. R. Anderson, and P. E. VanDorpe. 1984. *Stratigraphy of the Cherokee Group and Revision of Pennsylvanian Stratigraphic Nomenclature in Iowa.* Iowa Geological Survey Technical Information Series 12. Iowa City.

———, and B. J. Witzke. 1994. "The Mid-Cretaceous Boundary in the Western Interior Seaway, Central United States: Implications of Palynostratigraphy from the Type Dakota Formation." In *Perspectives on the Eastern Margin of the Cretaceous Western Interior Basin*, eds. G. W. Shurr, G. A. Ludvigson, and R. H. Hammond. Geological Society of America Special Paper 287. Boulder, Colo.

"Report on Geology of Iowa's Counties." 1906. Iowa Geological Survey, Iowa City.

Rhodes, R. S., II, and H. A. Semken Jr. 1986. "Quaternary Biostratigraphy and Paleoecology of Fossil Mammals from the Loess Hills Region of Western Iowa." *Proceedings of the Iowa Academy of Science* 93: 94–130.

Roosa, D. M., S. P. Esling, E. A. Bettis III, and J. C. Prior. 1984. *Natural History of the Lake Calvin Basin of Southeast Iowa.* Iowa Natural History Association Field Trip Guidebook 2. Iowa City.

———, J. C. Prior, D. D. Smith, R. M. Knutson, D. Henning, and E. A. Bettis III. 1983. *Natural History of the Upper Iowa Valley between Decorah and New Albin.* Iowa Natural History Association Field Trip Guidebook 1. Iowa City.

Rose, J. N. 1967. *The Fossils and Rocks of Eastern Iowa.* Iowa Geological Survey Educational Series 1. Iowa City.

Ross, C. A. 1964. "Early Silurian Graptolites from the Edgewood Formation of Iowa." *Journal of Paleontology* 38: 1107–1108.

Ruedemann, R. 1947. *Graptolites of North America.* Geological Society of America Memoir 19.

Ruhe, R. V. 1969. *Quaternary Landscapes in Iowa.* Iowa State University Press, Ames.

———, W. P. Dietz, T. E. Fenton, and G. F. Hall. 1968. *Iowan Drift Problem, Northeastern Iowa.* Iowa Geological Survey Report of Investigations 7. Iowa City.

———, and J. C. Prior. 1970. "Pleistocene Lake Calvin, Eastern Iowa." *Geological Society of America Bulletin* 81: 919–924.

Runkel, A. C. 1994. "Deposition of the Uppermost Cambrian (Croixan) Jordan Sandstone, and the Nature of the Cambro-Ordovician Boundary in the Upper Mississippi Valley." *Geological Society of America Bulletin* 106: 492–506.

Ryberg, W. 1990. "Mine Tunnels Offer Natural Storage Area." *Des Moines Sunday Register*, Oct. 28.

Sage, L. L. 1974. *A History of Iowa.* Iowa State University Press, Ames.

Salisbury, N. E., and J. C. Knox. 1969. *Glacial Landforms of the Big Kettle Locality, Dickinson County, Iowa.* Iowa State Advisory Board for Preserves, Development Series Report 6. Des Moines.

Savage, T. E. 1914. "The Relations of the Alexandrian Series to the Silurian Section of Iowa." *American Journal of Science* 38: 28–37.

Schermer, S. J., W. R. Green, and J. M. Collins. 1992. "A Brief Culture History of Iowa." In *Archaeology: An Activity Guide for Educators*, ed. S. J. Schermer. Office of the State Archaeologist, Iowa City.

Schmoker, J. W., and J. G. Palacas. 1990. "Porosity of Precambrian Sandstones in Lower Portion of the Eischeid #1 Well, Carroll County, Iowa." In *The Amoco M. G. Eischeid #1 Deep Petroleum Test, Carroll County, Iowa, Preliminary Investigations*, ed. R. R. Anderson. Geological Survey Bureau Special Report Series No. 2. Iowa City.

Schoewe, W. H. 1920. "The Origin and History of Extinct Lake Calvin." *Iowa Geological Survey Annual Report* 29: 49–222.

Scholle, P. A. 1978. *A Color Illustrated Guide to Carbonate Rock Constituents, Textures, Cements, and Porosities*. American Association of Petroleum Geologists Memoir 27. Tulsa, Okla.

Scholtes, W. H. 1955. "Properties and Classification of the Paha Loess-Derived Soils in Northeastern Iowa. *Iowa State College Journal of Science* 30: 163–209.

Schuldt, W. C. 1943. "Cambrian Strata of Northeastern Iowa." *Iowa Geological Survey Annual Report* 38: 379–422.

Schultze, H. P. 1992. "A New Long-Headed Dipnoan (Osteichthyes) from the Middle Devonian of Iowa, U.S.A." *Journal of Paleontology* 12: 42–58.

Schuster, R. D., R. R. Anderson, and S. B. Mukasa. 1988. "Rb/Sr and Sm/Nd Isotopic Analysis of a Keweenawan Gabbro from the Subsurface of Northeast Iowa." *Geological Society of America Abstracts with Programs* 20 (2): 128–129.

Schwert, D. P., and A. C. Ashworth. 1990. "Ice Age Beetles." *Natural History* 1: 10–14.

Schwieder, D. 1972. "The Last Pony Mine: The Closing of the New Gladstone." *Iowan* (winter): 29–33.

———, and R. Kraemer. 1973. *Iowa's Coal Mining Heritage*. Special Publication of the State Mining Board, Department of Mines and Minerals. Des Moines.

Scotese, C. R., L. M. Gahagan, and R. L. Larson. 1988. "Plate Reconstructions of the Cretaceous and Cenozoic Ocean Basins." *Tectonophysics* 155: 27–48.

Seigley, L. S., and D. J. Quade. 1996. "Gull Point State Park: A Glacial Legacy." *Iowa Geology* 21: 24–27.

Shaver, R. H. 1991. "A History of Study of Silurian Reefs in the Michigan Basin." In *Early Sedimentary Evolution of the Michigan Basin*, eds. P. A. Catacosinos and P. A. Daniels Jr. Geological Society of America Special Paper 256. Boulder, Colo.

———, and J. A. Sunderman. 1984. "Silurian Seascapes: Water Depth, Clinothems, Reef Geometry, and Other Motifs — A Critical Review of the Silurian Reef Model." *Geological Society of America Bulletin* 101: 939–951.

Sheehan, P. M. 1980. "Paleogeography and Marine Communities of the Silurian Carbonate Shelf in Utah and Nevada." In *Paleozoic Paleogeography of the West-Central United States*, eds. T. D. Fouch and E. R. Magathan. Rocky Mountain Section, Society of Economic Paleontologists and Mineralogists, Denver.

Shimek, B. 1896. "A Theory of the Loess." *Proceedings of the Iowa Academy of Science* 3: 82–89.

———. 1909. "Aftonian Sands and Gravels in Western Iowa." *Geological Society of America Bulletin* 20: 399–408.

———. 1910. "Geology of Harrison and Monona Counties." *Iowa Geological Survey Annual Report* 20: 271–485.

Shinn, E. A. 1968. "Practical Significance of Birdseye Structures in Carbonate Rocks." *Journal of Sedimentary Petrology* 38: 215–223.

Short, N. M. 1966. "Shock Processes in Geology." *Journal of Geological Education* 14: 149–166.

———, and T. E. Bunch. 1968. "A Worldwide Inventory of Features Characteristic of Rocks

Associated with Presumed Meteorite Impact Structures." In *Shock Metamorphism of Natural Materials*, eds. B. French and N. Short. Mono Book, Baltimore.

————, and D. P. Gold. 1996. "Petrography of Shocked Rocks from the Central Peak at the Manson Impact Structure." In *The Manson Impact Structure, Iowa: Anatomy of an Impact Crater*, eds. C. Koeberl and R. R. Anderson. Geological Society of America Special Paper 302. Boulder, Colo.

Shurr, G .W., G. A. Ludvigson, and R. H. Hammond, eds. 1994. *Perspectives on the Eastern Margin of the Cretaceous Western Interior Basin*. Geological Society of America Special Paper 287. Boulder, Colo.

Simon, J. A., and M. E. Hopkins. 1973. *Geology of Coal*. Illinois Geological Survey Report Series 1973H. Urbana.

Sims, P. K. 1985. *Precambrian Basement Map of the Northern Midcontinent, U.S.A.* U.S. Geological Survey Open-File Report 85-0604. Washington, D.C.

————, and G. B. Morey, eds. 1972. *Geology of Minnesota: A Centennial Volume*. Minnesota Geological Survey, St. Paul.

————, and Z. E. Peterman. 1986. "Early Proterozoic Central Plains Orogen: A Major Buried Structure in the North-Central United States." *Geology* 14: 488–491.

Sinotte, S. R. 1969. *The Fabulous Keokuk Geodes*. Wallace-Homestead, Des Moines.

Sloan, R. E. 1964. *The Cretaceous System in Minnesota*. Minnesota Geological Survey Report of Investigations 5. St. Paul.

Sloss, L. L. 1963. "Sequences in the Cratonic Interior of North America." *Geological Society of America Bulletin* 74: 93–114.

————. 1988. "Tectonic Evolution of the Craton in Phanerozoic Time." In *The Geology of North America*. Vol. D-2, *Sedimentary Cover-North American Craton*, ed. L. L. Sloss. Geological Society of America, Boulder, Colo.

Smith, R. K. 1967. "Mineralogy and Petrology of Silurian Bioherms of Eastern Iowa." M.S. thesis, University of Iowa, Iowa City.

Smith, T. A. 1971. "A Seismic Refraction Investigation of the Manson Disturbed Area." M.S. thesis, Iowa State University, Ames.

Southwick, D. L., G. B. Morey, and J. H. Mossler. 1986. "Fluvial Origin of the Lower Proterozoic Sioux Quartzite." *Geological Society of America Bulletin* 97: 1432–1441.

Stainbrook, M. A. 1935. "Stratigraphy of the Devonian of the Upper Mississippi Valley." In *Kansas Geological Society, Ninth Annual Field Conference Guidebook*.

————. 1937. "New Echinoderms from the Devonian Cedar Valley Formation of Iowa." *American Midland Naturalist* 18: 899–904.

————. 1938a. "*Atrypa* and *Stropheodonta* from the Cedar Valley Beds of Iowa." *Journal of Paleontology* 12: 229–256.

————. 1938b. "Pentameridae of the Cedar Valley Beds of Iowa." *American Midland Naturalist* 19: 723–739.

————. 1940a. "Orthid Brachiopods of the Cedar Valley Limestone of Iowa." *American Midland Naturalist* 23: 482–492.

————. 1940b. "*Prismatophyllum* in the Cedar Valley Beds of Iowa." *Journal of Paleontology* 14: 270–284.

————. 1941. "Biotic Analysis of Owen's Cedar Valley Limestones." *Pan-American Geologist* 75: 321–327.

————. 1942. "Brachiopoda of the Cedar Valley Beds of Iowa: Inarticulata, Rhynchonellacea, and Rostrospiracea." *Journal of Paleontology* 16: 604–619.

————. 1943a. "Spiriferacea of the Cedar Valley Limestone of Iowa." *Journal of Paleontology* 17: 417–450.

————. 1943b. "Strophomenacea of the Cedar Valley Limestone of Iowa." *Journal of Paleontology* 17: 39–59.

————. 1944. "The Devonian System of Iowa." In *Symposium on Devonian Stratigraphy*. Illinois Geological Survey Bulletin 68. Urbana.

————. 1945a. *Brachiopoda of the Independence Shale of Iowa*. Geological Society of America Memoir 14.

————. 1945b. "The Stratigraphy of the Independence Shale of Iowa." *American Journal of Science* 243: 66–83.

————. 1946. "Corals of the Independence Shale of Iowa." *Journal of Paleontology* 20: 401–427.

Stanley, S. M. 1993. *Exploring Earth and Life through Time*. W. H. Freeman, New York.

Steiner, M. B., and E. M. Shoemaker. 1996. "A Hypothesized Manson Impact Tsunami: Paleomagnetic and Stratigraphic Evidence in the Crow Creek Member." In *The Manson Impact Structure, Iowa: Anatomy of an Impact Crater*, eds. C. Koeberl and R. R. Anderson. Geological Society of America Special Paper 302. Boulder, Colo.

Steinhilber, W. L., and P. J. Horick. 1970. "Groundwater Resources of Iowa." In *Water Resources of Iowa*, ed. P. J. Horick. Iowa Academy of Science Special Publication. Iowa City.

Stewart, R. A. 1988. "Nature and Origin of Corrugated Ground Moraine of the Des Moines Lobe, Story County Iowa." *Geomorphology* 1: 111–130.

Stokes, W. L. 1973. *Essentials of Earth History*. 3rd ed. Prentice-Hall, Inglewood Cliffs, N.J.

Storrs, G. W. 1987. "An Ichthyofauna from the Subsurface Devonian of Northwestern Iowa and Its Biostratigraphic and Paleoecologic Significance." *Journal of Paleontology* 61: 363–374.

Strimple, H. L., and R. Boyt. 1965. "*Rhodocrinites beanei* New Species from the Hampton Formation (Mississippian) of Iowa." *Oklahoma Geology Notes* 25: 223–226.

————, and C. O. Levorson. 1969. "Catalogue of the Type Specimens of the Belanski Collection." *Bulletins of American Paleontology* 56: 259–271.

Svitil, K. A. 1993. "It's Alive, and It's a Graptolite." *Discover* 14 (7): 18–19.

Swade, J. W. 1985. *Conodont Distribution, Paleoecology, and Preliminary Biostratigraphy of the Upper Cherokee and Marmaton Groups (Upper Desmoinesian, Middle Pennsylvanian) from Two Cores in South-Central Iowa*. Iowa Geological Survey Technical Information Series 14. Iowa City.

Swisher, J. 1945. "The Rise and Fall of Buxton." *Palimpsest* 26: 179–192.

Tasch, P. L. 1955. "Paleoecologic Observations on the Orthoceratid Coquina Beds of the Maquoketa at Graf, Iowa." *Journal of Paleontology* 29: 510–518.

————. 1958. "The Internal Structure of *Diplograptus peosta* Hall." *Journal of Paleontology* 32: 1021–1025.

Templeton, J. S., and H. B. Willman. 1963. *Champlainian Series (Middle Ordovician) in Illinois*. Illinois Geological Survey Bulletin 89. Urbana.

Tester, A. C. 1931. "The Dakota Stage of the Type Locality." *Iowa Geological Survey Annual Report* 35: 197–332.

Thomas, A. O. 1924. "Echinoderms of the Iowa Devonian." *Iowa Geological Survey Annual Report* 29: 385–550.

Thomas, L. A. 1960. *Mississippian of North-Central Iowa*. Guidebook for the Twenty-fourth Annual Tri-State Geological Field Conference, Ames, Iowa.

Thompson, C. A. 1985. "Alluvial Aquifers." *Iowa Geology* 10: 8–11.

————. 1987. "Alluvial Aquifers: Northwest Iowa Summary." *Iowa Geology* 12: 20–21.

————. 1996. "Excelsior Fen Complex." In *Hogs, Bogs, and Logs: Quaternary Deposits and Environmental Geology of the Des Moines Lobe*, eds. E. A. Bettis III, D. J. Quade, and T. J. Kemmis. Geological Survey Bureau Guidebook 18. Iowa City.

————, E. A. Bettis III, and R. G. Baker. 1992. "Geology of Iowa Fens." *Journal of the Iowa Academy of Science* 99: 53–59.

———, and T. J. Kemmis. 1990. "Quaternary Geology and Aquifers of Iowa." In *Iowa's Principal Aquifers: A Review of Iowa Geology and Hydrogeologic Units*. Iowa Groundwater Association, Iowa City.

Tripp, R. B. 1959. "The Mineralogy of Warsaw Formation Geodes." *Proceedings of the Iowa Academy of Science* 66: 350–356.

Troeger, J. C. 1983. *From Rift to Drift: Iowa's Story in Stone*. Iowa State University Press, Ames.

Trowbridge, A. C. 1959. "The Mississippi River in Glacial Times." *Palimpsest* 40: 257–288.

———. 1966. *Glacial Drift in the "Driftless Area" of Northeast Iowa*. Iowa Geological Survey Report of Investigations 2. Iowa City.

Tucker, M., and V. P. Wright. 1990. *Carbonate Sedimentology*. Blackwell Scientific Publications, Oxford, England.

Twenhofel, W. H., G. O. Raasch, and F. T. Thwaites. 1935. "Cambrian Strata of Wisconsin." *Geological Society of America Bulletin* 46: 1687–1744.

Udden, J. A. 1900. "Geology of Pottawattamie County." *Iowa Geological Survey Annual Report* 11: 199–278.

———. 1903. "Geology of Mills and Fremont Counties." *Iowa Geological Survey Annual Report* 13: 123–133.

Upchurch, G. R., Jr., and D. L. Dilcher. 1990. *Cenomanian Angiosperm Leaf Megafossils, Dakota Formation, Rose Creek Locality, Jefferson County, Southeastern Nebraska*. U.S. Geological Survey Bulletin 1915.

VanDorpe, P. E. 1980. *A Bibliography of Pennsylvanian Geology and Coal in Iowa*. Geological Survey Miscellaneous Publication 21. Iowa City.

———, and M. R. Howes. 1986. "Mining Iowa's Coal Deposits." *Iowa Geology* 11: 12–16.

———, M. R. Howes, M. J. Miller, and S. J. Lenker. 1984. *Underground Mines and Related Subsidence Potential, What Cheer, Iowa*. Iowa Geological Survey Open-File Report 84-3. Iowa City.

Vandervelde, M. 1967. "Why the Town of Manson Is Happy." *Picture* magazine, *Des Moines Sunday Register*, Apr. 2.

Van Eck, O. J. 1965. "Geologic Setting." In *Coal Resources of Iowa*, eds. E. R. Landis and O. J. Van Eck. Iowa Geological Survey Technical Paper 4. Iowa City.

Van Schmus, W. R. 1979. "Geochronology of the Southern Wisconsin Rhyolites and Granites." *Geoscience Wisconsin* 2: 9–24.

———. 1980. "Chronology of Igneous Rocks Associated with the Penokean Orogeny in Wisconsin." In *Selected Studies of Archean Gneisses and Lower Proterozoic Rocks, Southern Canadian Shield*, eds. G. B. Morey and G. N. Hansen. *Geological Society of America Special Paper* 182. Boulder, Colo.

———, M. E. Bickford, R. R. Anderson, C. K. Shearer, J. J. Papike, and B. K. Nelson. 1989. "Quimby, Iowa, Scientific Drill Hole: Definition of Precambrian Crustal Features in Northwestern Iowa." *Geology* 17: 536–539.

———, M. E. Bickford, P. K. Sims, R. R. Anderson, C. K. Shearer, and S. B. Trevis. 1993. "Proterozoic Geology of the Western Midcontinent Basement." In *The Geology of North America*. Vol. C-2, *Precambrian Conterminous U.S.*, ed. J. C. Reed. Geological Society of America, Boulder, Colo.

———, M. E. Bickford, and I. Zietz. 1987. "Early Middle Proterozoic Provinces in the Central United States." In *Proterozoic Lithospheric Evolution*, ed. A. Kroner. American Geophysical Union Geodynamics Series 17. Washington, D.C.

———, M. E. Thurman, and A. E. Petermann. 1975. "Geology and Rb-Sr Chronology of Middle Precambrian Rocks in Eastern and Central Wisconsin." *Geological Society of America Bulletin* 86: 1255–1265.

———, and E. T. Wallin. 1990. "Age Relationships for Gabbro from the Amoco M. G.

Eischeid #1 Well." In *The Amoco M. G. Eischeid #1 Deep Petroleum Test, Carroll County, Iowa, Preliminary Investigations*, ed. R. R. Anderson. Geological Survey Bureau Special Report Series No. 2. Iowa City.

———, and E. T. Wallin. 1991. "Studies of the Precambrian Geology of Iowa: Part 3. Geochronologic Data for the Matlock Drill Holes." *Journal of the Iowa Academy of Science* 98: 182–188.

Van Tuyl, F. M. 1916. "The Geodes of the Keokuk Beds." *American Journal of Science* 42: 34–42.

———. 1925. "The Stratigraphy of the Mississippian Formations of Iowa." *Iowa Geological Survey Annual Report* 30: 33–374.

Van Zant, K. L. 1979. "Late-Glacial and Postglacial Pollen and Plant Macrofossils from Lake Okoboji, Northwestern Iowa." *Quaternary Research* 12: 358–380.

Wachsmuth C., and F. Springer. 1878. "Transition Forms in Crinoids and Description of Five New Species." *Philadelphia Academy of Natural Sciences Proceedings* 29: 224–266.

———, and F. Springer. 1890. "New Species of Crinoids and Blastoids from the Kinderhook Group of the Lower Carboniferous Rocks at Le Grand, Iowa." *Illinois Geological Survey Report* 8: 155–205.

———, and F. Springer. 1895. *Monograph of the Crinoidea Camerata of North America*. Memoirs of the Museum of Comparative Zoology, 3 vols. Harvard University and John Wilson and Son, Cambridge.

Walter, O. T. 1924. "Trilobites of Iowa and Some Related Paleozoic Forms." *Iowa Geological Survey Annual Report* 31: 167–389.

Walters, J. C. 1994. "Ice-Wedge Casts and Relict Polygonal Patterned Ground in Northeast Iowa, USA." *Permafrost and Periglacial Processes* 5: 269–282.

Wanless, H. R. 1970. "Late Paleozoic Deltas in the Central and Eastern United States." In *Deltaic Sedimentation: Modern and Ancient*, ed. J. P. Morgan. Society of Economic Paleontologists and Mineralogists Special Publication 15. Tulsa, Okla.

———, and F. P. Shepard. 1936. "Sea Level and Climatic Changes Related to Late Paleozoic Cycles." *Geological Society of America Bulletin* 47: 1177–1206.

———, J. B. Tubbs Jr., D. E. Gednetz, and J. L. Weiner. 1963. "Mapping Sedimentary Environments of Pennsylvanian Cycles." *Geological Society of America Bulletin* 74: 437–486.

———, and J. M. Weller. 1931. "Correlation and Extent of Pennsylvanian Cyclothems." *Geological Society of America Bulletin* 43: 1003–1016.

Weller, J. M. 1930. "Cyclical Sedimentation of the Pennsylvanian Period and Its Significance." *Journal of Geology* 38: 97–135.

———. 1958. "Cyclothems and Larger Sedimentation Cycles of the Pennsylvanian." *Journal of Geology* 66: 195–207.

White, C. A. 1870. *Report on the Geological Survey of the State of Iowa to the Thirteenth General Assembly, January, 1870, Containing Results of Examinations and Observations Made within the Years 1866, 1867, 1868, and 1869.* 2 vols. Mills and Company, Des Moines.

Wickham, J. T. 1979. *Pre-Illinoian Till Stratigraphy in the Quincy Area.* Illinois Geological Survey Guidebook 14. Urbana.

Willman, H. B., E. Atherton, T. C. Buschbach, C. Collinson, J. C. Frye, M. E. Hopkins, J. A. Lineback, and J. A. Simon. 1975. *Handbook of Illinois Stratigraphy.* Illinois Geological Survey Bulletin 95. Urbana.

Windom, K. E., K. E. Seifert, and R. R. Anderson. 1991a. "Studies of the Precambrian Geology of Iowa: Part 1. The Otter Creek Layered Igneous Complex." *Journal of the Iowa Academy of Science* 98: 170–177.

———, K. E. Seifert, and R. R. Anderson. 1991b. "Studies of the Precambrian Geology of Iowa: Part 2. The Matlock Keratophyre." *Journal of the Iowa Academy of Science* 98:

178–181.

Witzke, B. J. 1976. "Echinoderms of the Hopkinton Dolomite, Eastern Iowa." M.S. thesis, University of Iowa, Iowa City.

———. 1978. "Stratigraphy along the Plum River Fault Zone: A Guide to Late Ordovician through Middle Devonian Deposits of Eastern Iowa." In *Geology of East-Central Iowa*, ed. R. R. Anderson. Guidebook for the Forty-second Annual Tri-State Geological Field Conference. Iowa City.

———. 1980. "Middle and Upper Ordovician Paleogeography of the Region Bordering the Transcontinental Arch." In *Paleozoic Paleogeography of the West-Central United States*, eds. T. D. Fouch and E. R. Magathan. Rocky Mountain Section, Society of Economic Paleontologists and Mineralogists, Denver.

———. 1981a. "Cretaceous Vertebrate Fossils of Iowa and Nearby Areas of Nebraska, South Dakota, and Minnesota." In *Cretaceous Stratigraphy and Sedimentation in Northwest Iowa, Northeast Nebraska, and Southeast South Dakota*, eds. R. L. Brenner, R. F. Bretz, B. J. Bunker, D. L. Iles, G. A. Ludvigson, R. M. McKay, D. L. Whitley, and B. J. Witzke. Iowa Geological Survey Guidebook 4. Iowa City.

———. 1981b. *Silurian Stratigraphy of Eastern Linn and Western Jones Counties, Iowa*. Geological Society of Iowa Guidebook 35. Iowa City.

———. 1981c. "Stratigraphy, Depositional Environments, and Diagenesis of the Eastern Iowa Silurian Sequence." Ph.D. diss., University of Iowa, Iowa City.

———. 1983a. "Fossils: Evidence of Ancient Life in Iowa." *Iowa Geology* 8: 4–9.

———. 1983b. "Ordovician Galena Group in Iowa Subsurface." In *Ordovician Galena Group of the Upper Mississippi Valley: Deposition, Diagenesis, and Paleoecology,* ed. D. J. Delgado. Guidebook for the Thirteenth Annual Field Conference, Great Lakes Section, Society of Economic Paleontologists and Mineralogists. Dubuque, Iowa.

———. 1983c. "Silurian Benthic Invertebrate Associations of Eastern Iowa and Their Paleoenvironmental Significance." *Transactions of the Wisconsin Academy of Sciences, Arts, and Letters* 71: 21–47.

———. 1984. *Geology of the University of Iowa Campus Area*. Iowa Geological Survey Guidebook 7. Iowa City.

———. 1985. "Silurian System." In *The Plum River Fault Zone and the Structural and Stratigraphic Framework of Eastern Iowa*, eds. B. J. Bunker, G. A. Ludvigson, and B. J. Witzke. Iowa Geological Survey Technical Information Series 13. Iowa City.

———. 1987a. "Geodes: A Look at Iowa's State Rock." *Iowa Geology* 12: 8–9.

———. 1987b. "Models for Circulation Patterns in Epicontinental Seas Applied to Paleozoic Facies of the North American Craton." *Paleoceanography* 2: 229–248.

———. 1987c. "Pella Stratigraphy, Paleontology, and Correlation Problems of the 'St. Louis'–Pella Interval." In *Early Tetrapods, Stratigraphy, and Paleoenvironments of the Upper St. Louis Formation, Western Keokuk County, Iowa*, eds. R. M. McKay, B. J. Witzke, and M. P. McAdams. Geological Society of Iowa Guidebook 46. Iowa City.

———. 1987d. "Silurian Carbonate Mounds, Palisades-Kepler State Park, Iowa." In *Geological Society of America Centennial Field Guide — North-Central Section*, ed. D. L. Biggs. Geological Society of America, Boulder, Colo.

———. 1989. "Iowa's Ancient Seas." *Iowa Geology* 14: 4–8.

———. 1990a. "General Stratigraphy of the Phanerozoic and Keweenawan Sequence, M. G. Eischeid #1 Drillhole, Carroll Co., Iowa." In *The Amoco M. G. Eischeid #1 Deep Petroleum Test, Carroll County, Iowa, Preliminary Investigations*, ed. R. R. Anderson. Geological Survey Bureau Special Report Series No. 2. Iowa City.

———. 1990b. "Palaeoclimatic Constraints for Palaeozoic Palaeolatitudes of Laurentia and Euramerica." In *Palaeozoic Palaeogeography and Biogeography*, eds. W. S. McKerrow and C. R. Scotese. *Geological Society (London) Memoir* 12: 57–73.

————. 1991. "Salt and Pepper Sands of Western Iowa." *Iowa Geology* 16: 22–24.

————. 1992. *Silurian Stratigraphy and Carbonate Mound Facies of Eastern Iowa, Field Trip Guidebook to Silurian Exposures in Jones and Linn Counties.* Geological Survey Bureau Guidebook 11. Iowa City.

————. 1995a. "Bedrock Geology of Backbone State Park." In *The Natural History of Backbone State Park, Delaware County, Iowa*, ed. R. R. Anderson. Geological Society of Iowa Guidebook 61. Iowa City.

————. 1995b. "Geology of Backbone State Park." *Iowa Geology* 20: 22–25.

————. 1995c. "Stratigraphic Column of Iowa." Home Page. Geological Survey Bureau, Iowa City.

————. 1996. "Geologic Sources of Historic Stone Architecture in Iowa." *Iowa Geology* 21: 14–19.

————, and R. R. Anderson. 1988. "Iowa's Manson Crater: The Largest Impact Site in the U.S." *Iowa Geology* 13: 4–7.

————, and R. R. Anderson. 1996. "Sedimentary-Clast Breccias of the Manson Impact Structure." In *The Manson Impact Structure, Iowa: Anatomy of an Impact Crater*, eds. C. Koeberl and R. R. Anderson. Geological Society of America Special Paper 302. Boulder, Colo.

————, and B. J. Bunker. 1984. *Devonian Stratigraphy of North-Central Iowa.* Iowa Geological Survey Open-File Report 84-2. Iowa City.

————, and B. J. Bunker. 1985. *Stratigraphic Framework for the Devonian Aquifers in Floyd-Mitchell Counties, Iowa.* Iowa Geological Survey Open-File Report 85-2. Iowa City.

————, and B. J. Bunker. 1994. *Classic Geological Exposures, Old and New: Coralville Lake and Spillway, Devonian Fossil Gorge, Merrill A. Stainbrook Preserve, and Old State Quarry Preserve.* Geological Society of Iowa Guidebook 60. Iowa City.

————, and B. J. Bunker. 1995a. "Bedrock Geology in the Area of Osage Spring Park." In *Geology and Hydrogeology of Floyd-Mitchell Counties*, ed. B. J. Bunker. Geological Society of Iowa Guidebook 62. Iowa City.

————, and B. J. Bunker. 1995b. "Geology and Discussion of the Lithograph City Quarries." In *Geology and Hydrogeology of Floyd-Mitchell Counties*, ed. B. J. Bunker. Geological Society of Iowa Guidebook 62. Iowa City.

————, and B. J. Bunker. 1995c. "Type Section of Osage Springs Member." In *Geology and Hydrogeology of Floyd-Mitchell Counties*, ed. B. J. Bunker. Geological Society of Iowa Guidebook 62. Iowa City.

————, B. J. Bunker, and G. Klapper. 1985. "Devonian Stratigraphy in the Quad-Cities Area, Eastern Iowa-Northwestern Illinois." In *Devonian and Pennsylvanian Stratigraphy of the Quad-Cities Region Illinois-Iowa*, eds. W. R. Hammer, R. C. Anderson, and D. A. Schroeder. Guidebook for the Fifteenth Annual Field Conference, Great Lakes Section, Society of Economic Paleontologists and Mineralogists. Rock Island, Illinois.

————, B. J. Bunker, and F. S. Rogers. 1988. "Eifelian through Lower Frasnian Stratigraphy and Deposition in the Iowa Area, Central Midcontinent, U.S.A." In *Devonian of the World*, eds. N. J. McMillan, A. F. Embry, and D. J. Glass. Canadian Society of Petroleum Geologists Memoir 14. Calgary.

————, and B. F. Glenister. 1987a. "The Ordovician Sequence in the Guttenberg Area, Northeast Iowa." In *Geological Society of America Centennial Field Guide — North-Central Section*, ed. D. L. Biggs. Geological Society of America, Boulder, Colo.

————, and B. F. Glenister. 1987b. "Upper Ordovician Maquoketa Formation in the Graf Area, Eastern Iowa." In *Geological Society of America Centennial Field Guide — North-Central Section*, ed. D. L. Biggs. Geological Society of America, Boulder, Colo.

————, R. H. Hammond, and R. R. Anderson. 1996. "Deposition of the Crow Creek Member, Campanian, South Dakota and Nebraska." In *The Manson Impact Structure,*

Iowa: Anatomy of an Impact Crater, eds. C. Koeberl and R. R. Anderson. Geological Society of America Special Paper 302. Boulder, Colo.

———, and M. E. Johnson. 1992. "Silurian Brachiopod and Related Benthic Communities from Carbonate Platform and Mound Environments of Iowa and Surrounding Areas." In *Report for Project Ecostratigraphy*. Cambridge University Press, Cambridge, England.

———, and D. R. Kolata. 1988. "Changing Structural and Depositional Patterns, Ordovician Champlainian and Cincinnatian Series of Iowa-Illinois." In *New Perspectives on the Paleozoic History of the Upper Mississippi Valley: An Examination of the Plum River Fault Zone*, eds. G. A. Ludvigson and B. J. Bunker. Geological Survey Bureau Guidebook 8. Iowa City.

———, and G. A. Ludvigson. 1982. *Cretaceous Stratigraphy and Depositional Systems in Guthrie County, Iowa*. Geological Society of Iowa Guidebook 38. Iowa City.

———, and G. A. Ludvigson. 1987. "Cretaceous Exposures, Big Sioux River Valley, North of Sioux City, Iowa." In *Geological Society of America Centennial Field Guide — North-Central Section*, ed. D. L. Biggs. Geological Society of America, Boulder, Colo.

———, and G. A. Ludvigson. 1990. "Petrographic and Stratigraphic Comparisons of Sub-Till and Inter-Till Alluvial Units in Western Iowa: Implications for the Development of the Missouri River Drainage." In *Holocene Alluvial Stratigraphy and Selected Aspects of the Quaternary History of Western Iowa*, ed. E. A. Bettis III. Geological Survey Bureau Guidebook 9. Iowa City.

———, and G. A. Ludvigson. 1994. "The Dakota Formation in Iowa and the Type Area." In *Perspectives on the Eastern Margin of the Cretaceous Western Interior Basin*, eds. G. W. Shurr, G. A. Ludvigson, and R. H. Hammond. Geological Society of America Special Paper 287. Boulder, Colo.

———, and G. A. Ludvigson. 1996a. "Coarse-grained Eastern Facies." In *Mid-Cretaceous Fluvial Deposits of the Eastern Margin, Western Interior Basin: Nishnabotna Member, Dakota Formation, Field Guide to the Cretaceous of Guthrie County*, eds. B. J. Witzke and G. A. Ludvigson. Geological Survey Guidebook 17. Iowa City.

———, and G. A. Ludvigson, eds. 1996b. *Mid-Cretaceous Fluvial Deposits of the Eastern Margin, Western Interior Basin: Nishnabotna Member, Dakota Formation, Field Guide to the Cretaceous of Guthrie County*. Geological Survey Bureau Guidebook 17. Iowa City.

———, G. A. Ludvigson, R. L. Brenner, and R. M. Joeckel. 1996. "Regional Dakota Sedimentation." In *Mid-Cretaceous Fluvial Deposits of the Eastern Margin, Western Interior Basin: Nishnabotna Member, Dakota Formation, Field Guide to the Cretaceous of Guthrie County*, eds. B. J. Witzke and G. A. Ludvigson. Geological Survey Guidebook 17. Iowa City.

———, G. A. Ludvigson, J. Poppe, and R. L. Ravn. 1983. "Cretaceous Paleogeography along the Eastern Margin of the Western Interior Seaway, Iowa, Southern Minnesota, and Eastern Nebraska and South Dakota." In *Mesozoic Paleogeography of West-Central United States*, eds. M. Reynolds and E. Dolly. Rocky Mountain Section, Society of Economic Paleontologists and Mineralogists, Denver.

———, G. A. Ludvigson, R. L. Ravn, R. M. Joeckel, and R. L. Brenner. 1996. "Regional Sedimentation of Coarse-Grained Fluvial Facies, Basal Dakota Formation, Albian, Western Iowa and Eastern Nebraska." *Geological Society of America Abstracts with Programs* 28 (6): 71.

———, and R. M. McKay. 1987. "Cambrian and Ordovician Stratigraphy in the Lansing Area, Northeastern Iowa." In *Geological Society of America Centennial Field Guide — North-Central Section*, ed. D. L. Biggs. Geological Society of America, Boulder, Colo.

———, R. M. McKay, B. J. Bunker, and F. J. Woodson. 1990. *Stratigraphy and Paleoenvironments of Mississippian Strata in Keokuk and Washington Counties, Southeast Iowa*. Geological Survey Bureau Guidebook 10. Iowa City.

———, R. L. Ravn, G. A. Ludvigson, R. J. Joeckel, and R. L. Brenner. 1996. "Age and Correlation of the Nishnabotna Member." In *Mid-Cretaceous Fluvial Deposits of the Eastern Margin, Western Interior Basin: Nishnabotna Member, Dakota Formation, Field Guide to the Cretaceous of Guthrie County*, eds. B. J. Witzke and G. A. Ludvigson. Geological Survey Guidebook 17. Iowa City.

———, and H. L. Strimple. 1981. "Early Silurian Camerate Crinoids of Eastern Iowa." *Proceedings of the Iowa Academy of Science* 88: 101–137.

Wolf, R. C. 1983. *Fossils of Iowa: Field Guide to Paleozoic Deposits*. Iowa State University Press, Ames.

———. 1991. *Iowa's State Parks, also Forests, Recreation Areas, and Preserves*. Iowa State University Press, Ames.

Woodson, F. J., ed. 1989a. *An Excursion to the Historic Gilmore City Quarries*. Geological Society of Iowa Guidebook 50. Iowa City.

———. 1989b. "A Short History of the Quarry Industry at Gilmore City." In *An Excursion to the Historic Gilmore City Quarries*, ed. F. J. Woodson. Geological Society of Iowa Guidebook 50. Iowa City.

———, and B. J. Bunker. 1989. "Lithostratigraphic Framework of Kinderhookian and Early Osagean (Mississippian) Strata, North-Central Iowa." In *An Excursion to the Historic Gilmore City Quarries*, ed. F. J. Woodson. Geological Society of Iowa Guidebook 50. Iowa City.

Yoho, W. H. 1967. *Preliminary Report on Basement Rocks of Iowa*. Iowa Geological Survey Report of Investigations 3. Iowa City.

Zawistoski, A. N., E. P. Kvale, and G. A. Ludvigson. 1996. "Upper Albian (Basal Dakota Fm.) Tidal Rhythmites, Eastern Nebraska." *Geological Society of America Abstracts with Programs* 28 (6): 73.

Zawistowski, S. J. 1971. "Biostromes in the Rapid Member of the Cedar Valley Limestone (Devonian) in East-Central Iowa." Master's thesis, University of Iowa, Iowa City.

Ziegler, A. M. 1965. "Silurian Marine Communities and Their Environmental Significance." *Nature* 207: 270–272.

———, A. J. Boucot, and R. P. Sheldon. 1966. "Silurian Pentameroid Brachiopods Preserved in Position of Growth." *Journal of Paleontology* 40: 1032–1036.

Index

Illustrations are indicated by italicized numbers.

Bur Oak Books / Natural History